T0328293

URBAN GEOMORPHOLOGY

URBAN GEOMORPHOLOGY

Landforms and Processes in Cities

Edited by

MARY J. THORNBUSH
University of Oxford, Oxford, United Kingdom

CASEY D. ALLEN
The University of the West Indies, Cave Hill Campus, Barbados

Elsevier
Radarweg 29, PO Box 211, 1000 AE Amsterdam, Netherlands
The Boulevard, Langford Lane, Kidlington, Oxford OX5 1GB, United Kingdom
50 Hampshire Street, 5th Floor, Cambridge, MA 02139, United States

Library of Congress Cataloging-in-Publication Data
A catalog record for this book is available from the Library of Congress

British Library Cataloguing-in-Publication Data
A catalogue record for this book is available from the British Library

ISBN: 978-0-12-811951-8

For information on all Elsevier publications visit our website at
https://www.elsevier.com/books-and-journals

Publisher: Candice Janco
Acquisition Editor: Amy Shapiro
Editorial Project Manager: Tasha Frank
Production Project Manager: Bharatwaj Varatharajan
Designer: Matthew Limbert

Typeset by Thomson Digital

Contents

Contributors ix
Preface xi

1. Introduction
CASEY D. ALLEN, MARY J. THORNBUSH

1.1 Introduction 1
References 5

SECTION I
PALEOGEOMORPHOLOGY AND ARCHAEOGEOMORPHOLOGY

2. Interactions between Geomorphology and Urban Evolution since Neolithic Times in a Mediterranean City
JOANA M. PETRUS, MAURICI RUIZ, JOAN ESTRANY

2.1 Introduction 10
2.2 The Geography of Palma, a Mediterranean City 11
2.3 Urban Evolution and Geomorphological Processes since the Talayotic Period (BC 3000–Present) 18
2.4 Land Use as the Crucial Change of Urban Geomorphology in the 20th Century 29
2.5 Concluding Remarks 31
References 32

3. Geotourism Development in an Urban Area Based on the Local Geological Heritage (Pruszków, Central Mazovia, Poland)
MARIA GÓRSKA-ZABIELSKA, RYSZARD ZABIELSKI

3.1 Introduction 37
3.2 Georesources of Pruszków and its Surroundings 39
3.3 Relief and Deposits 40
3.4 Water 43
3.5 Erratics 43
3.6 Stones in an Open Urban Space 49
3.7 Final Remarks 51
References 53

4. Anthropogeomorphological Metamorphosis of an Urban Area in the Postglacial Landscape: A Case Study of Poznań City
ZBIGNIEW ZWOLIŃSKI, IWONA HILDEBRANDT-RADKE, MAŁGORZATA MAZUREK, MIROSŁAW MAKOHONIENKO

4.1 Introduction 55
4.2 Geological and Sedimentological Setting 60
4.3 Geomorphological Setting 64
4.4 Anthropogenic Changes in Morphological Landscapes 66
4.5 Urban Geosites 71
4.6 Conclusion 73
References 74

SECTION II

ANTHROPOGEOMORPHOLOGY

5. Urban Stream Geomorphology and Salmon Repatriation in Lower Vernon Creek, British Columbia (Canada)

ALEXANDER MACDUFF, BERNARD O. BAUER

5.1 Introduction 81
5.2 Methods 84
5.3 Results 87
5.4 Discussion 96
5.5 Summary and Conclusions 97
References 98

6. Landform Change Due to Airport Building

EDYTA PIJET-MIGOŃ, PIOTR MIGOŃ

6.1 Introduction 101
6.2 Types of Geomorphic Change 103
6.3 Conclusions 110
References 111

SECTION III

LANDSCAPE INFLUENCES ON URBAN GROWTH

7. Environmental Contamination by Technogenic Deposits in the Urban Area of Araguaína, Brazil

CARLOS A. MACHADO, SILVIO C. RODRIGUES

7.1 Introduction 115
7.2 Methodology 116
7.3 Technogenic Deposits (TDs) and Soil Contamination 117
7.4 Soil Contamination by TDs in the Urban Area of Araguaína 119
7.5 Conclusion 125
References 125

8. Transforming the Physical Geography of a City: An Example of Johannesburg, South Africa

JASPER KNIGHT

8.1 Introduction 129
8.2 The Physical Environment of the City 130
8.3 Development of the City 135
8.4 Discussion: Challenges of the City Today 139
8.5 Conclusions and Future Outlook 143
References 144

9. When Urban Design Meets Fluvial Geomorphology: A Case Study in Chile

PAULINA ESPINOSA, JESÚS HORACIO, ALFREDO OLLERO, BRUNO DE MEULDER, EDILIA JAQUE, MARÍA DOLORES MUÑOZ

9.1 Introduction 150
9.2 Objective and Methods 151
9.3 An Interdisciplinary Dialogue 151
9.4 Study Area Characterization 154
9.5 Design Exercise: Creating Scenarios 161
9.6 Discussion Around Feasibility Issues 169
9.7 Conclusions 169
References 171

SECTION IV

DEVELOPING GEOMORPHOLOGICAL HAZARDS DURING THE ANTHROPOCENE

10. Urban Geomorphology of an Arid City: Case Study of Phoenix, Arizona

ARA JEONG, SUET YI CHEUNG, IAN J. WALKER, RONALD I. DORN

10.1 Sonoran Desert Setting of the Phoenix Metropolitan Area 177

10.2 Common Desert Geomorphic Processes in the Phoenix Metropolitan Area 179
10.3 Desert Geomorphic Hazards 195
10.4 Summary Perspective on Human Influences on the Arid Geomorphic System in the Urbanizing Sonoran Desert 200
References 201

11. Bivouacs of the Anthropocene: Urbanization, Landforms, and Hazards in Mountainous Regions

KEVIN GAMACHE, JOHN R. GIARDINO, PANSHU ZHAO, REBECCA HARPER OWENS

11.1 Introduction 206
11.2 Study Area 207
11.3 The New Awareness of the Critical Zone 208
11.4 Mining Town Development 210
11.5 Geomorphic Processes 212
11.6 Location, Location, Location: A Planner's Dream 221
11.7 Predicting Urban Suitability in the San Juan Mountains 223
11.8 Results 226
11.9 Conclusion 226
References 228

12. Pokhara (Central Nepal): A Dramatic Yet Geomorphologically Active Environment Versus a Dynamic, Rapidly Developing City

MONIQUE FORT, BASANTA R. ADHIKARI, BHAWAT RIMAL

12.1 Introduction 232
12.2 Pokhara City in Its Valley: A Long, Dramatic, and Complex History 232
12.3 A Tourist City with Major Attractions Related to Its Geomorphology 241
12.4 Potential Threats: Natural Hazards and Risks 248
12.5 Conclusions 255
References 257

SECTION V

URBAN STONE DECAY: CULTURAL STONE AND ITS SUSTAINABILITY IN THE BUILT ENVIRONMENT

13. Urban Stone Decay and Sustainable Built Environment in the Niger River Basin

OLUMIDE ONAFESO, ADEYEMI OLUSOLA

13.1 Introduction 261
13.2 Decay of Clay Sandstones and Mudstones 262
13.3 Evidence of Rock decay Consequent to Urban Stone Decay 264
13.4 Warm Wet Climates of the River Niger Basin Region 269
13.5 Prevailing Atmospheric Pollution of the Urban-Built Environment 270
13.6 Conclusion 273
References 274

14. A Geologic Assessment of Historic Saint Elizabeth of Hungary Church Using the Cultural Stone Stability Index, Denver, Colorado

CASEY D. ALLEN, STACY ESTER, KAELIN M. GROOM, RODERICK SCHUBERT, CAROLYN HAGELE, DANA OLOF, MELISSA JAMES

14.1 Introduction and Background 278
14.2 Methods: Basics of the Cultural Stone Stability Index 281
14.3 Saint Elizabeth's CSSI Analysis 283
14.4 Implications and Conclusion 298
References 301

15. Photographic Technique Used in a Photometric Approach to Assess the Weathering of Pavement Slabs in Toronto (Ontario, Canada)

MARY J. THORNBUSH

15.1 Introduction 303
15.2 A New Method 305
15.3 Results with Discussion 310
15.4 Conclusions 313
References 313

16. Conclusion

MARY J. THORNBUSH, CASEY D. ALLEN

16.1 Introduction 317
16.2 Future Studies 318
16.3 Conclusions 319
References 320

Index 321

Contributors

Basanta R. Adhikari Civil Engineering Department, Institute of Engineering, Tribhuvan University, Kirtipur, Nepal

Casey D. Allen The University of the West Indies, Cave Hill Campus, Barbados

Bernard O. Bauer University of British Columbia, Kelowna, BC, Canada

Suet Yi Cheung Arizona State University, Tempe, AZ, United States

Bruno De Meulder University of Leuven, Heverlee (Leuven), Belgium

Ronald I. Dorn Arizona State University, Tempe, AZ, United States

Paulina Espinosa University of Leuven, Heverlee (Leuven), Belgium

Stacy Ester University of Colorado Denver, Denver, CO, United States

Joan Estrany Department of Geography, University of the Balearic Islands, Palma, Mallorca, Spain; Institute of Agro-Environmental and Water Economy Research–INAGEA, University of the Balearic Islands, Palma, Spain

Monique Fort Département de Géographie, Université Paris-Diderot-SPC, Paris Cedex 13, France

Maria Górska-Zabielska Jan Kochanowski University, Kielce, Poland

Kevin Gamache Texas A&M University, College Station, TX, United States

John R. Giardino Texas A&M University, College Station, TX, United States

Kaelin M. Groom Arizona State University, Tempe, AZ, United States

Carolyn Hagele University of Colorado Denver, Denver, CO, United States

Iwona Hildebrandt-Radke Institute of Geoecology and Geoinformation, Adam Mickiewicz University in Poznań, Poznań, Poland

Jesús Horacio University of Concepcion, Concepción, Chile; University of Santiago de Compostela, Galicia, Spain

Melissa James University of Colorado Denver, Denver, CO, United States

Edilia Jaque University of Concepcion, Concepción, Chile

Ara Jeong Arizona State University, Tempe, AZ, United States

Jasper Knight University of the Witwatersrand, Johannesburg, South Africa

Alexander MacDuff University of British Columbia, Kelowna, BC, Canada

Carlos A. Machado Tocantins Federal University (UFT), Araguaína, Brazil

Mirosław Makohonienko Institute of Geoecology and Geoinformation, Adam Mickiewicz University in Poznań, Poznań, Poland

Małgorzata Mazurek Institute of Geoecology and Geoinformation, Adam Mickiewicz University in Poznań, Poznań, Poland

Piotr Migoń Institute of Geography and Regional Development, University of Wrocław, Wrocław, Poland

María Dolores Muñoz University of Concepcion, Concepción, Chile

Alfredo Ollero University of Zaragoza, Zaragoza, Spain

Dana Olof University of Colorado Denver, Denver, CO, United States

Adeyemi Olusola University of Ibadan, Ibadan, Nigeria

Olumide Onafeso Olabisi Onabanjo University, Ago Iwoye, Ogun State, Nigeria

Rebecca Harper Owens Texas A&M University, College Station, TX, United States

Joana M. Petrus Department of Geography, University of the Balearic Islands, Palma, Mallorca, Spain

Edyta Pijet-Migoń Institute of Tourism, Wrocław School of Banking, Wrocław, Poland

Bhawat Rimal Institute of Remote Sensing and Digital, Earth (RADI), CAS, Beijing, China

Silvio C. Rodrigues Uberlândia Federal University (UFU), Uberlândia, Brazil

Maurici Ruiz GIS and Remote-Sensing Service, University of the Balearic Islands, Palma, Mallorca, Spain

Roderick Schubert University of Colorado Denver, Denver, CO, United States

Mary J. Thornbush University of Oxford, Oxford, United Kingdom

Ian J. Walker Arizona State University, Tempe, AZ, United States

Ryszard Zabielski Polish Geological Institute-National Research Institute, Warsaw, Poland

Panshu Zhao Texas A&M University, College Station, TX, United States

Zbigniew Zwoliński Institute of Geoecology and Geoinformation, Adam Mickiewicz University in Poznań, Poznań, Poland

Preface

The human influence on altering landscapes is faster and more dramatic than most natural processes. It is, therefore, crucial to explore geomorphology from the perspective of environments that have been transformed by human activity and occupation. Urban geomorphology is an essential part of anthropogenic geomorphology (anthropogeomorphology), as the built environment represents the quintessential human-altered or anthropogenic landscape. *Urban Geomorphology: Landforms and Processes in Cities* examines human impacts on landscapes through processes and landforms created over time through development and urbanization.

The volume is organized into five main sections: Paleogeomorphology and archaeogeomorphology, Anthropogeomorphology, Landscape influences on urban growth, Developing geomorphological hazards during the Anthropocene, and Urban stone decay. In order, these sections represent:

1. The temporal aspect, with the hope that lessons can be learned from the past
2. The core focus of this book, anthropogeomorphology, including studies from all parts of the world
3. How landscapes influence urban growth, with a specific focus on less developed countries (LDCs) and their environmental constraints
4. Geomorphological hazards during the perceived Anthropocene, which includes modern hazards, such as sinkholes and mass wasting events
5. Urban stone decay (rock weathering) focusing on cultural stone and sustainability in the built environment

Importantly, the volume contributes to the development of a "human geomorphology" that is linked to the environmental past and landscape change. The focal point, however, is the more recent past, with increasing human alterations of landscapes engulfing natural landscapes and transforming them to urbanscapes and, in the process, producing potentially irreparable damage to the Earth surface.

To address the subsequent issues of the human transformation of natural landscapes and the environmental impacts and geomorphological hazards that environmental change can encompass, this volume adopts a multidisciplinary perspective. This approach remains appropriate for audiences from a range of disciplines and occupations—from geologists, conservationists, and land-use planners to architects, developers, and environmental management professionals. *Urban Geomorphology* not only transcends disciplines, it also covers varied spatial-temporal frameworks and presents a diverse set of strategies and possible solutions to human impacts and geomorphological hazards within urban landscapes.

Mary J. Thornbush and Casey D. Allen

CHAPTER

1

Introduction

Casey D. Allen, Mary J. Thornbush***

*The University of the West Indies, Cave Hill Campus, Barbados;
**University of Oxford, Oxford, United Kingdom

OUTLINE

1.1 Introduction 1
References 5

1.1 INTRODUCTION

Coined by Coates (1976) as "...the study of [humans] as a physical process of change whereby [s/he] metamorphoses a more natural terrain to an anthropogene cityscape," and expanded upon later by the author (Coates, 1984), urban geomorphology as a specific concept has been around since at least the 1960s (Xizhi, 1988). Put another way, urban geomorphology centers on the pursuit of understanding the impacts that landforms—and the inherent processes that give rise to them—can have on urban areas, and vice versa (Coates, 1976; Cooke, 1976; Thornbush, 2015). Even before these earlier works, and continuing still today, different components of urban geomorphology (e.g., fluvial regimes, landslides and hazards, and seismic activity) have been researched extensively, including regional studies, such as Gupta's (1987) review of Singapore. Still, while more research in geomorphology is being focused on examining human impacts on landscapes, this is often done in the context of attempting to understand human-environment interactions during the Anthropocene (e.g., Goudie and Viles, 2016). Recent attention has focused specifically on cities, where human activity has historically been concentrated.

This volume addresses similar themes tied to human impacts on landscapes through occupation (urbanization) and development, but more through the lens of anthropogenic geomorphology or "anthropogeomorphology" (i.e., the intersection of geomorphology, and what Coates, 1984 called *anthropogene*—the human-created landscape, or city). Although most of the research in this volume occurs during the perceived Anthropocene epoch, the focus centers on the built environment, and particularly as it applies to land clearance, conservation

Urban Geomorphology. http://dx.doi.org/10.1016/B978-0-12-811951-8.00001-1

issues, pollution, decay and erosion, urban climate, and anthropogenic climate change, and more. These topics shed more light on the human transformation of natural landscapes and the environmental impacts and geomorphological hazards that landscape change can encompass for cities.

Although the topic of urban geomorphology occurs in numerous articles, it is not always noted as such, and no (known) focused compendium yet exists to address the topic specifically. This volume rectifies the gap in knowledge, bringing together specialists from around the world who conduct groundbreaking research in urban geomorphology, showcasing and highlighting current research trends and directions in this neglected, but important, area of study. Overall, the volume focuses on the built environment as the specific location of concentrated human impacts and change and not just in large cities, like metropolises or megalopolises, but smaller urbanized areas too. It takes a cross-disciplinary approach that is international in scope, highlighting case studies from around the globe. The volume further contributes to developing a "human geomorphology"—*anthropogeomorphology*—that Coates (1976) and Cooke (1976) envisioned, where people are considered agents of environmental history and landscape change.

For the researcher wanting to approach landscape from a holistic human-environmental perspective, this volume can serve as a port of first call to assess the diversity involved in urban geomorphology and anthropogeomorphological studies. It is particularly well-suited to mature audiences of researchers, from graduate level students to professionals, although the content is also accessible and useful for advanced undergraduate level courses/students. Its interdisciplinary approach also appeals to audiences from a range of disciplines and professions, such as conservationists and land-use planners as well as architects and developers. This volume's research not only informs research across disciplines, but also encompasses varied spatial-temporal frameworks, presenting a diverse set of approaches and potential solutions to human impacts and geomorphological hazards within urban landscapes.

Specifically, this volume uses five overarching sections focused on urban geomorphology, each containing case studies centered on a specific urban region. Some of these are rather well-studied cities, such as Johannesburg (South Africa), Phoenix (Arizona, USA), and Toronto (Canada), while others are more broad-reaching in both location and scope: the Niger River basin (Africa), Poland, and airports around the world. Regardless of site setting and topic, however, each chapter retains a focus on the urban environment and how different geomorphic agents interact with that locale—its anthropogeomorphology.

Beginning with ancient and historical geomorphology, the first section begins with a discussion of changes in the sometimes rapidly developing Mediterranean city of Palma (on the island of Mallorca) since ancient times. Built on an alluvial complex, as many cities were/are, the nearby river and estuary provided a natural harbor. As populations increased—and without regard for potential future hazards—the city expanded, forcing anthropogenic changes to the river, harbor, and alluvial cover itself. Petrus et al. (2018) expound upon the historicity of those events, drawing conclusions from archaeological evidence and archival records, connecting their on-the-ground findings with modern 3-D renderings. In the next chapter, Górska-Zabielska and Zabielski (2018) outline the potential of abiotic tourism in Pruszków, Poland, focusing specifically on the lack of awareness among municipalities when it comes to geotourism. Although the city offers several museums and other tourist attractions, the rich geodiversity has not yet been included in that economic sector. They argue that, if done responsibly, georesources too could be useful in promoting tourism. Then, Zwoliński et al. (2018) highlight the Polish lowland city of Poznań—situated currently and historically along the banks of the Warta River—and its modification of the landscape over time. Their aim rests in highlighting the area's complex geological and morphological characteristics—created primarily during the most recent glaciation—and discussing the city's geomorphic changes through the ages.

Moving forward in time, the next section centers on current anthropogeomorphology and, more specifically, addressing the effects that urban development can have on stream and steam biota as well as the influence that airports can have on changing the landscape. In the former case, MacDuff and Bauer (2018) take the reader to British Columbia (Canada), where, due to intensive engineering projects in the 1950s, salmon were removed from the Okanagan basin. Although their study area around the city of Vernon has also been modified, they use field data and hydraulic model simulations to demonstrate the viability of salmon reintroduction to the basin. Taking on a broader topic, Pijet-Migoń and Migoń (2018) utilize examples from around the world to showcase how airports modify geomorphic landscapes. They review several instances where swamps get dredged, wetlands altered, ground leveled, and even artificial islands are built, all to increase airport land area and keep up with the growing demand for air travel, even if such modifications are not necessarily visually impressive.

The third section links urban expansion and the overarching landscape by examining the role that landforms can play in mitigating and/or exacerbating geomorphic change. For example, Machado and Rodrigues (2018) utilize satellite imagery to identify "technogenic deposits" (e.g., improper disposal of household, industrial, and civil construction waste), and then assess those areas for potential soil contamination. Their assessment sheds light on a growing worldwide problem that increases in tandem with population growth. Following a related thread, Knight (2018) discusses Johannesburg's (South Africa) changing physical landscape as it pertains to the city's rich mining history and apartheid movement. As extractive processes increased, various features were restructured—from waterways to mountains—to make room for a burgeoning populace (exacerbated by apartheid) that resulted in irregular sprawl and development as well as landscape modification. Knight showcases several examples of how the city has used greening and memory to construct the modern landscape and how these efforts have modified the physical landscape. Extending Knight's examples, in the final chapter of this section, Espinosa et al. (2018) offer new solutions to urban design that encompasses the physical landscape. Using the densely populated city of Concepción (Chile), they examine a highly modified riverine environment that is known to be flood-prone since modern urban development began in the latter part of the 20th century. Instead of trying to modify the landscape to fit the geomorphology (or vice versa) as many modern efforts do, Espinosa et al. argue for combining them, incorporating the physical landscape into the design process, and showcase one such model.

Building on these case studies, the next section tackles a topic of interest to many geomorphologists: hazards. While sole volumes exist on the topic, this section presents highly focused examples of *potential* hazards in cities as related specifically to their geomorphology. Hazards, we know, only become such when people are involved—otherwise, the so-called hazard represents a natural event. Still, often people influence, or even create the (potential) occurrence of hazards, sometimes through a lack of knowledge about the landscape and other times by political means. In any case, the effect of hazards on urban environments, especially when it comes to reading the landscape for signs of past hazard activity, should not be overlooked. The fourth section addresses this topic with three interesting case studies. In the first instance, Jeong et al. (2018) use the Sonoran Desert city of Phoenix (Arizona, USA) to highlight the impact that understanding paleohazards can have on planning. As one of the nation's fastest growing urban areas, and hosting a sparsely vegetated desert landscape, the Phoenix area allows for studying/learning from ancient geomorphological hazards, such as large rockfall and landslide events, paleofloods, and debris flows.

Taking the reader to urban areas in "mountain towns" of Colorado's (USA) San Juan Range, Gamache et al. (2018) discuss the intersection of the critical zone and mining in the region, both past and present. Their review discusses the anthropogenic building and landform modification related to mining towns and bedroom communities (e.g., seasonally populated towns) and landform change across different environments—from glacial and periglacial to fluvial and mass-wasting areas—to predict the suitability of more permanent urban settlements in the San Juan Mountains. Keeping the mountain theme, Fort et al. (2018) focus on Nepal's second largest city, Pokhara, located just south of the Annapurna Range (elevations above 8000 m). While Pokhara rests on a broad plain, it remains surrounded by an incredible geomorphology, including lakes, caves, gorges, and scenic glaciated mountains, making it a highly touristed area. Its beautiful landscape, however, belies an ominous geology. Influenced by Himalayan tectonics as well as monsoonal climate events, hazards abound in the region and Fort et al. (2018) discuss these in detail, noting the devastating events that pose a serious threat to Pokhara's future economic well-being.

The final section centers on the often under-appreciated subfield of stone decay (weathering). This seemingly simple process triggers the beginning of landscape evolution and change—without rocks decaying, mountains would never change, valleys would remain the same, and rivers would stay their course. To this end, the fifth section offers a look into stone decay patterns and assessment in North America and Africa. Here, Onafeso and Olusola (2018) discuss and gather evidence of ongoing decay of ancient structures in the warm and wet environment of the Niger River basin, finding increasing rates of algal growth among Neolithic archaeological sites as atmospheric pollution increases. Additionally, they incorporate their evidence into a GIS database and offer further insight into the decay trends and patterns of the region. Then, Allen et al. (2018) offer a case study in an assessment technique used previously on an ancient urban area (Petra, Jordan) to evaluate the basic geologic stability of a historic building in Denver, Colorado (USA). As the first study using the Cultural Stone Stability Index (modeled after the successful Rock Art Stability Index; Dorn et al., 2008), this chapter offers the geomorphologist a way to quickly and efficiently quantify building decay noninvasively. In the volume's last chapter, Thornbush (2018) utilizes photogeomorpholgy to assess the decay of sidewalks in downtown Toronto (Canada). Specifically, she uses integrated digital photography and image processing in an outdoor setting to gain a 3-D perspective of sidewalk pavement decay. Her findings, based on quantitative photography, represent the first use of this technique on a horizontal urban surface.

As the case studies demonstrate, this volume takes the approach of developing anthropogeomorphology as a complement to the focus on physical landscapes (and physical geomorphology) that exists in geomorphology as a discipline at large. More specifically, it remains focused on the urban environment explicitly, rather than a particular people or landform process. This offers an opportunity to showcase recent research on a breadth of topics that have emerged on human impacts converging with urbanization and development of not just simply human-made creations, human influences, or human activity, but the effects that these have on landforms specifically, and vice versa—as landforms and their processes can also influence people in several ways. In the end, although the volume does include these concepts, they represent only pieces of a larger whole, and within the context of urban environments exclusively, because more than half of the world's population now lives in urban areas, with that factor increasing to nearly two-thirds by 2030 (United Nations, 2016). Regardless of locale and scale, however, the topics covered in this volume focus on cities as landscapes that have been severely and perhaps even irrevocably altered by people.

References

Allen, C.D., Ester, S., Groom, K.M., Schubert, R., Hagele, C., Olof, D., James, M., 2018. A geologic assessment of historic Saint Elizabeth of Hungary Church using the Cultural Stone Stability Index, Denver, Colorado. In: Thornbush, M.J., Allen, C.D. (Eds.), Urban Geomorphology: Landforms and Processes in Cities. Elsevier, San Diego, CA, Chapter 14.

Coates, D.R., 1976. Urban Geomorphology. Geological Society of America, Boulder, CO.

Coates, D.R., 1984. Urban geomorphology. Applied Geology. Springer, New York, pp. 601–605.

Cooke, R., 1976. Urban Geomorphology. Geogr. J. 142, 59–65.

Dorn, R.I., Whitley, D.S., Cerveny, N.V., Gordon, S.J., Allen, C.D., Gutbrod, E., 2008. The Rock Art Stability Index: a new strategy for maximizing the sustainability of rock art as a heritage resource. Herit. Manage. 1, 37–70.

Espinosa, P., Horacio, J., Ollero, A., de Meulder, B., Jaque, E., Muñoz, M.D., 2018. When urban design meets fluvial geomorphology: a case study in Chile. In: Thornbush, M.J., Allen, C.D. (Eds.), Urban Geomorphology: Landforms and Processes in Cities. Elsevier, San Diego, CA, Chapter 9.

Fort, M., Adhikhari, B.R., 2018. Pokhara (central Nepal): a dramatic yet geomorphologically active environment vs. a dynamic, rapidly developing city. In: Thornbush, M.J., Allen, C.D. (Eds.), Urban Geomorphology: Landforms and Processes in Cities. Elsevier, San Diego, CA, Chapter 12.

Gamache, K., Giardino, J.R., Zhao, P., Owens, R.H., 2018. Bivouacs of the Anthropocene: urbanization, landforms and hazards in mountainous regions. In: Thornbush, M.J., Allen, C.D. (Eds.), Urban Geomorphology: Landforms and Processes in Cities. Elsevier, San Diego, CA, Chapter 11.

Górska-Zabielska, M., Zabielski, R., 2018. Geotourism development in an urban area. Based on the local geological heritage (Pruszków, central Mazovia, Poland). In: Thornbush, M.J., Allen, C.D. (Eds.), Urban Geomorphology: Landforms and Processes in Cities. Elsevier, San Diego, CA, Chapter 3.

Goudie, A.S., Viles, H.A., 2016. Geomorphology in the Anthropocene. Cambridge University Press, Cambridge, UK.

Gupta, A., 1987. Urban geomorphology in the humid tropics: the Singapore case. International Geomorphology 1986. Proc. 1st conference 1, 303–317.

Jeong, A.S.Y., Walker, I.J., Dorn, R.I., 2018. Urban geomorphology of an arid city: case study of Phoenix, Arizona. In: Thornbush, M.J., Allen, C.D. (Eds.), Urban Geomorphology: Landforms and Processes in Cities. Elsevier, San Diego, CA, Chapter 10.

Knight, J., 2018. Transforming the physical geography of a city: an example of Johannesburg, South Africa. In: Thornbush, M.J., Allen, C.D. (Eds.), Urban Geomorphology: Landforms and Processes in Cities. Elsevier, San Diego, CA, Chapter 8.

MacDuff, A., Bauer, B.O., 2018. Urban stream geomorphology and salmon repatriation in Lower Vernon Creek, British Columbia (Canada). In: Thornbush, M.J., Allen, C.D. (Eds.), Urban Geomorphology: Landforms, Processes in Cities. Elsevier, San Diego, CA, Chapter 5.

Machado, C.A., Rodrigues, S.C., 2018. Environmental contamination by technogenic deposits in the urban area of Araguaína, Brazil. In: Thornbush, M.J., Allen, C.D. (Eds.), Urban Geomorphology: Landforms and Processes in Cities. Elsevier, San Diego, CA, Chapter 7.

Onafeso, O., Olusola, A., 2018. Urban stone decay and sustainable built environment in the Niger River basin. In: Thornbush, M.J., Allen, C.D. (Eds.), Urban Geomorphology: Landforms and Processes in Cities. Elsevier, San Diego, CA, Chapter 13.

Petrus, J.M., Ruiz, M., Estrany, J., 2018. Interactions among geomorphology and urban evolution since the antiquity in a Mediterranean city. In: Thornbush, M.J., Allen, C.D. (Eds.), Urban Geomorphology: Landforms and Processes in Cities. Elsevier, San Diego, CA, Chapter 2.

Pijet-Migoń, E., Migoń, P., 2018. Landform change due to airport building. In: Thornbush, M.J., Allen, C.D. (Eds.), Urban Geomorphology: Landforms and Processes in Cities. Elsevier, San Diego, CA, Chapter 6.

Thornbush, M., 2015. Geography, urban geomorphology and sustainability. Area 47, 350–353.

Thornbush, M.J., 2018. Photographic technique used in a photometric approach to assess the weathering of pavement slabs in Toronto (Ontario, Canada). In: Thornbush, M.J., Allen, C.D. (Eds.), Urban Geomorphology: Landforms and Processes in Cities. Elsevier, San Diego, CA, Chapter 15.

United Nations, 2016. The World's Cities in 2016—Data Booklet (ST/ESA/SER.A/392). Department of Economic and Social Affairs, Population Division. Available from: http://www.un.org/en/development/desa/population/publications/pdf/urbanization/the_worlds_cities_in_2016_data_booklet.pdf.

Xizhi, D., 1988. A brief discussion of urban geomorphology. J. Mount. Res. 6 (2), 65–72.

Zwoliński, Z., Hildebrandt-Radke, I., Mazurek, M., Makohonienko, M., 2018. Anthropogeomorphological metamorphosis of urban area on post-glacial landscape: case study Poznań City. In: Thornbush, M.J., Allen, C.D. (Eds.), Urban Geomorphology: Landforms and Processes in Cities. Elsevier, San Diego, CA, Chapter 4.

SECTION I

PALEOGEOMORPHOLOGY AND ARCHAEOGEOMORPHOLOGY

2 *Interactions between Geomorphology and Urban Evolution Since Neolithic Times in a Mediterranean City* 9

3 *Geotourism Development in an Urban Area Based on the Local Geological Heritage (Pruszków, Central Mazovia, Poland)* 37

4 *Anthropogeomorphological Metamorphosis of an Urban Area in the Postglacial Landscape: A Case Study of Poznań City* 55

CHAPTER

2

Interactions between Geomorphology and Urban Evolution Since Neolithic Times in a Mediterranean City

Joana M. Petrus, Maurici Ruiz**, Joan Estrany*,†*

*Department of Geography, University of the Balearic Islands, Palma, Mallorca, Spain; **GIS and Remote-Sensing Service, University of the Balearic Islands, Palma, Mallorca, Spain; †Institute of Agro-Environmental and Water Economy Research–INAGEA, University of the Balearic Islands, Palma, Spain

OUTLINE

2.1 Introduction 10

2.2 The Geography of Palma, a Mediterranean City 11

2.3 Urban Evolution and Geomorphological Processes Since the Talayotic Period (BC 3000 Present) 18

2.3.1 Talayotic Period (BC 3000–550) 18

2.3.2 Roman and Late Ancient Age (2nd Century BC to 6th Century AD) 24

2.3.3 Islamic Period (10–13th Centuries AD) 25

2.3.4 Late Middle Age (13–15th Centuries) 27

2.3.5 Modern Age (16–18th Centuries) 28

2.3.6 Contemporary Age (19–20th Centuries) 28

2.4 Land Use as the Crucial Change of Urban Geomorphology in the 20th Century 29

2.5 Concluding Remarks 31

References 32

Urban Geomorphology. http://dx.doi.org/10.1016/B978-0-12-811951-8.00002-3

2.1 INTRODUCTION

Human-environment relationships become more complex and intense as societies progress to a higher level of technological development that provides the possibility of further transforming their environment (Raudsepp-Hearne et al., 2010). The combined effect on the planet of human activity since the end of the 19th century to present—with more than 7 billion inhabitants—is obviously greater in quantitative terms than in earlier times when the population was 1 million inhabitants in BC 10 000, 50 million in the year BC 1000, and 200 million in 1 AD (Goudie, 2013). However, the ecological footprint for human activity has also been locally and regionally very intense in certain areas where anthropogenic influence has existed for thousands of years. Because of this, Crutzen and Stoermer (2000) proposed the use of the term Anthropocene for the current geological epoch based on major and still growing impacts of human activities on earth and atmosphere. Despite the scientific community's disagreement on when the Anthropocene began (Cearreta, 2015), "anthropogenesis" as a process can be traced back to the beginning of the Holocene (ca 11 500 years), when agriculture was stationary in the Mesopotamia region and then began to progressively extend to Europe around BC 6000 (Zalasiewicz et al., 2011).

Since their appearance, agriculture and livestock systems have transformed natural ecosystems into cultural landscapes that have been shaped and managed by human activities. In the Mediterranean, these transformations took place over long periods of time, during different stages in which deforestation and soil degradation led to erosion and greater aridity (Mazoyer and Roudart, 2006); and were also conditioned by the Mediterranean hydrologic cycle, with intensely seasonal characteristics of rainfall and a very strong coupling between climate and vegetation cover. This combination of features is also related to the occurrence of flash floods, another common feature in the region, affected by the closeness of mountains to the coastline, low vegetation cover, and low infiltration capacities of soils (Wainwright and Thornes, 2004). The uninterrupted history of plant cover reduction in the Mediterranean has enhanced the flood risk, causing an acceleration of sediment transfer from the headwaters of catchments to downstream deposition areas, such as alluvial fans or deltas, that enabled the creation of new areas suitable for cultivation.

Since the Neolithic period, distribution patterns of human settlements have been related to fluvial systems, with many often located on middle reaches as well as at river mouths. In the absence of a chronological study sequencing the appearance of human settlements along the Mediterranean, the establishment of a relationship with coastline variations also affected by climate oscillations during the Holocene—especially in the late Holocene—is problematic because only some evidence can be considered for specific periods and coastal sites (Delile et al., 2015; López Castro, 2016; Romero Recio, 1996). Carayon (2008) confirmed the existence of at least 183 Phoenician port enclaves that were well known as cities between the 2nd century BC and 1st century AD (e.g., Carthage, Troy, Tire, Sidon, Cartagena, Melilla, Tarifa, Ibiza, and Palermo). Nowadays, their vestiges are located inland, far from the coast, or have even altogether disappeared due to silting processes promoted by fluvial activity and modification of the coastline.

The occupation of coastal areas by ports or urban agglomerations is very frequent throughout the Mediterranean basin, as it is in other periods and societies (Flaux et al., 2017; Giaime et al., 2017; Morhange et al., 2003), which clearly establishes a relationship between the development of coastal urban areas—either cities or commercial colonies—and the occupation of estuaries, bays, and river mouths. The location of archaeological sites with respect to the current coastline, and absolute dating of nearby deposits, enables successive progradation and retrogradation stages along the coastline to be related to climatic oscillations or tectonic processes at the regional scale. Still, it is just as important to note that archaeological vestiges of coastal occupation correlated most often with the maximum sea level reached in the late Holocene (Laborel et al., 1994; Morhange et al., 2003; Morhange and Marriner, 2011).

The Phoenician, Greek, and Roman urban settlements during the first millennium on the Mediterranean coast followed a similar location pattern. Here, settlements tended to be located on promontories, headlands, or coastal peninsulas, since proximity to the coast and altitude guaranteed maritime control of defensible and safe sites. Settlements were also located at river mouths due to easy access to water supply and the protection of maritime traffic in estuaries. In addition, these ports facilitated inland access, where agricultural and extractive activities were initiated by the local population (Pounds, 1976). Mediterranean cities of Phoenician, Greek, or Roman origin, subsequently transformed into large cities, are examples of settlements consolidated over millennia due to their location, availability of resources, and historical evolution. Therefore, they remain excellent case studies for analysis of anthropogenic activity as an external modeling agent of natural systems and also represent an opportunity to gain a better understanding of the evolutionary, geomorphic, and ecological dynamics affecting human activity.

Palma, the capital of the island of Mallorca, the largest in the Balearic Archipelago (western Mediterranean), was founded by the Romans in BC 123 (García Riaza, 2003, p. 73). It follows the prototypical location pattern of Mediterranean ancient cities: elevated position above sea level (ASL) with good visibility, close to the coast and an estuary, and close to freshwater springs. In addition, its surrounding area is a sheltered low coastal plain eased maritime traffic and a wooded hinterland provided for natural resource extraction. The study of the urban transformation of Palma, its land use, and how these have shaped natural systems throughout the last two millennia, represents an opportunity to analyze human-environment interactions and the role that geomorphology has played historically in such Mediterranean coastal cities founded in Roman times.

2.2 THE GEOGRAPHY OF PALMA, A MEDITERRANEAN CITY

The Mediterranean Sea covers approximately 2.5 million km^2. The geological history of the region was the breakup of the supercontinent Pangea (250 Ma), generating the Thethys Sea, which is the ancestor of the present-day Mediterranean Sea. The Mediterranean is also the westernmost part of the Alpine-Himalayan orogenic belt that stretches from Spain to New Zealand (Mather, 2009, p. 5). The collision of Africa and Eurasia generated the Mediterranean Sea (45 Ma), and the Mediterranean's Quaternary geodynamics have generated an area that includes a complex mixture of plate subduction at varying ages and stages.

The Mediterranean Sea contains 165 inhabited isles that are >10 km^2, Mallorca being the seventh largest (Fig. 2.1A; 3640 km^2). Its geology and geomorphology (Silva et al., 2005, p. 1), are characterized by a basin-and-range topographical configuration that resulted from Late Miocene to Early Pleistocene faulting (Silva et al., 2005, p. 1-3). The mountain areas (i.e., Tramuntana Range, along the northwest coast, Central Ranges in the central part of the island,

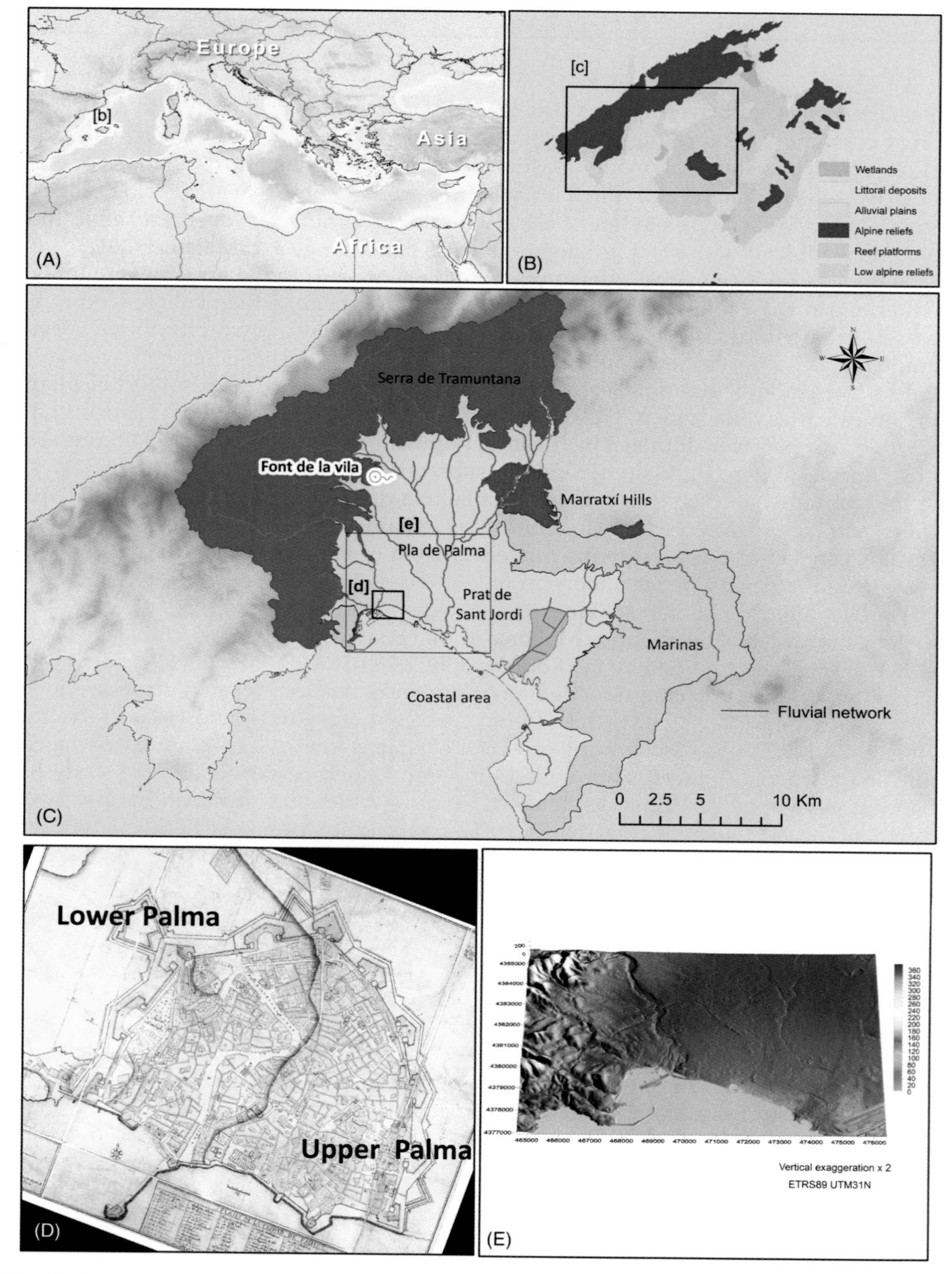

FIGURE 2.1 (A) Location of Mallorca Island within the Mediterranean Sea; (B) relief units of Mallorca Island defined by geology and geomorphology; (C) topography, fluvial network, and relief units within the Palma basin, including the location of *Sa Font de la Vila*; (D) map of the city of Palma, ca 1726—attributed to Gerónimo Canobes (Centro Geográfico del Ejército)—where the *Palma Alta* and *Palma Baixa* are separated by a scarp generated by a fault; (E) digital elevation model of the western side of Palma Bay, where the city was founded and has expanded during the last 2000 years (z values were exaggerated doubled, whilst x and y coordinates are in m within the UTM 31N grid zone).

and Llevant Ranges along the eastern coast; Fig. 2.1B) correspond to uplifted blocks of the Alpine fold belt. They present semihorst geometry and run from northeast to southwest. Late Miocene, Pliocene, and Pleistocene deposits overlap the folded Mesozoic to Middle Miocene rocks, constructing near-horizontal platforms around the ranges and filling down-dropped areas. Flat platform areas extend to the south (Marina de Llucmajor), in the center (Es Pla), and to the east (Marina de Santanyí) of the island (Jenkins et al., 1990).

The city of Palma is located on Mallorca's south coast on the Bay of Palma, which is the sea outlet of the Palma hydrographic basin (533 km^2). It is a semicircular depression that is open to the sea (south) through this plain by a 33-km coastline and delimited by the Tramuntana Range (northwest) and Miocene tabular reliefs (northeast-east) of the Marina de Llucmajor. Within this geological and geomorphological background, five relief units are described in the Palma basin (Fig. 2.1C):

1. Tramuntana Range. The alpine process of compression in a northwesterly direction gave rise to this range's successive folds and thrusts, aligned in a northeasterly/southwesterly direction and stacked toward the northwest. Lithological differences demonstrate that the cliffs and massifs are comprised of hard limestone rocks, while much softer materials have settled at the base, such as clays or calcarenites—materials characteristic of mountain slopes and valley bottoms. This alternation between hard and clayish materials is also important because it explains the emergence of water through abundant springs. In addition, fluvial systems make use of soft materials found on the base of the limestone massifs to generate wide longitudinal valleys through which the drainage network is organized.
2. The Marina is a reefal Upper Miocene tabular platform composed of calcarenites, calcisiltites, and terra-rossa postreef sediments, occupying the eastern slopes of the basin and the western part of Palma city attached to the Tramuntana Range. Streams have incised the Miocene platform since the Pleistocene through striking (Silva et al., 2005), forming canyons as an interdependent activity of tectonics, karst, and fluvial activity (Estrany and Grimalt, 2014), mainly in the Sant Jordi, Jueus, and Son Verí Rivers, located in the eastern part of the Palma basin.
3. Pla de Palma is a flat and subsided depression that occupies the central part of the basin, where the city was founded. Specifically, it is an alluvial plain constituted by a juxtaposition of Quaternary alluvial fans of low-gradient slope that are incised by several streams (Grimalt and Rodríguez-Perea, 1994).
4. Marratxí Hills. Formed by well-developed intervening antiform-like reliefs occupying the northeastern slopes of the Palma basin (Giménez, 2003).
5. The coastal area comprises the southern boundary of the hydrographic basin. The coastline alternates between sandy beaches and rocky coasts. The last foothills of the Tramuntana Range produce a west coast that is high and rocky, with some significant "*calas*," such as Cala Major or Portopí. Transition to the eastern coast is characterized by Pleistocene to Holocene alluvial and littoral deposits that appear alternately in the sector where Palma is located. The main streams of this sector are—from west to east—the Sa Riera, Na Bàrbara, and Gros Rivers, oriented in the downstream part from northeast to southwest, following the main fault strikes. At the mouths of these streams, estuaries or deltas that were present during the last two millennia have now disappeared due to urbanization.

On the eastern coast, sand beaches and a gentle rocky shore alternate with well-developed Pleistocene-to-Holocene littoral deposits that enclose an onshore wetland at the distal sector of the alluvial fans where subsidence processes dominate (Silva et al., 2005).

Further explanation of the Pla de Palma as a relief unit is required to better understand its historical urban evolution and interaction with geomorphic processes. As a combination of fluvial and neotectonic processes, three relief subunits can be described at the geological microscale in the Palma area (Fig. 2.1C–D):

1. Palma Alta (in Catalan Alta means Upper). A promontory approximately 13 m ASL mostly composed of alternating alluvial and littoral Plio-Quaternary deposits.
2. Palma Baixa (in Catalan Baixa means Lower). Sa Riera is the river adjacent to the Palma Alta. The stream and its estuary, with a large floodplain, fit within a fault in which the primitive harbor was also established. This fault is included in a system of normal NNE-SSW faults, with the main one being the Palma fault (Silva et al., 2005).
3. Prat de Sant Jordi. Located 3 km southeast of the city of Palma, it was a wetland on 50 m-thick Quaternary alluvial deposits. Sediment deposition and sea-level oscillations during the last two millennia as well as neotectonics (Silva et al., 2005) affected this wetland. Specifically, the highest global sea level of the past 110 000 years probably occurred during the Medieval Warm Period of 1100–1200 AD, when warm conditions similar to today's climate caused the sea level to rise 12–21 cm higher than its present level (Grinsted et al., 2010, p.486). In the 19th century, it was drained and divided into plots for agricultural and livestock farming, finally being modified intensely by urban expansion during the second half of the 20th century.

The hydrology of catchments in the Palma basin as shaped by Mediterranean climate is strongly influenced by the seasonal distribution of precipitation, catchment geology, vegetation type and extent, the geomorphology of the slope and channel systems, and anthropogenic modifications. The hydrological regime is ephemeral or intermittent, with flash floods occurring during intense rainfall events. Throughout Quaternary glacial periods, the coastal zone was displaced seaward onto the continental shelf, with the river cutting to new base levels (Rose and Meng, 1999). Meanwhile, during interglacial periods with high sea levels like current conditions, coastal valleys were drowned to form lagoons or estuaries. These lagoons subsequently began filling with sediment from the land and sea. This cut-and-fill cycle has been repeated many times and, consequently, most coastal valleys now contain remnants of previous interglacial deposits in addition to those deposited over the Holocene period (the last 10 000 years). The most recent phase of sedimentation was initiated near-simultaneously throughout the western Mediterranean region toward the end of the Postglacial Marine Transgression approximately 7000–8000 years ago (Fleming et al., 1998). Since the Postglacial Marine Transgression, infilling rates have varied depending on the sediment load carried by the rivers and the hydrodynamic conditions that transport sediment into fluvial outlets to the sea.

Paleohydrological studies show that most characteristic drainage catchments in the southern regions of the western Mediterranean presented a combination of distinctive morphometric and anthropogenic elements, such as steep slopes, thin and permeable soils, scarce vegetation cover, and significant human transformation (Machado et al., 2011). These parameters—combined with scarce and irregular precipitation—converted ephemeral regimes into fluvial systems characterized by flash floods and variable erosion/sedimentation processes (Calle et al., 2017; Camarassa and Segura, 2001 ; Ruiz et al., 2014). These morphometric parameters describe the physical geographic characteristics of fluvial systems, helping to understand the underlying geomorphic influences involved in landscape evolution. Thus, a series of morphometric parameters (Fig. 2.2) obtained for Palma's catchments provide an essential description of the physical and topographic features of waterways as a simple diagnosis of the natural sensitivity of individual catchments (Haines et al., 2006). Previously, the fluvial drainage network was digitally obtained through ArcGIS's 3D Analyst, Hydrology, and Spatial Analysis toolsets (ESRI, 2017). However, the massive direct transformation of many channels required a more accurate analysis of the fluvial drainage network arrangement. Therefore, photographs and topographic maps (contour lines) had to be interpreted to define the final fluvial drainage network. The morphometric analysis of Palma basin catchments showed wide variability in their parameters, ranging from 1.6 to 205.5 km^2 in the catchment area, with a range from up to 90 to 1065 m in altitude, and 1.3–15.3% in gradient slope (Fig. 2.2). The catchments can be classified by their locations on the coast. The eastern coast has six small catchments with an average surface area of 4.0 ±1.7 km^2 that cover 23.8 km^2 (4.5% of the Palma basin). Three catchments constitute the central coast with an average surface area of 100.8 ±52.7 km^2 and cover 302.3 km^2 (56.7% of the Palma basin). The eastern coast contains three catchments with an average surface area of 69.0 ±42.1 km^2 that cover 207.9 km^2 (38.8% of the Palma basin).

Torrentiality and clinometric variables were the parameters most closely related to geological settings. The western catchments are located in abrupt alpine reliefs characterized by the highest values of torrentiality (average 25 ±5) and clinometry (average 10.1 ±1.9%) due to the presence of impervious materials in the lower parts (i.e., Dogger and Oligocene marls); and slope gradients >30% in the upper parts, where Lias dolomites and limestone predominate. The central coast contains the largest catchments of the Palma basin, with very steep headwaters in the Tramuntana Range and lower gradient slopes in the middle and downstream parts over the alluvial fans, where torrentiality (average 14 ±4) and clinometry (3.7 ±0.2%) are subsequently moderate. However, the significant Quaternary depositional process facilitates the transport capacity of these fluvial systems, whose headwater parts are characterized by high clinometry and connectivity between the slopes and channels. On the western coast, rivers incised into the carbonate Miocene platform, forming canyons in the upper and middle parts of their catchments. The hardness of calcarenites and tabular reliefs ensures that torrentiality (2 ±1) and clinometry (1.8 ±0.4) are very low over these pervious materials of the Upper Miocene carbonate platform.

Palma basin's climate is classified on the Emberger climate scale (Guijarro, 1986) as Mediterranean cool hyperhumid at the Tramuntana Range's headwaters and hot semiarid along the coastal area where Palma is located. The average annual rainfall is 710 mm, exceeding 1000 mm at the headwaters of Tramuntana Range and decreasing to 450 mm on the coast, heavily influenced by the orographic effect. The average annual temperature is 18°C in Palma-Portopí (0 m ASL), although at the headwaters it does not exceed 12°C. In the upper part of the Palma basin, oaks (*Quercion ilicis* species) predominate on north-facing aspects, and scrubland or garrigue (*Oleo ceratonion* species) on south-facing ones. The Pla de Palma is characterized by agricultural use, together with extensive urban use areas, while intensive urban uses dominate in the lower part to its mouth.

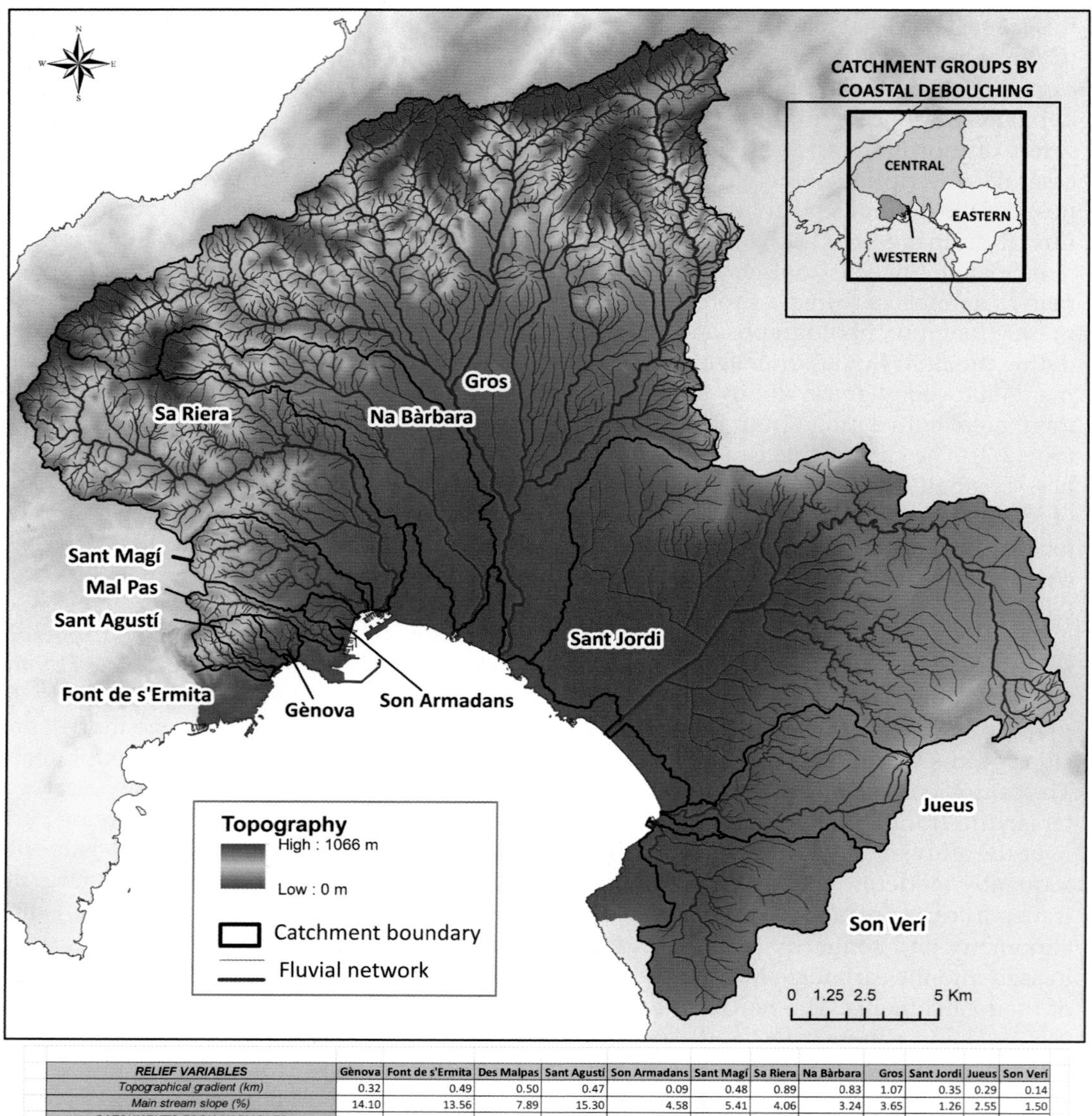

RELIEF VARIABLES	Gènova	Font de s'Ermita	Des Malpas	Sant Agustí	Son Armadans	Sant Magí	Sa Riera	Na Bàrbara	Gros	Sant Jordi	Jueus	Son Verí
Topographical gradient (km)	0.32	0.49	0.50	0.47	0.09	0.48	0.89	0.83	1.07	0.35	0.29	0.14
Main stream slope (%)	14.10	13.56	7.89	15.30	4.58	5.41	4.06	3.24	3.65	1.26	2.55	1.50
CATCHMENT'S FORM VARIABLES												
Area (km^2)	1.56	2.14	3.98	2.60	1.56	11.96	58.25	38.57	205.50	153.18	24.12	29.79
Perimeter (km)	6.02	8.58	14.85	8.84	6.26	17.99	45.42	43.67	99.53	70.75	27.22	33.78
Main stream length (km)	2.27	3.58	6.34	3.07	1.97	8.78	21.91	25.62	29.19	27.44	11.20	9.32
Elongation Ratio	0.44	0.38	0.40	0.54	0.50	0.86	1.69	0.96	4.47	3.54	1.37	2.03
Articulation Index	5.20	9.43	15.88	5.72	3.91	10.15	12.99	26.82	6.54	7.75	8.19	4.60
FLUVIAL INTENSITY VARIABLES												
Drainage density	4.15	4.41	4.13	4.02	2.64	2.76	3.08	2.45	3.37	1.54	2.02	1.24
Torrenciality	29.25	41.23	32.12	30.95	6.76	12.22	14.84	5.60	20.56	2.12	3.09	1.21

FIGURE 2.2 **Map of Palma basin showing the topography, studied rivers, and their related catchments.** Inset: (Map) the catchment groups based on the coastal area where it emerges; and (table) the main geometric and morphometric parameters used to examine the influence of catchments on the urban geomorphology.

Among Mediterranean islands, Mallorca is fourth in terms of population (861 430 inhabitants), after Sicily, Sardinia, and Cyprus. Palma, with 402 949 inhabitants (2016), is the Mediterranean's second most populated insular city after Palermo (673 735 inhabitants; ISTAT, 2017). The evolution of land use over the last 60 years shows a divergent pattern caused by important socioeconomic changes. On the one hand, farmland in marginal areas has been gradually abandoned, particularly in the Tramuntana Range, leading to a process of afforestation. On the other hand, there has been an expansion of urban uses in the middle and lower parts of the Pla de Palma. The long-term analysis of the urban evolution of Palma (i.e., 2000 years) requires the examination of two key variables: population and the area occupied by the city. In this long period, the population followed an upward trend that has become exponential since 1950 (Fig. 2.3). Historical documents allow the Roman colony to be calculated at ca 3000 people. However, from the beginning of the Christian era until practically the Muslim conquest, there are no reliable demographic data (Mas Florit, 2013, p. 42 and 216; Rosselló Bordoy, 1973, p. 86). Similar uncertainties are derived from the surface area, although archaeological findings established the first Roman settlement (1 ha) and the first fortification (7 ha; Alcántara Peña, 1882).

From the 2nd century BC to the beginning of the 20th century, the urban geography of Palma is linked to the succession of three main walled enclosures: (1) Roman (BC 1); (2) Islamic

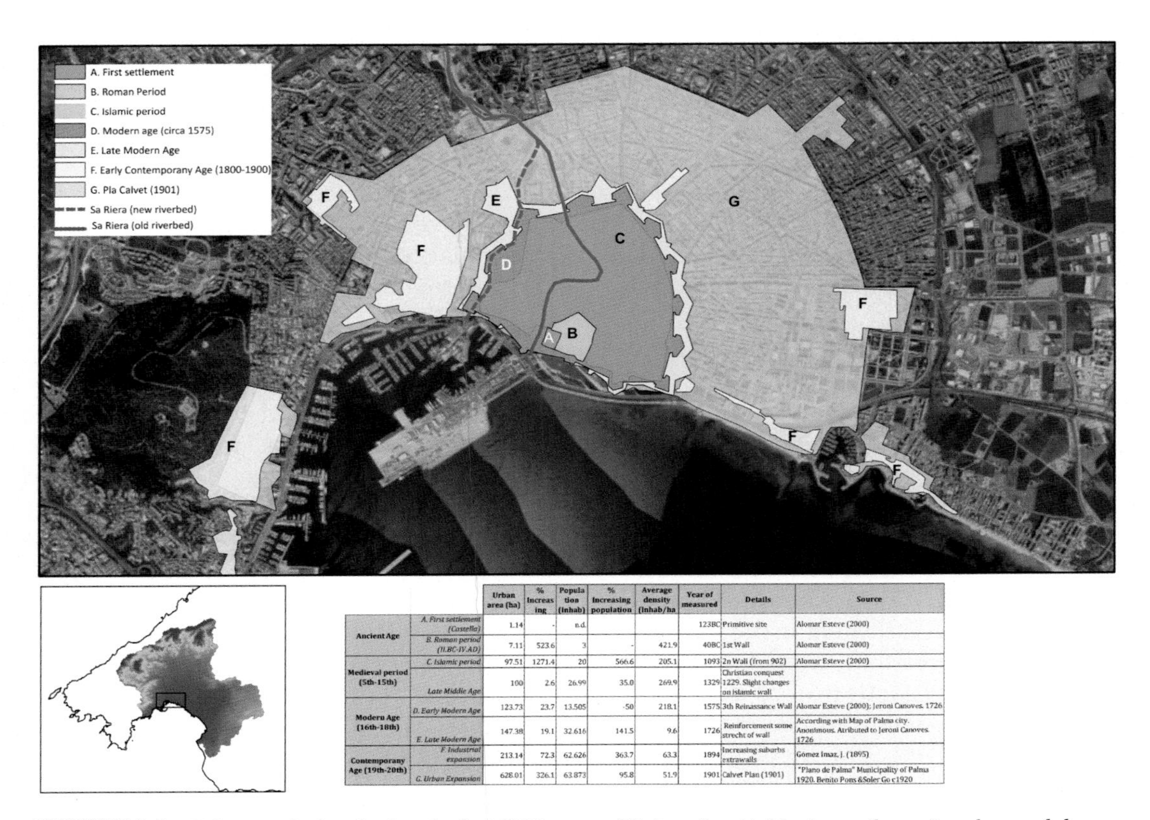

		Urban area (ha)	% Increasing	Population (Inhab)	% Increasing population	Average density (Inhab/ha	Year of measured	Details	Source
Ancient Age	*A. First settlement (Castella)*	1.14	-	n.d.			123BC	Primitive site	Alomar Esteve (2000)
	B. Roman period (II.BC-IV.AD)	7.11	523.6	3	-	421.9	40BC	1st Wall	Alomar Esteve (2000)
Medieval period (5th-15th)	*C. Islamic period*	97.51	1271.4	20	566.6	205.1	1093	2n Wall (from 902)	Alomar Esteve (2000)
	Late Middle Age	100	2.6	26.99	35.0	269.9	1329	Christian conquest 1229. Slight changes on islamic wall	
Moderu Age (16th-18th)	*D. Early Modern Age*	123.73	23.7	13.505	-50	218.1	1575	3th Reinassance Wall	Alomar Esteve (2000); Jeroni Canoves. 1726
	E. Late Modern Age	147.38	19.1	32.616	141.5	9.6	1726	Reinforcement some strecht of wall	According with Map of Palma city. Anonimous. Atributed to Jeroni Canoves. 1726
Contemporany Age (19th-20th)	*F. Industrial expansion*	213.14	72.3	62.626	363.7	63.3	1894	Increasing suburbs extrawalls	Gómez Imaz, J. (1895)
	G. Urban Expansion	628.01	326.1	63.873	95.8	51.9	1901	Calvet Plan (1901)	"Plano de Palma" Municipality of Palma 1920. Benito Pons &Soler Go c1920

FIGURE 2.3 **Urban evolution during the last 2000 years of Palma.** Inset table shows the main urban and demographic parameters during each historical stage.

(late 11th century); and (3) Renaissance (1575)—although some authors identify up to five (García Delgado, 2000; Tous Meliá, 2004). Between the Roman and Islamic enclosures, Palma increased its surface area exponentially, acquiring toward the end of the 11th century before the Pisano-Genoese invasion a surface area of around 100 ha, at a time when Barcelona, for example, did not cover more than 42 ha (Tous Meliá, 2004). At the end of the Islamic period, the population was ca 25 000 inhabitants, a figure that would not occur again until the 17th century because of recurrent epidemics, poor harvests, and inefficient distribution of land (Juan Vidal, 2007, p. 24). As a result, between the 14 and 16th centuries, the population reached an estimated minimum of 8650 inhabitants in 1531, during the well-known Medieval demographic stagnation (Juan Vidal, 1977, p. 57). In the modern age, the population increased constantly due to suburban development outside the wall, reaching 13 505 inhabitants (Sevillano Colom, 1974, p. 250). Figure 2.3 reveals that, during the 19th century, total population and urban extension grew exponentially due to the change from an agrarian to industrial economy and to improved hygienic conditions that reduced epidemic mortality (Salvà, 1981, p. 84).

2.3 URBAN EVOLUTION AND GEOMORPHOLOGICAL PROCESSES SINCE THE TALAYOTIC PERIOD (BC 3000–PRESENT)

It is essential to adopt a holistic vision that integrates human activity into a long-term perspective for understanding how human societies adapted to environmental changes in the past (Sivapalan et al., 2016). By analyzing the urban evolution of Palma and its relationship with the environment, for example, it is possible to establish a sequence of six historical stages based on: (1) socioeconomic organization; (2) urban interventions; and (3) significant geomorphological processes (Table 2.1). These stages have in common: (1) the continuity in the location of the city that has expanded from its initial core to its present size; and (2) its special relationship with water as a resource and risk throughout its history.

2.3.1 Talayotic Period (BC 3000–550)

The human presence of Mallorca prior to the Roman colonization did not consolidate into any type of permanent settlements that could be considered urban (Riera Frau, 1998). Accordingly, the areas with the highest concentration of population during the pre-Talayotic and Talayotic periods (BC 3000–550) in the Palma basin are in archaeological sites located at the headwaters of the Sa Riera, Coanegra, and Na Bàrbara catchments (Gili Suriñach, 1995). The number of settlements in the middle reaches is considerably lower, except for the Sant Jordi catchment (Fig. 2.2). Urban transformation in the later millennia does not verify the existence of a megalithic settlement in the primitive urban nucleus of Palma.

Other cultures that expanded from the eastern to western Mediterranean were present in the Balearic Islands toward BC 1000. Thus, at the end of the Talayotic period, the Phoenician presence—mainly for commercial reasons—is noted in archaeological records at coastal strategic enclaves of Mallorca that were also occupied by the Talayotic community for funerary use. These enclaves are currently small islands near the coast of Palma (Fig. 2.1C). Here, archaeological evidence has been useful to infer changes in the coastline related to erosion processes and/or variations in sea level. Thus, the area surrounding these islets contain

TABLE 2.1 Main Urban Actions and Environmental Events Occurring during a Sequence of Six Historical Stages Since the Neolithic (3000 BC) and Their Relationship with Key Historical Dates, Demography, and Urban Expansion Evolution

BC/AD	Stage	Historical period	Year	Main urban actions	Environmental events	Key historical dates	Urban population	Urban extension (ha)
AD	VI	Contemporary age	2000				402 949 (2016)	628
				2006 Urban plan modification	2007–2008 Severe floods in Sa Riera, Na Barbara and Gros		375 776 (2005)	
							333 485 (2000)	
			1900	1985 Urban Development Plan	1974 Severe floods in Jueus and Son Veri rivers	1983 First Statute of Autonomy of the Balearic Islands	297 042 (1975)	
				1973 Urban Plan by Ribas-Piera	1962 Extreme floods affecting all rivers in catchment area (Sa Riera, Na Barbara, Gros, Sant Magi, Mal Pas, Jueus, Son yen)	1962 Air passengers over 1 million	157 131 (1960)	
				1954 Extension of Palma airport				
				1941 Urban plan by G. Alomar	1942 Extreme flood in Sa Riera, Na Barbara, Gros	1937 Industrial railway to Genova neighborhood	117 188 (1940)	
						1936 Spanish Civil War	78 363 (1920)	
				1937 Limestone quarry Genova	1934 Extreme floods in Sa Riera, Sant Magi, Aigua Dolca and Mal Pas	1931 Urban rail line to harbor		
				1902 Royal Order to demolish the walls	1933 Extreme floods in Sa Riera, Na Barbara, Gros	1918 Spanish flu epidemic	68 416 (1910)	
				1901 Urban expansion plan by B. Calvet	1902 Floods		63 837 (1900)	
			1800	1873 Demolition of stretch of wall	1887 Landslide	1872 Beginning of the railroad	62 626 (1897)	213
				Opening of shipyard		1870 Yellow fever epidemic	54 421 (1875)	

(Continued)

TABLE 2.1 Main Urban Actions and Environmental Events Occurring during a Sequence of Six Historical Stages since the Neolithic (3000 BC) and Their Relationship with Key Historical Dates, Demography, and Urban Expansion Evolution (*cont.*)

BC/AD	Stage	Historical period	Year	Main urban actions	Environmental events	Key historical dates	Urban population	Urban extension (ha)
				1869 Expansion Santa Catalina neighborhood	1851 Earthquake/landslide	1865 Cholera epidemic		
				1856 Royal Order bans construction nearer than 400 m to wall	1850 Severe floods			
				1850 Improvement of diverted riverbed	1850 End of Little Ice Age			
				1835–1868 Drying of Prat de Sant Jordi wetlands	1842 Floods	1835 MendizAbal disentailment	51 871 (1857)	
					1827, 1835 Landslides	1820-1821 Yellow fever epidemic	41 094 (1838)	
						1812 Public lighting (oil system)		
					1808 Floods	1801 Introduction of maize crop		
	V	Modern age	1700	1778 Began stone paving of streets	1763 Severe floods	1788 First discussion about public lighting (RESFC)	31 437 (1797)	
				1767 Second attempt at drying Prat de Sant Jordi wetlands		1785 Mining development plans (RESFC)		
				1715 Reinforcement of maritime stretch of wall	1750 Extreme floods. Overflow on the former river-bed	1778 Royal Economic Society of Friends of the Country (RESFC)	32 616 (1746)	147
					1734 Severe floods			
					1733 1783 Landslides			
			1600	1656 Quarantine zone	1635 Extreme floods. Overflow on the former river-bed	1652 Black Death epidemic	25 988 (1635)	

BC/AD	Stage	Historical period	Year	Main urban actions	Environmental events	Key historical dates	Urban population	Urban extension (ha)
				1613 Approval of Sa Riera river-bed diversion	1618 Extreme floods. Overflow on the former Sa Riera river-bed	1606 Clashes between Canamunt and Canavall bandit groups		
			1500	1575 End of last fortification	1501–1502; 1507; 1516 severe droughts	1521 Rebellion of the Guilds	13 505 (1573)	
				1560 Start of last fortification			9235 (1566)	
				Third wall (Renaissance)			8650 (1531)	129
							10 445 (1524)	
	IV	Late Middle Age	1400	Expansion of crop fields	1495 Heavy snowfalls	1492 Discovery of America		
					1491–1493 Severe drought			
					1490 Extreme floods			
					1469/1487 Cold waves	1475 Black Death epidemic	15 785 (1475)	
				Urban demographic loss	1454–1455–1456; 1470; 1473; 1478; 1484–1485; severe droughts	Late Middle Ages crisis	10 275 (1444)	
					1408; 1444 Extreme floods	1465 Black Death epidemic	13 990 (1421)	
				Great commercial movement. Palma attracts merchants from all over the Mediterranean	1403 Extreme, catastrophic floods	1453 Surrender of Constantinople		
						1450 Civil war. Foreign rebellion		
						1439 Vallseca's nautical map		

(Continued)

TABLE 2.1 Main Urban Actions and Environmental Events Occurring during a Sequence of Six Historical Stages since the Neolithic (3000 BC) and their Relationship with Key Historical Dates, Demography, and Urban Expansion Evolution (*cont.*)

BC/AD	Stage	Historical period	Year	Main urban actions	Environmental events	Key historical dates	Urban population	Urban extension (ha)
			1300	1321 King Sanc's first attempt to dry Prat de Sant Jordi wetlands		1383 Black Death epidemic		
						1375 Mallorcan School of Cartography. Catalan Atlas		
						1375 Black Death epidemic		
						1349 End of Kingdom of Mallorca		
				1306 Began La Seu Cathedral building		1348 Black Death epidemic	20 620 (1343)	
				Acquisition of land for new parish church of Sant Nicolau (1302)	Beginning of Little Ice Age		26 990 (1329)	
			1200	1248 Division into four Parishes.		1276 Beginning of Kingdom of Mallorca		
				Construction of public hydraulic system		1229 Christian conquest		
	III	Islamic period	1100	New urban layout	Medieval Warm Period	1115 Pisan raid		100
			1000	Second wall (Islamic). Sa Riera River is within the wall				
			900	Madina Mayurqa		902 Arab conquest		
	II	Late ancient age Dark Ages	800					
			700					

			600	Possible Byzantine reinforcement of the wall		534 Byzantine conquest		
			500	First wall (Roman)		459 Vandal conquest		
		Roman period	400			123 Roman conquest of Mallorca		
			300					
			200					
			100					7
			1					1
BC			50–70	Roman settlement in Palma				
			123	Primitive settlement Castella				
	I	Talayotic period	200	Phoenician commercial enclaves				
			800					
			5000	Postalayotic				
			10000	Prototalayotic				

submerged archaeological remains, such as wharves or piers, indicating the existence of submerged channels and progradation processes of the coastline (Guerrero, 1981, 1989).

2.3.2 Roman and Late Ancient Age (2nd Century BC to 6th Century AD)

Documentary evidence dates the Roman conquest of Mallorca to BC 123 (García Riaza, 2003). Classical sources indicate the existence of five Roman cities: Pollentia, Palma, Guium, Tuccis, and Bocchorum (Arribas, 1983; García Riaza, 2003). Pollentia and Palma were the most important cities during this period. While Pollentia has been located precisely and is, thus, widely studied from both archaeological and paleogeographic points-of-view (Cau Ontiveros and Chávez, 2003; Giaime et al., 2017), the location of Palma is somewhat diffuse, as the continuous superposition of new urban strata has impeded the extensive excavation of the Roman city. The archaeological data show an initial location of Palma (Fig. 2.3), defended by a coastal cliff and a wall built probably in 2 AD (Rosselló Bordoy, 1973). This first walled enclosure would define an urban area of about 7 ha as the beginning of the current city, although it was also possible that there was a port and a suburb outside the enclosure (Cau Ontiveros and Chávez, 2003; Rosselló Bordoy, 1983).

The selection of a high flat promontory was optimal because of its good visibility over the wide, open bay to the southwest, nearby harbors, freshwater sources (streams and springs), and easy defense (García Delgado, 2000). In addition, it was well-connected to Pollentia, located 48 km (30 miles) northeast in the current Alcúdia Bay, at only 7 h walking distance, and also very close to a wetland (Scheidel, 2013).

At the foot of the western slope of the promontory, a natural harbor was located within the estuary of the Sa Riera River. In the 1st century BC, the estuary opened 800 m upstream from the mouth of the river to the sea, enabling the location of the old port (Alomar Esteve, 2000). Figure 2.4 illustrates the topographical reconstruction by García Delgado (2000), where the coastline would be set more inland from its current position. In the most downstream part of the Sa Riera, the river incised a pronounced meander. On its left margin, the topography is steep and higher (about 18–20 m ASL), while on its right bank a floodplain was generated attached to La Sang Hill. The current street of Sant Jaume functioned as a flood channel at high-water stages, cutting off the meander to connect directly with the estuary and temporarily leaving the interior area as an oxbow lake. On the left bank, downstream of the described meander, deposition processes generated a flat area, where a port suburb and some characteristic Roman buildings (e.g., a theater) were likely located. Archaeological probes conducted in the area established sediment 4 m deeper than the current topographical street level. This sediment consists mainly of clays and sands of alluvial material as well as a wall corresponding to the base of a platform that could be the orchestra of the Roman theater (Moranta Jaume, 1997). This indicates that toward the end of the 1st century BC, the water level was at least 4 m lower than at present. Considering that it was navigable, the base level of the river would be at a height of at least 9 m (García Delgado, 2000).

At the contact between the southern boundary of the Tramuntana Range and the Pla de Palma, the Sa Font de la Vila flows (Fig. 2.1C). This is a freshwater spring that discharges an annual average water volume of 5 million m^3. In this same area, other important springs are located, such as Font de na Bastera and Font del Mestre Pere (Mateos Ruiz and González Casasnovas, 2009, p. 109). The left margin of Sa Riera River is part of an 8 km alluvial lobe with a gentle

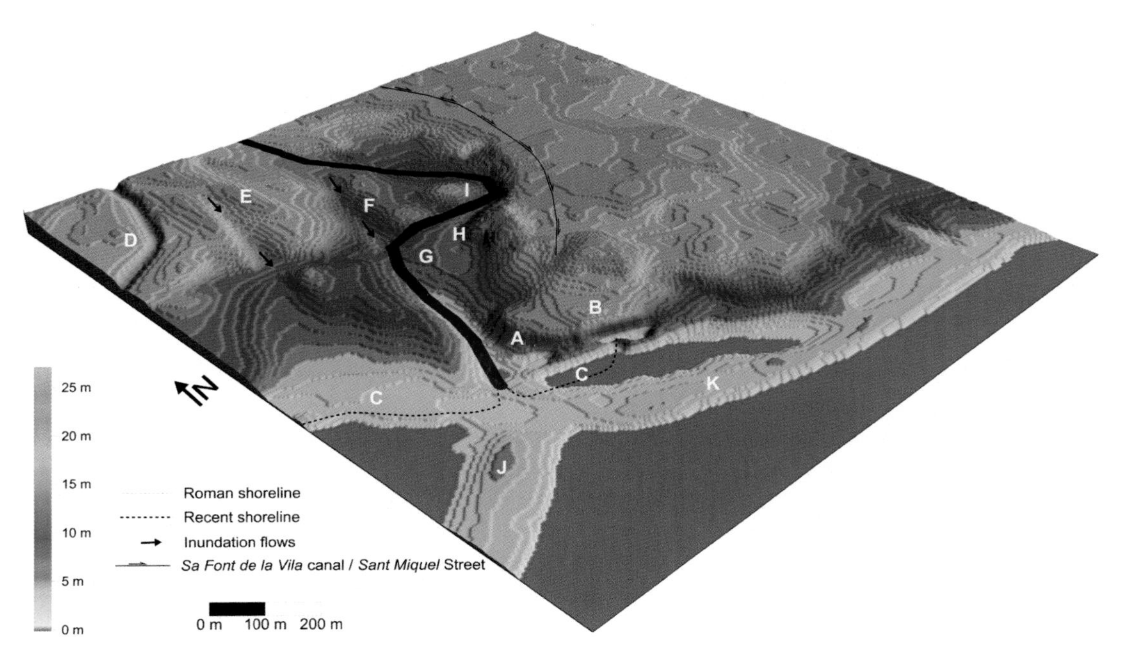

FIGURE 2.4 Geomorphological reconstruction of Palma since the Roman Age. A: *Almudaina* Castel; B: Cathedral; C: Recent shoreline accretion; D: *Sa Riera* River deviation; E: *La Sang* Hill; F: *Sant Jaume* Street; G: Roman theater location; H: *Brossa* Street; I: Meander of *Sa Riera* River; J: Harbor. *Source: García Delgado, C., 2000. Las raíces de Palma. Olañeta (Ed.), Palma.*

slope of ~2% from the apex at the Tramuntana Range to the coastal cliff promontory, where the Roman city was located. This relief allowed the canalization of water in the 10th century by the Arabs. However, some authors believe that it was probably known and used in earlier times because the Roman city of Palma was characterized by surface-water drainage and sewers. This is visible today in the presence of many houses in the lower part of the city still having filtering wells around 15 m deep for utilizing this waterflow (García Delgado, 2000).

2.3.3 Islamic Period (10–13th Centuries AD)

The period from the beginning of the Christian era to the Arab domination (starting in 902 AD) has been little studied, although some authors argue that improvements were probably made in the city's fortification during that timeframe (Alcántara Peña, 1882; Cau Ontiveros and Chávez, 2003). The hypothesis of depopulation has not been proven (Fontanals, 1982), as the Islamic settlement on Madina Mayurqa "was not unplanned but also based on the preexistence of a city" (Cau Ontiveros and Chávez, 2003, p. 46).

The Islamic period represented a prosperity period for the city of Palma, which reached 25 000 inhabitants—a figure not seen again until 1635 (Riera Frau, 1998; Riera-Frau and Roman-Quetgles, 2014; Rosselló Bordoy, 2007). In 1115, the Pisano-Genoese invasion destroyed the city and forced its reconstruction and densification, as did the Almoravid immigration processes. The Pisan chronicle of these events and the Llibre del Repartiment (in Catalan,

"Book of Land Distribution")—written in the 13th century and containing donations and distribution of land made by the King to those who participated in the Christian conquest of Mallorca—explain the urban structure of the city, its walls, and some key functional elements.

The most significant urban transformation was the construction of the second fortified enclosure, expanding the city from 6–7 ha to almost 100 ha and incorporating the river into the fortified city. This exposed its surrounding, recently created suburbs to the risk of flooding. The construction of Madina Mayurqa used the southern and western layout of the existing Roman wall, continuing to the sea cliff ca 700 m to the east. Additionally, a new perimeter was built providing the Madina with a structure that was not modified until the 16th century (Tous Melià, 2004). The new city was divided into two large areas separated by Sa Riera River (Al-Saqqiyya, Fig. 2.3) and connected to inland Mallorca and its port area through different gates. In terms of geomorphology, the Azequia gate ("Channel gate," named Santa Margalida in the Christian epoch), was located at the highest point of the Palma Alta, where the water of the Sa Font de la Vila (current Sant Miquel Street; Fig. 2.4) was already channeled. The access gate of Sa Riera to the city (Barbolet, Plegadissa in Christian times) is also significant. The final reach of the Sa Riera River was still navigable, although the progressive alluvial sedimentation created muddy areas that are preserved in the toponymy of Christian times (Port Fangós, Catalan for "Muddy Harbor").

The human occupation of Palma and its surroundings is closely related to the use and management of water through elements, such as the aforementioned *exechin* (meander) of Sa Riera, which eroded the base of the alluvial promontory, the *enemir* (spring of the *emir* or *font de la vila*, the toponym in Catalan), which provided freshwater to the city, and the *dar as-sin'a* (shipyard) built at the foot of the *castella* (Hort del Rei, in Catalan: the King's Garden). The neighborhood of Madina Mayurca hosted the highest urban density, but outside the wall there was a periphery of great economic dynamism in agricultural and commercial activities. The wetland known as Qatín (Prat de Sant Jordi, in Christian times, from the Arabic *qtn*, meaning "cotton"), located 3 km east of Palma, etimologically suggests that was used for grazing and cultivation of cotton during a period with a climate most favorable to this crop (Medieval Warm Optimum, Rosselló Bordoy, 2007).

The Madina Mayurca, with a population of at least 25 000 inhabitants, was one of the most populated cities of Al-Andalus, with significant economic activity, some of which was highly polluting. Many of these activities started inside the city, located near ditches, springs, or the river, which was often used as a spillway. Wool carders or tanners were located in streets near Sa Riera River, such as Carrer de Peraires ("Wool Carders Street") or Costa d'en Brossa. These were later displaced to the periphery or moved outside the city when urban density increased. Dyeing and tannery were also moved to the southeast of the city (Sa Calatrava) during this time, as the location was optimal—between the sea and a zone of orchards—favored by the land breeze that displaced the unpleasant smells toward the sea. Similarly, the marine breeze moved these smells away from inhabited spaces. This location's function was maintained during the Industrial Revolution of the 19th century until its abandonment with the socioeconomic metamorphosis experienced during the second half of the 20th century (Bernat Roca, 1997, 2012; Escartín Bisbal, 2001; Gutiérrez Lloret, 1987).

The availability of water from both the Sa Font de la Vila and Sa Riera Rivers and the Islamic urban tradition, facilitated the proliferation of orchards and gardens in the city (Riera-Frau and Roman-Quetgles, 2014). These urban green areas favored the internal recirculation of water and were useful as absorption zones during the floods of Sa Riera River. The use of water is, therefore, a key element explaining the urban and rural dynamics of Madina Mayurca, which changed in the following centuries due to three key facts: (1) the diversion of Sa Riera River outside the wall enclosure; (2) the drying of the Prat de Sant Jordi; and (3) the demolition at the beginning of the 20th century of walls that had defined the city limits for nearly 2000 years.

2.3.4 Late Middle Age (13–15th Centuries)

In 1229, King Jaume I of Aragon conquered Madina Mayurqa (as "Ciutat de Mallorca" the new name in Catalan), commencing the region's Christian period. This rule did not introduce important changes in the second fortification, but the dynamics of land use and forms of land tenure determined new environmental relationships. While the Islamic city based its organization on public and widespread access to water for intensive agricultural uses, which encouraged small properties—the new Christian society established a system of land distribution that favored the appearance of big landowners, who combined the cultivation of cereals, pasture, irrigation lands, and hunting. Thus, as the Prat de Sant Jordi (the Muslim *Qatín*) was an unhealthy and disease-prone place, King Sanç proposed in 1321 that it should be drained and dessicated, a project that was again reviewed in 1767 and finally completed in 1846 (Rosselló, 1989).

Within the city, the most significant modification was the construction of churches and the initial division of the city into eight portions, half of which corresponded to the Jaume I and the remaining four to magnates who contributed to the conquest: Nuno Sanç, the Bishop of Barcelona, the Viscount of Bearn, and the Count of Empúries. The division resulted in four parishes, two in the Palma Alta (Santa Eulàlia and Sant Miquel) and two in the Palma Baixa (Santa Creu and Sant Jaume). In 1302, as a split-off from Santa Eulàlia, Sant Nicolau was added as a parish. Each parish, except for Sant Nicolau, was also given a hinterland outside the walls, again using the eight parts division. The distribution of houses, orchards, workplaces, mills, and bakeries varied according to the portions.

Some of the most devastating and well-documented flash floods occurred during this period. Several authors have compiled the main anomalies and catastrophic events recorded in the 15 and 16th centuries (Barceló Crespí, 1991; Grimalt Gelabert, 1989). Climatological data from the 15th century illustrate patterns of highly irregular weather with periods of drought alternating with periods of torrential rain. These events were recorded due to the great social commotion caused by disturbance of the agricultural cycle, especially in rain-fed cereal crops, which led to expensive imports to meet domestic demand. During this period, severe droughts were documented (1454–1456, 1470, 1473, 1478, and 1484–1485) as well as cold waves in 1469 and 1487, in addition to large snowfalls in 1495. Catastrophic floods in 1403, 1408, 1444, and 1490 are also documented (Barceló Crespí, 1991, p. 135). The 1403 flood (October 14th, known as the "year of the flood"), was the most catastrophic, causing 5000 deaths (Damians and Manté, 1997; Grimalt Gelabert and Rosselló Geli, 2010; Rosselló, 1985). The flash flood generated a great wave that entered the city by breaking down the wall (>3 m) and acquired greater flow velocity in its descent toward the main interior meander (>4 m) (Grimalt Gelabert, Rosselló Geli, 2010). The flood demolished and damaged at least 2000 buildings and eight of the bridges that crossed the river in the city. At its mouth, the flood flowed over the 10 m-high wall, which retained the water until its collapse delivered the water column to the sea. Corroboration of the devastating flood effects brought on by extreme weather events is found in comparison to demographic data for 1364 and 1421 that shows an overall decrease of 7965 inhabitants (−36%). The most affected parishes were those located in the Palma Baixa (Sant Jaume, Sant Nicolau, and Santa Creu), with a relative population decrease of 45% (Grimalt Gelabert and Rosselló Geli, 2010). However, the parishes located in the Palma Alta (Santa Eulàlia and Sant Miquel) only experienced a population decrease of 33%.

2.3.5 Modern Age (16–18th Centuries)

The old Muslim wall enclosure, clearly insufficient to defend the city against advances in artillery, was reinforced and enlarged between 1560 and 1575, especially in its western sector, where a hornwork was built. The most important urban action of this period was the diversion of the Sa Riera River, with the opening of a new moat in the wall in its western sector (Tous Melià, 2003, p. 91; Zaforteza Musoles, 1987, p. 96-101). In December 1613, the project was approved and construction was carried out in 1614 and 1615. However, the new channel created by the diversion did not have enough capacity to contain the severe flash floods of 1618, 1635, and 1734 and significant damage occurred once again, especially because the old streambed continued as a sewage and wastewater collector (still today this artificial channel is integrated into the urban sewer network) (Grimalt Gelabert, 1989, p. 23) (Grimalt Gelabert, 1989, p. 25). In the 18th century, the population of Palma declined due to infectious diseases, such as tuberculosis, compounding the river's effluent discharge by serving as crematoria for the clothes and belongings of those affected. Fajarnés-Tur (1900), for example, noted that the Mal Pas River, a short 4 km from Palma (Fig. 2.2), functioned as the main crematorium.

The urban morphology of Palma in the 18th century reveals growing industrial and commercial activity linked to new urbanization in the 17th century (Bejarano, 1993). The walls represented not only a line of defense, but also an economic border for the distribution of land use because occupation outside the walls depended on the distance to the wall as a security military ring. The only settlement adjacent to the wall was Santa Catalina, located west of the city, where fishing (Llonja) and grinding (flour mills of Es Jonquet) activities have been located since the 14th century. Nevertheless, due to its proximity to the Sa Riera River, this settlement was also affected by several catastrophic floods.

2.3.6 Contemporary Age (19–20th Centuries)

From the beginning of the Christian era, the relationship between the city and the alluvial plain of Pla de Palma extended beyond the Sa Riera River. However, most historians of 19th century Mallorcan economy conclude that Palma was a city with a specific industrialization based on its own business and factories model (Manera and Petrus, 1991). As in the rest of Europe's industrial cities, demographic growth and industrialization made the urban ecosystem more sensitive. The main evidence for this new environmental crisis was the frequent epidemic crises (e.g., typhus, smallpox, and diphtheria), whose sources of infection were mainly in the port area (Santa Catalina), probably due to the arrival of infected populations from merchant vessels that contributed to overcrowding and the lack of sanitary and hygienic measures for the entire city (Escartín Bisbal, 2001). Palma also undertook two major projects during this period: the demolition of the walls and draining and desiccating of Prat de Sant

Jordi (another area prone to the spread of epidemic diseases), both of which enabled urban expansion to occur under better hygienic conditions.

During the 19th century, the impact of seismic activity was also seen—although to a lesser extent than floods—in the urban evolution of the city of Palma. The most powerful recorded earthquake (VIII on the MSK scale) occurred on May 15th, 1851, and caused considerable damage to many buildings and structures in Palma, including the cathedral. Another seismic effect was the change in groundwater circulation, increasing the flow of Palma's source and the temperature of hydrogen sulfide emanation in the thermal source of Campos (Giménez, 2003, p. 81; Bouvy, 1851). The aftershocks of this earthquake lasted until August 1852. The seismicity is related to NE-SW faults that give the central zone of the island and the Palma basin a higher probability of seismic incidence (Giménez, 2003, p. 79). Although they cannot be considered frequent, in the last 400 years three earthquakes were documented with intensities higher than VI on the MSK scale (Mezcua and Martínez Solares, 1983).

2.4 LAND USE AS THE CRUCIAL CHANGE OF URBAN GEOMORPHOLOGY IN THE 20TH CENTURY

Geomorphological processes determined and restricted the urban evolution of Palma throughout its long history. However, during the last century to present, urban sprawl and transport infrastructure grew haphazardly around the city. The role played by territorial development was significantly the opposite of natural dynamics, with a lack of respect for the conservation of natural landscapes and intensive (although perhaps necessary) agricultural use.

A spatial analysis of land-use changes over the past 60 years using Corine Land Cover (1990, 2000, 2006, and 2012), with the support of land-use data from 1956 and 1973 extracted from Pons Esteva (2003), showed that urbanized land in the catchments of the Palma basin increased exponentially during the 20th century (Fig. 2.5). Urbanized areas grew from 1380 ha in 1956 (4% of the Palma basin) to 8175 ha in 2012 (15%). The largest increase of urban areas was between 1973 and 1995, involving 3941 ha. The catchments located on the central coast were the most affected, with a total of 4211 ha being transformed to urbanized land. This increase was especially significant in the Gros and Na Bàrbara catchments as a result of the first expansion crown of Palma into areas under special sensitivity to flooding.

Urban sprawl was the main driver of the catchment's hydrological alteration, in which change in soil permeability conditions led to an acceleration of the runoff response to rainfall. Impact on the fluvial network, however, was not only caused by the change in land use, but also by specific transformation, recanalization, burial, and diversion of the streams. Additionally, the ephemeral regime of the fluvial network and the infrequency of catastrophic flood events limited society's perception of flooding hazard. This situation generated irreversible geomorphological changes with consequences on hydrological function, such as quantity and dynamics of flow, connection to groundwater, and changes in hillslope-channel connectivity (EEA, 2016).

Finally, environmental and territorial planning tools did not properly consider natural hazards, with direct consequences related to the increase in damage caused by flood events. The Palma basin displays a lack of integrated planning against flood risk from ancient times

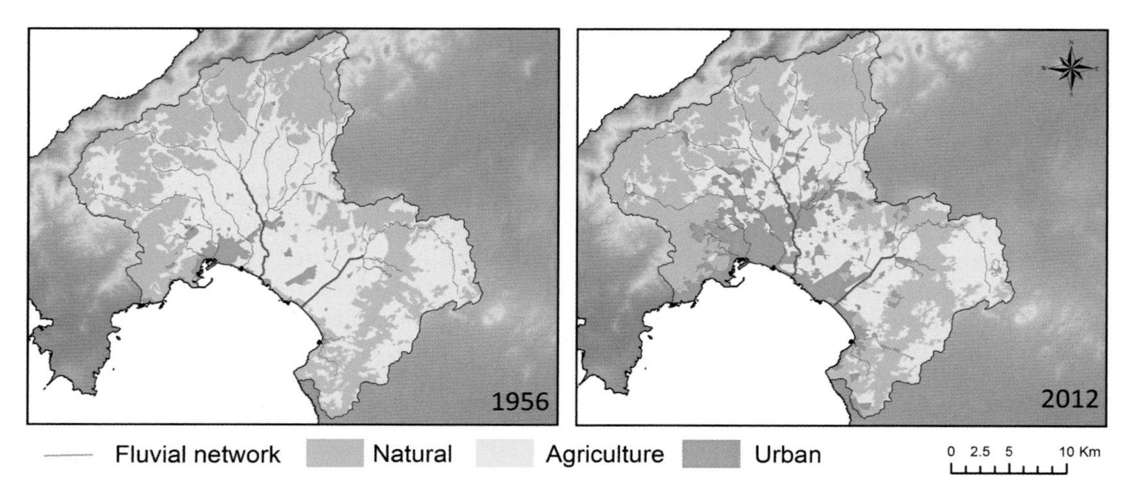

FIGURE 2.5 **Land use evolution of the Palma basin between 1956 and 2012.**

to the present. Urban development in the last century was the main factor responsible for roads and buildings being massively exposed to flood risks (Burby and French, 1981). The damage caused by flood episodes in the last century was significant in terms of diseases, but also caused significant economic losses. Despite this historical failure to plan for natural hazards, in the last 20 years public administrations have begun to integrate this planning at an all-island scale, making it possible to obtain the level of exposure to flooding by overlapping the flood areas of the Territorial Plan of Mallorca (http://www.conselldemallorca.net) and land use. The results showed an exponential increase of the areas under flood risk caused by the sprawl of urban use/expansion across all catchments (1956: 7 ha; 1973: 23 ha; 1995: 179; 2000: 195 ha; and 292 ha in 2012). The period of greatest increase coincides with the stage of greatest growth in urbanized areas (i.e., 1973–1995). The catchments with high exposure to flood risk are those located on the central coast, that is, on the Gros and Na Bàrbara Rivers.

Agricultural uses experienced a reduction during the last century, of 9159 ha for the entire Palma basin, that were inversely proportional to the increase in urbanized land. However, this decrease illustrated a slowdown in relation to the changes of urban use. This is because from 1940s to the 1970s, the Spanish postwar autarkic period increased autochthonous agricultural activity. Still, during the last ca 40 years, the losses of cultivated areas were not only based on their transformation to urbanized areas, but also on the abandonment of marginal agricultural land and its progressive afforestation (García-Comendador et al., 2017, p. 2255). The catchments located on the central coast received the most impacted due to their low relief and proximity to Palma, with a loss of 6033 ha. Catchments located on the western coasts, however, saw a greater reduction of cultivated land, such as Sant Agustí (49%) and Sa Riera (31.3%), due to a combination of urbanization and afforestation. In terms of the latter process, natural areas experienced a significant increase from 22 510 ha (42%) in 1956 to 24 874 ha (47%) in 2012. Most of these areas were under agricultural use managed by historical terracing. Their abandonment could promote the collapse of soil conservation structures leading to land degradation (Calsamiglia et al., 2017).

Land-use transformation can also be linked to global change in recent decades (Rodriguez-Lloveras et al., 2016). The Mediterranean region is very sensitive to climate change, where observational data forecast a substantial change toward higher temperatures and more extreme precipitation regimes (IPCC, 2013). The increase in erosive processes should also be noted, specifically at the headwaters of the Palma basin due to gradient slope and terrace abandonment. Anthropogenic transformation in the last century also had a significant impact on coastal dynamics. The expansion of Palma's port and construction of harbors near the city, for example, have modified sediment transport in the Bay of Palma, where dune systems located in the surrounding coastal area were buried by the construction of tourism resorts, meaning beach nourishment must now be carried out on a regular basis.

2.5 CONCLUDING REMARKS

The interface between environment and humans influenced urban settlements historically, although people have not acted like other animal species, which usually reach a point of equilibrium within their ecological niche. Consequently, human have increased their population and exceeded the carrying capacity of their occupied territory—the theory on which the population-resources dilemma is based (Malthus, 1826). The exponential increase of Palma's population, a city located in a physical territory that is limited (i.e., an island), requires dynamic analysis of its human and physical variables in order to decipher the sensitivity of the urban ecosystem and its limiting natural factors.

Still, overall, it can be assumed that carrying capacity is elastic against population variations because this capacity cannot be determined endogenously or exogenously, as the city maintains economic, cultural, technological, and human exchanges and relationships with other places (Cohen, 1995, 1998; Stutz, 2014). On a temporal scale, a long-term perspective is key to explaining demographic and urban evolution, also providing information on how the territory is constructed and transformed. The comprehension of urban ecosystems—influenced by their inherent (urban) geomorphology—can add to the understanding of the change and governance of ecosystems in a world that is increasingly dominated by people.

Human impacts on the environment at more local scales (Goudie, 2017), such as in the case study given here, is a good example of combining geomorphology and historical geography and the appearance of interactions and feedbacks that moves beyond unidirectional cause-and-effect (human impacts), representing one of the main features of Earth and social systems science (Harden, 2014; Stutz, 2014). There are still important gaps in knowledge of the impact of feedback loops within geomorphic systems under significant human disturbance, where geomorphic systems receiving impacts in turn disturb humans.

The study of human-natural systems focusing on ecological phenomena are creating emerging concepts for social-ecological systems, in recognition of the many diffuse feedbacks of both local- and global-scale processes (Chin et al., 2014). The case study of Palma, where anthropogenic influences date back to the Roman Age, sheds some light on how interscale feedback processes between Earth and social systems are very intense in a high-energy and sensitive environment, such as that found in the Mediterranean region. When seen through to the end, the approach of integrating historical geography and geomorphology—especially in urban areas—is useful for better understanding the resilience of social-ecological systems influenced by the impact of global change.

References

Alcántara Peña, P., 1882. Antiguos Recintos Fortificados De Palma. Editorial Mallorquina de Francisco Pons, Palma.

Alomar Esteve, G., 2000. La Reforma de Palma (1950). In: Vaquer, P.P. (Ed.), Hacia la renovación de una ciudad a través de un proceso de revolucíon creativa. Colegio de Arquitectos de Baleares, Palma, Mallorca.

Arribas, A., 1983. La Romanització de les Illes Balears, Inaugural lecture of the 1983–84 academic year. University of the Balearic Islands, Palma, Mallorca.

Barceló Crespí, M., 1991. Per a una aproximació a la climatologia de la Mallorca baixmedieval a través de textos històrics. Bolletí de la Societat Arqueològica Lul·liana XLVII, 123–140.

Bejarano, E., 1993. Estructura urbana de la Palma preindustrial en el siglo XVIII según la localización de actividades de transformación y abasto. Bolletí de la Societat Arqueològica Lul·liana 49, 333–352.

Bernat Roca, M., 1997. Feudalisme i infraestructura artesanal: De Madina Mayurqa a Ciutat de Mallorca (1270–1315). Bolletí de la Societat Arqueològica Lul·liana LIII, 27–70.

Bernat Roca, M., 2012. Onofre Rodríguez contra la Universitat: sobre tints i noves murades (Ciutat de Mallorca 1576–1622). Memòries de la Reial Acadèmia Mallorquina d'Estudis Genealoògics, Heràldics i Històrics 22, 99–118.

Bouvy, P., 1851. Sobre el terremoto ocurrido en la isla de Mallorca el 15 de Mayo último. Revista Minera II (26) 356–375.

Burby, R.J., French, S.P., 1981. Coping with floods: the land use management paradox. J. Am. Plan. Assoc. 47, 289–300. doi: 10.1080/01944368108976511.

Calle, M., Alho, P., Benito, G., 2017. Channel dynamics and geomorphic resilience in an ephemeral Mediterranean river affected by gravel mining. Geomorphology 285, 333–346. doi: 10.1016/j.geomorph.2017.02.026.

Calsamiglia, A., Fortesa, J., García-Comendador, J., Lucas-Borja, M.E., Calvo-Cases, A., Estrany, J., 2018. Spatial patterns of sediment connectivity in terraced lands: anthropogenic controls of catchment sensitivity. Land Degrad. Dev. 29, 1198–1210.doi: 10.1002/ldr.2840.

Camarassa, A.M., Segura, F., 2001. Flood events in Mediterranean ephemeral streams (ramblas) in Valencia region, Spain. CATENA 45, 229–249. doi: 10.1016/S0341-8162(01)00146-1.

Carayon, N., 2008. Les ports pheniciens et puniques: geomorphologie et infraestrcutures. Texte – Revue de Critique et de Theorie Litteraire 1, 1372, These de Doctorat, Available from: https://tel.archives-ouvertes.fr/tel-00283210/document.

Cau Ontiveros, M.Á., Chávez, M.E., 2003. El fenómeno urbano en Mallorca en época romana: los ejemplos de Pollentia y Palma. Mayurqa 29, 27–49.

Cearreta, A., 2015. La definición geológica del Antropoceno según el Athropocene working group (AWG). Enseñanza de las Ciencias de la Tierra 23, 263–271.

Chin, A., Florsheim, J.L., Wohl, E., Collins, B.D., 2014. Feedbacks in human–landscape systems. Environ. Manag. 53, 28–41. doi: 10.1007/s00267-013-0031-y, Available from: http://link.springer.com/10.1007/s00267-013-0031-y.

Cohen, J., 1995. Population growth and earth's human carrying capacity. Science 269 (5222), 341–346.

Cohen, J., 1998. How Many People Can the Earth Support? W.W Norton & Company, New York.

Crutzen, P.J., Stoermer, E., 2000. The "Anthropocene". Global Change Newslett. 41 (17), 18.

Damians, Manté, A., 1997. Ayguat en Mallorca en 1403. Bolletí de la Societat Arqueològica Lul·liana VIII, 289–290.

Delile, H., Abichou, A., Gadhoum, A., Goiran, J.-P., Pleuger, E., Monchambert, J.-Y., Jerbania, I., Ben, Ghozzi, F., 2015. The geoarchaeology of utica, tunisia: the paleogeography of the Mejerda Delta and hypotheses concerning the location of the ancient harbor. Geoarchaeology 30, 291–306.

EEA (European Economic Area), 2016. Flood risks and environmental vulnerability: exploring the synergies between floodplain restoration, water policies and thematic policies.

Escartín Bisbal, J., 2001. La Ciutat Amuntegada. Edicions Documenta Balear, Palma, Mallorca.

ESRI (Environmental Systems Research Institute), 2017. ArcGIS Desktop. Release 10.4. Environmental Systems Research Institute, Redlands, CA.

Estrany, J., Grimalt, M., 2014. Catchment controls and human disturbances on the geomorphology of small Mediterranean estuarine systems. Estuarine Coastal Shelf Sci. 150, 230–241. doi: 10.1016/j.ecss.2014.03.021.

Fajarnés-Tur, E., 1900. Mortalidad de Tisis En Palma de Mallorca Durante El Siglo XVIII. IX Congreso Internacional de Medicina E Higiene, 52. Tipografía de los hijos de Juan Colomar, Palma, Mallorca.

Flaux, C., El-Assal, M., Shaalan, C., Marriner, N., Morhange, C., Torab, M., Goiran, J.-P., Empereur, J.-Y., 2017. Geoarchaeology of Portus Mareoticus: Ancient Alexandria's lake harbour (Nile Delta, Egypt). J. Archaeol. Sci. Rep.doi: 10.1016/j.jasrep.2017.05.012.

Fleming, K., Johnston, P., Zwartz, D., Yokoyama, Y., Lambeck, K., Chappell, J., 1998. Refining the eustatic sea-level curve since the Last Glacial Maximum using far- and intermediate-field sites. Earth Planet. Sci. Lett. 163, 327–342.

Fontanals R., 1982. Una hipòtesi sobre la situació de Palma alta. In: Estudis de prehistòria, d'història de Mayurqa i d'història de Mallorca dedicats a Guillem Rosselló i Bordoy, Palma de Mallorca, 189–194.

García-Comendador, J., Fortesa, J., Calsamiglia, A., Calvo-Cases, A., Estrany, J., 2017. Post-fire hydrological response and suspended sediment transport of a terraced Mediterranean catchment. Earth Surf. Proc. Landf.doi: 10.1002/esp.4181.

García Delgado, C., 2000. Las raíces de Palma. Olañeta (Ed.), Palma.

García Riaza, E., 2003. Mallorca y su diversidad estatutaria. Mayurqa 29, 71–83.

Giaime, M., Morhange, C., Cau Ontiveros MÁ, Fornós, J.J., Vacchi, M., Marriner, N., 2017. In search of Pollentia's southern harbour: geoarchaeological evidence from the Bay of Alcúdia (Mallorca, Spain). Palaeogeogr. Palaeoclimatol. Palaeoecolo. 466, 184–201. doi: 10.1016/j.palaeo.2016.11.023.

Gili Suriñach, S., 1995. Territorialidades de la Prehistoria Reciente Mallorquina. Universitat Autònoma de Barcelona (UAB).

Giménez, J., 2003. New data on the post-Neogene tectonic activity of Mallorca Island. Geogaceta 33, 79–82.

Goudie, A.S., 2013. The Human Impact on the Natural Environment: Past, Present and Future. Seventh ed. Blackwell, W. (ed), Wileyl-Blackwell. A John Wiley & Sons, Ltd., Publication.

Goudie, A.S., 2017. The integration of human and physical geography revisited. Canad. Geogr./Le Géographe canadien 61, 19–27. doi: 10.1111/cag.12315.

Grimalt, M., Rodríguez-Perea, A., 1994. Unidades morfológicas del llano de Palma (Mallorca), 403–411

Grimalt Gelabert, M., 1989. Les inundacions històriques a Sa Riera. Territoris 42, 19–42.

Grimalt Gelabert, M., Rosselló Geli, J., 2010. Reconstruction of the 1403 Flood of Palma (Mallorca) From Historical Data. HyMeX Workshop, Minorca, Spain.

Grinsted, A., Moore, J.C., Jevrejeva, S., 2010. Reconstructing sea level from paleo and projected temperatures 200 to 2100 ad. Climat. Dyn. 34, 461–472. doi: 10.1007/s00382-008-0507-2.

Guerrero, V., 1981. Los asentamientos humanos sobe los islotes costeros de Mallorca. Bolletí de la Societat Arqueològica Lul·liana XXXVIII, 192–231.

Guerrero, A.V., 1989. Puntos de escala y embarcaderos púnicos en Mallorca: Illot d'En Sales. Bolletí de la Societat Arqueològica Lul·liana ILV, 27–38.

Guijarro, J.A., 1986. Contribución a la Bioclimatología de Baleares. Universitat de les Illes Balears.

Gutiérrez Lloret, S., 1987. Elementos del urbanismo de la capital de Mallorca: funcionalidad espacial. Trabajos del Museo de Mallorca 41, 206–224.

Haines, P.E., Tomlinson, R.B., Thom, B.G., 2006. Morphometric assessment of intermittently open/closed coastal lagoons in New South Wales, Australia. Estuarine Coastal Shelf Sci. 67, 321–332. doi: 10.1016/j.ecss.2005.12.001.

Harden, C.P., 2014. The human-landscape system: challenges for geomorphologists. Phys. Geogr. 35, 76–89. doi: 10.1016/j.ecss.2005.12.001.

IPCC (Intergovernmental Panel on Climate Change), 2013. Climate change 2013, the physical science basis. In: Stocker, T.F., et al. (Ed), Contribution of Working Group I to the Fifth Assessment Report of the Intergovernmental Panel on Climate Change, Cambridge and New York, 1522.

ISTAT (Italian National Institute of Statistics), 2017. Resident population on 1st January. All municipalities, 'Palermo'. Data extracted on 09 January from http://www.istat.it/en/.

Jenkins, H.C., Sellwood, B.W., Pomar, L., 1990. A Field Excursion Guide to the Island of Mallorca. In: Lister, C.J. (Ed.), The Geologist's Association, London.

Juan Vidal, J. (2007): "Mallorca en la segunda mitad del siglo XV. Panorama socioeconómico" Medicina Balear, Palma p.22–32

Juan Vidal, J., 1977. Notas sobre la población y la vida urbana de la Mallorca Moderna. Mayurqa 17, 57–62.

Laborel, J., Morhange, C., Lafont, R., Le Campion, J., Laborel- Deguen, F., Sartoretto, S., 1994. Biological evidence of sea-level rise during the last 4500 years, on the rocky coasts of continental southwestern France and Corsica. Marine Geol. 120, 203–223.

López Castro, J.L., 2016. Los puertos fenicios en la antigüedad. De oriente a la península ibérica y norte de África. In: López Ballesta, J.M., Asensio, Sebastián, R., Ros Sala, M.M., Gianfrotta, P.A., García Charton, J.A. (Eds.), Los Puertos Del Mediterráneo: Contactos, Multiculturalidad E Intercambios: Estrategias Socieconómicas, Políticas y Ecológicas. Universidad Popular de Mazarrón, Mazarrón, Spain, 17–26.

Machado, M.J., Benito, G., Barriendo, M., Rodrigo, E.S., 2011. 500 years of rainfall variability and extreme hydrological events in southeastern Spain drylands. J. Arid Environ. 75, 1244–1253. doi: 10.1016/j.jaridenv.2011.02.002.

Malthus, T.R., 1826. An Essay on the Principle of Population; or a view of its past and present effects on human happiness; with an enquiry into our prospects respecting the future removal or mitigation of th evils which it occasions, Sixth ed. John Murray, Arbemarle Street: London.

Manera, C., Petrus, J.M., 1991. Del taller a la Fàbrica. El Procés d'Industrialització de Mallorca. Ajuntament de Palma. Regidoria de Cultura, Palma, Mallorca.

Mas Florit, C., 2013. El poblamiento de Mallorca durante la Antigüedad tardía: la transformación del mundo rural (ca. 300-902/903 AC). PhD Thesis. Universitat de Barcelona.

Mateos Ruiz, R.M., González Casasnovas, C., 2009. Los caminos del agua en las Islas Baleares-Acuíferos y manantiales. Instituto Geológico y Minero de España (IGME) & Consejería de Medio Ambiente. Govern de les Illes Balears, Palma, Mallorca.

Mazoyer, M., Roudart, L., 2006. A History of World Agriculture: From Neolithic Age to Current Crisis. Earthscan. International Institute for Environment and Development, London.

Mezcua, J., Martínez Solares, J.M., 1983. Sismicidad del área íbero-magrebí. Presidencia del Gobierno. Instituto Geográfico Nacional.

Moranta Jaume, L., 1997. El teatro romano de Palma. Una hipótesis y sus primeras comprobaciones.

Morhange, C., Blanc, F., Schmitt-Mercury, S., Bourcier, P., Oberlin, C., Prone, A., Vivent, D., Hesnard, A., 2003. Stratigraphy of late-Holocene deposits of the ancient harbour of Marseilles, southern France. HolocenE 13, 593–604.

Mather, A., 2009. Tectonic setting and landscape development. In: Woodward, J.C. (Ed.), The Physical Geography of the Mediterranean. Oxford University Press, Oxford, UK, 5–32.

Morhange, C., Marriner, N., 2011. Hazards in the coastal mediterranean: A geoarchaeological approach. In: Martini, I.P., Chesworth, W. (Eds.), Landscapes and Societies: Selected Cases. Springer, Netherlands, 223–234.

Pons Esteva, A., 2003. Evolució dels usos del sòl a les illes Balears. Territoris, 129–145, 1956–2000.

Pounds, N.J.G., 1976. An Historical Geography of Europe 450 B.C.-A.D. 1330. First pub. Cambridge University Press, London-New York-Melbourne.

Raudsepp-Hearne, C., Peterson, G.D., Tengö, M., Bennett, E.M., Holland, T., Benessaiah, K., MacDonald, G.K., Pfeifer, L., 2010. Untangling the environmentalist's paradox: why is human well-being increasing as ecosystem services degrade? BioScience 60, 576–589. doi: 10.1525/bio.2010.60.8.4.

Riera-Frau, M.M., Roman-Quetgles, J., 2014. Jardines, Huertos y Espacios cultivados en las islas orientales del Al-Andalus: estudio de la Madina Mayurqa. Butlletí de la Societat Arqueològica Lul·liana 70, 35–49.

Riera Frau, M.M., 1998. La ciudad islámica en las Islas Baleares. In: García Arenal, M. (Ed.), Genése de la ville islamique en al-Andalus et au Magrheb occidental, Cressiers, Pierre. Madrid.

Rodriguez-Lloveras, X., Buytaert, W., Benito, G., 2016. Land use can offset climate change induced increases in erosion in Mediterranean watersheds. CATENA 143, 244–255. doi: 10.1016/j.catena.2016.04.012.

Romero Recio, M., 1996. Los puertos fenicios y púnicos, 105–135.

Rose, J., Meng, X., 1999. River Activity in Small Catchments over the Last 140 ka, North-east Mallorca, Spain. In: Brown, A.G., Quine, T.A. (Eds.), Fluvial Processes and Environmental Change. John Wiley & Sons Ltd, Chichester, pp. 91–102.

Rosselló, R., 1985. El diluvi de l'any 1403. Elements 4, 15–16.

Rosselló, R., 1989. El pla de Sant Jordi. Notes històriques (segles XIII-XVI). Bolletí de la Societat Arqueològica Lul·liana 45, 91–103.

Rosselló Bordoy, G., 1973. Los siglos oscuros de Mallorca. Mayurqa X, 77–99.

Rosselló Bordoy, G., 1983. Palma romana: nuevos enfoques a su problemática. In: Pollentia y la romanización de las Baleares, Symposium de Arquelologia. Alcudia, 141–155.

Rosselló Bordoy, G., 2007. El islam en las Baleares. Mallorca musulmana. Remembranza de Nunyo Sanç y el Repartiment de Mallorca. Servicio de Palma.

Ruiz, J.M., Carmona, P., Pérez Cueva, A., 2014. Flood frequency and seasonality of the Jucar and Turia Mediterranean rivers (Spain) during the "Little Ice Age". Méditerranée, 121–130. doi: 10.4000/mediterranee.7208.

Salvà Tomàs, P. 1981. La dinámica de la población de las islas Baleares en el último tercio del siglo XIX (1878-1900). Treballs de Geografia 38, 77–139.

Scheidel, W., 2013. The shape of Roman World.

Sevillano Colom, F., 1974. La demografía de Mallorca a través del impuesto del morabatín. Siglos XIV, XV y XVI. Boletín de la Sociedad Arqueológica Luliana XXXIV(820-821), 233–273.

Silva, P.G., Goy, J.L., Zazo, C., Giménez, J., Fornós, J., Cabero, A., Bardají, T., Mateos, R., González-Hernández, F.M., Hillaire-Marcel, C.B.G., 2005. Mallorca Island: geomorphological evolution and Neotectonics. Desir, G., Gutiérrez, F.G.M. (Eds.), Field Trip Guides of Sixth International Conference on Geomorphology, II, University of Zaragoza, 433–472.

Sivapalan, M., Troy, T.J., Srinivasan, V., Kleidon, A., Gerten, D., Montanar, A.G. (Eds.), 2016. Predictions Under Change: Water, Earth, and Biota in the Anthropocene. Hydrology and Earth System Sciences and Earth System Dynamics Copernicus Publications HESS/ESD.

Stutz, A.J., 2014. Modeling the pre-industrial roots of modern super-exponential population growth. PLoS ONE 9 doi: 10.1371/journal.pone.0105291.

Tous Melià, J., 2003. Palma a través de la cartografía (1596-1902). Palma de Mallorca: Ajuntament de Palma.

Tous Meliá, J., 2004. La evolución urbana de Palma, una visión iconográfica. Biblio 3W: revista bibliografica de geografía y ciencias sociales IX.

Wainwright, J., Thornes, J.B., 2004. Environmental issues in the Mediterranean. In: Routledge (Ed.), Processes and Perspectives From the Past and Present. first ed. Routledge, London.

Zaforteza Musoles, D.. 1953–1988. La Ciudad de Mallorca. Ensayo histórico toponímico. Five-volume work. Antigua Imprenta Soler at 1953, 1954, 1957 and 1960. Facsimile edition 1987.

Zalasiewicz, J., Williams, M., Haywood, A., Ellis, M., 2011. The Anthropocene: a new epoch of geological time? Philos. Trans. A Math Phys. Eng. Sci. 369, 835–841. doi: 10.1098/rsta.2010.0339.

CHAPTER

3

Geotourism Development in an Urban Area based on the Local Geological Heritage (Pruszków, Central Mazovia, Poland)

Maria Górska-Zabielska, Ryszard Zabielski***

***Jan Kochanowski University, Kielce, Poland; **Polish Geological Institute-National Research Institute, Warsaw, Poland**

OUTLINE

3.1 Introduction 37

3.2 Georesources of Pruszków and its Surroundings 39

3.3 Relief and Deposits 40

3.4 Water 43

3.5 Erratics 43

3.6 Stones in an Open Urban Space 49

3.7 Final Remarks 51

References 53

3.1 INTRODUCTION

Pruszków is district town in central Mazovia (central Poland; Fig. 3.1) that currently serves as a bedroom community for Poland's capital, Warsaw. Today, it is a very quiet town without heavy industry, but this was not always so. The historical and cultural heritage of the region, reaching back to ancient times, is linked to the consequences of the Industrial Revolution in the 19th century and the tragic events of World War II. Museum exhibits at places like the

Urban Geomorphology. http://dx.doi.org/10.1016/B978-0-12-811951-8.00003-5

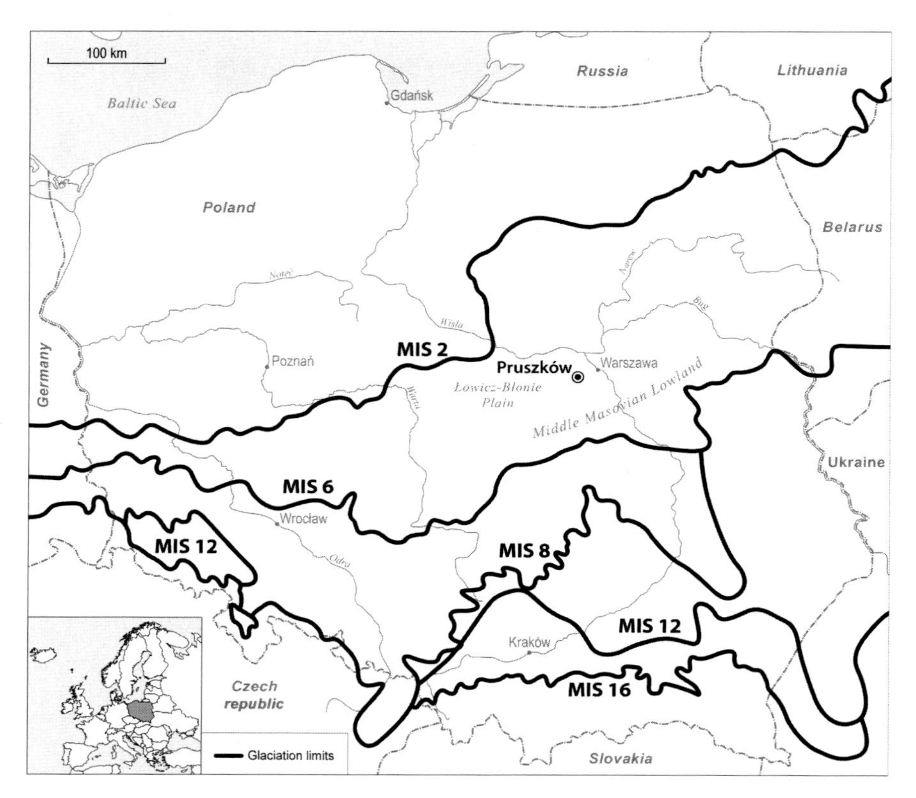

FIGURE 3.1 **Location of Pruszków in relation to the extents of Pleistocene glaciations in Poland.** Ice limit data from Marks et al. (2016). Dates based on INQUA, Bern (2011). Lisiecki and Raymo (2005): MIS 2, Last Glacial Maximum, 29–14 ka BP; MIS 6, 191–130 ka BP; MIS 8, 300–243 ka BP; MIS 12, 478–424 ka BP; MIS 16, 676–621 ka BP.

Museum of Ancient Mazovian Metallurgy (http://en.mshm.pl/) and Dulag 121 Museum (http://dulag121.pl/?lang.en) showcase these times, even as their remnants are being lost. For example, while the quarter of the old Majewski Pencil Factory (http://st-majewski.pl/en/) is awaiting revitalization, factories of the Association of Polish Mechanics from America (in which machine tools were designed and produced) are currently being demolished. Still, they are at least worthy of being commemorated, but who gets to make decisions about what is commemorated and what should be destroyed?

In Pruszków, not only are there objects of cultural heritage in need of preservation, but also elements of natural heritage needing attention. Due to their location in the urban area and insufficient knowledge of municipal authorities and residents, however, these have remained unnoticed.

Among all the natural and cultural resources of a region (Kowalczyk, 2000), there are those that arouse curiosity and are interesting for tourists. Lijewski et al. (2002) have called them "tourist values." In order to distinguish these abiotic resources of the natural environment, which are valuable and attractive for tourists, there is a proposal to call them "geotourist values." As a formality, it should be noted that abiotic natural resources or "georesources" include, among others, bedrock (including outcrops and petrographic types of rocks), soil, relief, surface, and underground water, weather, and local climate (Dowling, 2013; Kożuchowski, 2005; Palacio-Prieto, 2015).

Georesources occur not only in nonurban areas. Valuable examples of geological heritage (or "geosites") may also occur within towns (del Monte et al., 2013; Dowling, 2013; Lollino et al., 2015; Migoń, 2012; Pica et al., 2015; Rubinowski and Wójcik, 1978). Palacio-Prieto (2015) defined urban geosites as places representing geological or geomorphological values that formed as a result of geological processes or were produced by humans, but that show a close relationship with geology. According to Palacio-Prieto (2015), these can be, among others, buildings constructed of natural rocks. Migoń (2012) and Reynard (2008) indicated that erratics integrated into developed urban space can also be such objects.

Geovalues are of geotouristic interest (Hose, 1995; Migoń, 2012; Newsome et al., 2006, 2010; Słomka and Kicińska-Świderska, 2004), including urban geotourism (del Lama et al., 2015; del Monte et al., 2013; Lollino et al., 2015; Pica et al., 2015, 2016; Rodrigues et al., 2011). Geotourism is a new idea of a branch of tourism that appeared around 15 years ago and combines sightseeing and qualified tourism. It is a part of "cognitive tourism" that is based on the exploration of geological objects and processes that provide aesthetic experiences (Słomka and Kicińska-Świderska, 2004). Geotourism is also an economic activity because it offers geoproducts (Reynard et al., 2015) that bring real financial benefits to people involved in any way with its implementation.

The purpose of this chapter is to promote the geomorphological and geological heritage of Pruszków that will give geotourism a chance to develop in this small Polish town. The authors hope that the chapter contributes to an increase in the knowledge of inanimate nature occurring in the urban area. When the inhabitants acquire knowledge of geotourist objects in their town, tourism servicing can result in a real economic development of the town (Gordon, 2012). Photographs were taken by Maria Górska-Zabielska (2011, 2015) and are used here for illustration (Górska-Zabielska 2016).

3.2 GEORESOURCES OF PRUSZKÓW AND ITS SURROUNDINGS

There are some objects of abiotic heritage in Pruszków and its surroundings, including old glacial relief, Pliocene clays, till, erratics and stones, bog iron, and water. They occur in different combinations (Table 3.1). Their location is shown in Fig. 3.2.

The Pruszków District is located in the Łowicz-Błonie Plain, belonging to the Middle Mazovian Lowland (Kondracki, 2013) in central Poland (Fig. 3.1). The almost flat (90–100 m ASL) plain is built of ground moraine that was left by the Scandinavian ice sheet during the Warthian stadial of the Odranian glaciation (MIS 6; Fig. 3.1), that is about 185–130 ka BP (Mojski, 2005). Ground moraine is composed of till—deposits that are characterized by a low permeability—a feature that is especially valuable in agriculture. Pedogenesis has transformed till into sapric gleysol. Between Pruszków and Grodzisk Mazowiecki, these soils are used for growing vegetables. Agricultural lands within the limits of Pruszków are located in the northern part of the town (Fig. 3.2A). They belonged to the historical Production Complex of Horticulture that was established by Piotr Ferdynand Hoser around 1898 (District Register of Monuments, 2009). Vast open areas of the Łowicz-Błonie

TABLE 3.1 Objects of Abiotic Heritage in Pruszków

Symbol	Explanation
A	Pedogenesis has transformed till into sapric gleysol used in agriculture
B	Historical palace of Count and Countess Antoni and Jadwiga Potulicki, the last owners of Pruszków
C	Palace of Helenówek and former fish ponds
D	Old tenements built of brick in the so-called "Quarter of Millionaires"
E	"Count's clay pit ponds" in the city district of Ostoja
F	Fish ponds in Count Potulicki's Park are former "Count's clay pit ponds"
G	Small ponds in Żwirowisko Park are former "Count's clay pit ponds"
H	Museum of Ancient Mazovian Metallurgy
I	Local bathing pool in the Mazovia Park of culture and recreation is a former "Count's clay pit ponds"
J	Komorów Reservoir was formed by damming the Utrata River
K	Historic hydrotechnical systems located along the Utrata River valley are evidence of the old fishponds belonging to Potulicki
L	Water intake from the Oligocene aquifer from a depth of 244 m in Jasna St.
M	Water intake from the Oligocene aquifer from a depth of 245 m in Lipowa St.
N	Water intake from the Oligocene aquifer from a depth of 244 m in Żbikowska St.
O	Water intake from the quaternary aquifer from a depth of 29.5 m in Prusa St.
P	Gabions in Żwirowisko park (former "Count's clay pit ponds") have aesthetic, decorative, and stabilizing function
R	Gabions also create a unique fence on the private property in Podhalańska St. No. 10
S	The outer walls of the District Authority Office building are covered with Novabrik elevation brick made from a mixture of broken granite, marble, and mica
T	Most of the *matzhevas* in the Jewish cemetery are made of so-called Kunów sandstone (lower Jurassic)

Plain that developed as meadows and cultivated fields are the foreground for the palace and park complexes and highlight their architectural and composition values (Lewin and Korzeń, 2008). The historical palace of Count and Countess Antoni and Jadwiga Potulicki is located in the very heart of Pruszków (Fig. 3.2B). In the nearby villages of Helenówek (Fig. 3.2C) and Pęcice, there are two other mansions: the Representative Center of the Ministry of National Defense and private old Polish Manor, respectively.

3.3 RELIEF AND DEPOSITS

Activity of the last ice sheet in the region resulted in, among other things, the intensive glaciotectonic push and squeeze processes that moved the underlying Pliocene clays from their original position (Kowalczyk and Nowicki, 2007). The clays, commonly called loam, occur on the ground surface or at a shallow depth in many places in Pruszków and the surrounding areas. The good quality of these clays, used as raw material for the construction of industrial and residential buildings, was recognized by Jonas Abramson, Szulim Ditman, and Count Antoni Potulicki, who established the company and managed a quite

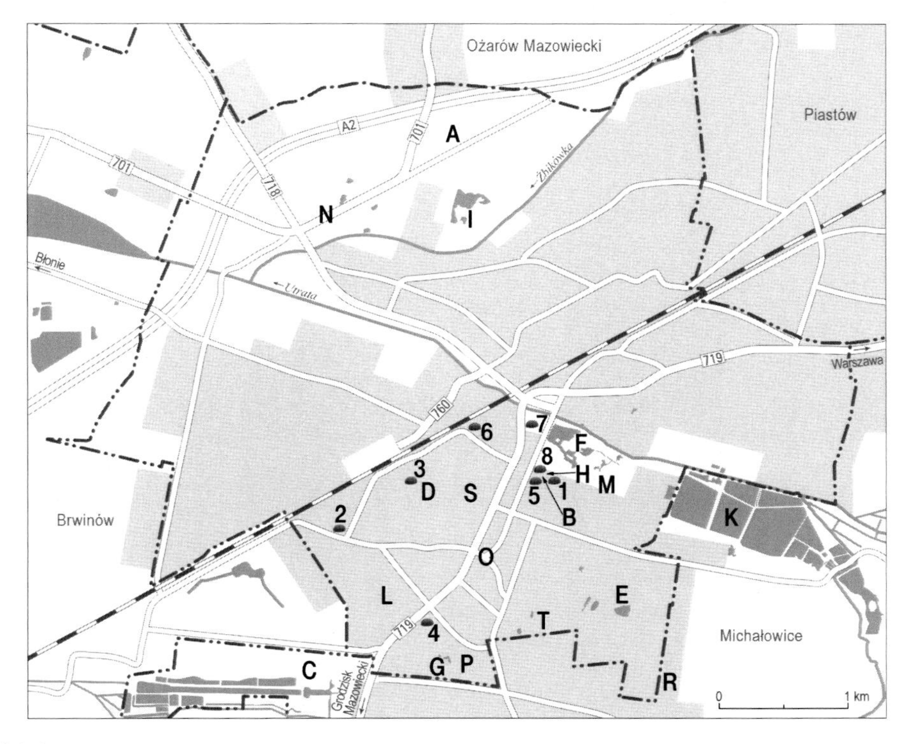

FIGURE 3.2 **Objects of abiotic heritage in Pruszków.** A–T, for an explanation. See Table 3.2; 1–8, erratics—for an explanation see Table 3.2.

prosperous brickyard in the years 1878–1938 (Kaleta, 2010). Today, we find in the town some remains of the construction boom of the late 19 and early 20th centuries. These are the old tenements that are built of brick on Steel Street or Pencil Street in the so-called "Quarter of Millionaires" (Fig. 3.2D) (Krzyczkowski, 2009) and on several other streets, for example, Brick, Brickyard, Ceramic, and Gravel Streets. Traces of old mining pits are visible as the depressions (commonly known as "Count's clay pit ponds") occurring in Ostoja (Fig. 3.2E), a quarter of Pruszków (Kaleta, 2010), and ponds in the Potulicki Park (Fig. 3.2F) in the town center (Bielawski, 2009). Small ponds in the so-called pits, which occur between the church dedicated to Our Lady of Perpetual Help and the Municipal Kindergarten No. 13 in the southern part of the city (Fig. 3.2G), are also evidence of clay mining extraction in this part of Pruszków.

Bog iron ore was another raw material exploited near Pruszków and Brwinów (Fig. 3.3A). This is a sedimentary rock with low iron content that occurs in mires and other wetlands (Mazurek, 2011; Ratajczak and Rzepa, 2011). The formation of bog iron ore is shown schematically in Fig.3.3B. From the 2nd century BC to the 4th century AD, these natural resources became the basis for the development of a large center of production and processing of iron (Tomczak, 2007; Woyda 2002, 2006). The results of archaeological excavations are exhibited in the recently renovated Museum of Ancient Mazovian Metallurgy (Fig. 3.2H; http://en.mshm.pl/).

(A)

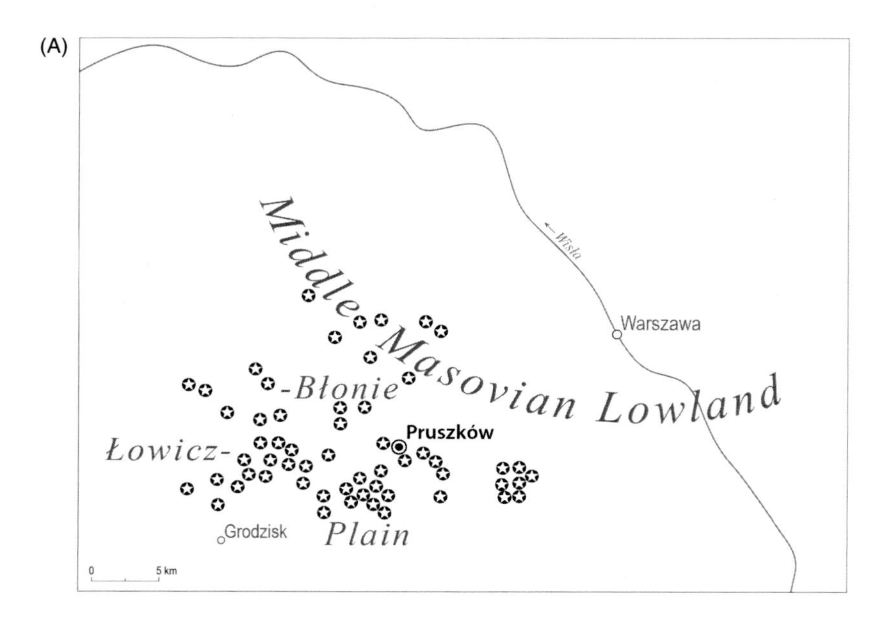

(B)

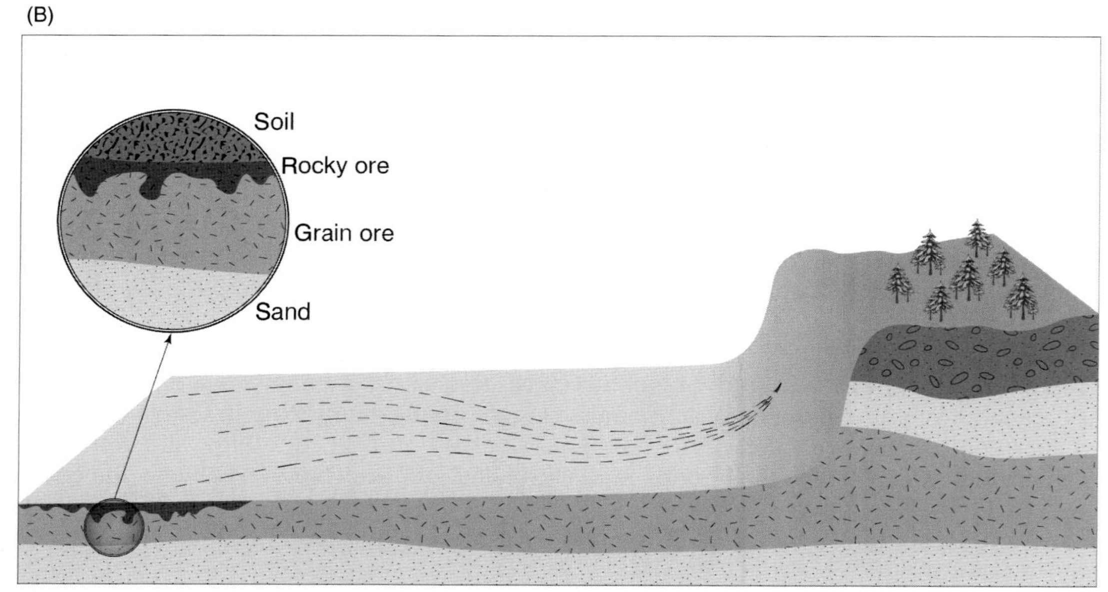

FIGURE 3.3 (A) Ancient Mazovian Metallurgy—archaeological sites are marked. (B) Formation of bog iron ore. *Source: From Rutkowski, M., 2001. Żelazne łąki (in Polish) (Iron meadows). Wiedza i Życie 5. Available from: http://archiwum.wiz.pl/2001/01050500.asp.*

3.4 WATER

Water is a natural resource that is underestimated by the inhabitants of the Mazovia region. In the Łowicz-Błonie Plain, water occurs mainly in numerous rivers and streams. This area is drained by the Utrata River (Fig. 3.2) and its tributaries (Żbikówka, Regułka, Raszynka, and Zimna Woda). These rivers, shifting their courses form one bank to the other, have formed classic meanders and waterlogged areas. Some meanders have turned into oxbow lakes. All of these forms are the effects of lateral erosion of the rivers. Fluvial relief, being a natural composition element of parks, constitutes great tourism potential for this region.

In the immediate vicinity of Pruszków, there are no lakes, but a local bathing pool (Fig. 3.2I) was opened in the northern part of the town in Mazovia Park of Culture and Recreation in 2014. Its popular name, "clay pit ponds," indicates that the Pliocene clays have been formerly exploited there. After recent modernization, the park is an ideal place for the residents of Pruszków and the surrounding area to relax.

Another water body, the Komorów Reservoir (Table 3.1J), which is adjacent to Pruszków from the southeast, is already located in Commune Michałowice (Fig. 3.2). This reservoir was formed by damming the Utrata River between the villages of Komorów-Wieś and Pęcice. Today, it is a storage reservoir for the river. The natural water resources in Pruszków and immediate surroundings, which have been transformed by humans, have an undoubted geotouristic value that is appreciated by the inhabitants of Pruszków and the surrounding area who go there for walks. The Komorów Reservoir is also appreciated by anglers. The newly built playground for children, located at the north side of the reservoir, is an additional attraction for the whole family.

Water occurs in historic hydrotechnical systems located along the Utrata River valley, that is, in the old fishponds belonging to Potulicki (Fig. 3.4F and K in Fig. 3.2 and Table 3.1), in the ponds belonging to the Pęcice Estate, to Tworkowski (Jakubowski, 2009; Lewin and Korzeń 2008; Skwara and Pruszków, 2002), and in the aforementioned "Count's clay pit ponds." The ponds in Potulicki Park (Fig. 3.2F) are the most impressive of them. The landscape value of these ponds was appreciated in 1963, when they were entered together with the palace into the Register of Monuments. All of these objects were created due to a favorable natural relief and the adaptation of the Utrata River oxbows and old mining pits of the brickyard (Bielawski, 2009).

Resources of confined groundwater are exploited in four water intakes for the needs of Pruszków residents. Groundwater from the Oligocene Aquifer is drawn from a depth of 244 m (Fig. 3.2L), 245 m (Fig. 3.2M), and 238 m (Fig. 3.2N) below the ground surface on Jasna, Lipowa, and Żbikowska Streets, respectively. Groundwater from the Quaternary Aquifer is drawn from a depth of 29.5 m (Fig. 3.2O) below the ground surface on Prus Street (Kowalczyk et al., 2007).

3.5 ERRATICS

In the natural environment of the town, in addition to those mentioned earlier, there are many objects of unquestionable importance as geotouristic resources. These are erratics that are scattered in the square greenery, which lie on the sides of the avenues and appear as obelisks in city parks. Unfortunately, they have not attracted the attention of city authorities

FIGURE 3.4 **The largest of the four ponds in the Potulicki Park with the new (2011) element in the form of a fountaim.** *Source: Photo by Górska-Zabielska (2011).*

and residents of Pruszków. Very few people see erratics at all and even fewer are conscious of their scientific and educational value, not to mention their importance for conservation and geotourism. If the authorities made an effort to showcase the beauty of these boulders by protecting them, installing information kiosks near them, and creating a network of paths around them, then the erratics would attract the attention of inhabitants and tourists. Then, they would no longer merely be natural objects of beauty and have value for geotourism.

Erratics are traces of the last ice sheet that covered the Mazovia region. These rock fragments were transported by the Scandinavian ice sheet during its advance some 215–210 and 130–125 ka BP (=MIS 6; Marks et al., 2016; Mojski, 2005). The authors know of at least eight large erratics located within the city limits of Pruszków (Table 3.2).

The largest erratic is made from granite (Fig. 3.5A; No. 1 in Fig. 3.2 and Table 3.2); according to the information board, it was excavated during the construction works on Grunwald Street in 1995. The occurrence of this erratic in surface deposits indicates that it was brought to this region by the Scandinavian ice sheet during the later part of the Odranian glaciations (=Warthian stadial, MIS 6; Marks et al., 2016). As is indicated by its mineral composition, it was eroded from the bedrock in the Swedish Småland region. The boulder is mainly composed of gray-reddish, brick-red, or brownish-reddish feldspar crystals 1–10 mm long that are accompanied by white or light yellow plagioclase crystals 1–3 mm long and single crystals of blue quartz (typical of Småland) with a diameter of not more than a few mm (Czubla et al., 2006). The size of the boulder can also indirectly indicate that it was transported by an ice sheet from that region. This granite is a monument of inanimate nature protected by law in accordance with the Nature Conservation Act of 2004. It is the only one of its kind in the Pruszków District.

The second largest erratic in the town is a piece of granite-gneiss (No. 2 in Fig. 3.2 and Table 3.2). It is accompanied by three smaller erratics. All of the boulders were dug up in a

TABLE 3.2 List of the Eight Largest Erratics in Pruszków

Petrographic type, origin, age, figure no.	Length (m)	Width (m)	Height (m)	Circumference (m)	Volume (m^3)	Weight (t)	Location in the town
1. Småland granite from southeastern Sweden; 1.75–1.5 Ga (Fig. 3.5A)	3.1	2.3	1.3	8.0	4.64	12.8	Museum of Ancient Mazovian Metallurgy 52°09′51.8″ N 20°48′32.3″ E
2. Granite-gneiss from the Baltic Shield; 1.96–1.75 Ga	1.85	1.95	0.95	6	1.79	4.93	Secondary and Sport School Complex in Pruszków, 2 Gomuliński St. 52°09′39.5″ N 20°47′04.6″ E
3. Gaize from the bedrock of Gdańsk Bay or Lower Vistula Valley; 145–66 Ga (Fig. 3.5B)	2.05	1.7	1.15	6.0	2.1	4.19	In front of the building of Social Insurance Institution, Steel St. 52°09′51.3″ N 20°47′38.0″ E
4. Småland granite from southeastern Sweden; 1.75–1.5 Ga	2.15	1.3	1.0	5.6	1.46	4.0	Southern part of the town 52°09′14.5″ N 20°47′37.6″ E
5. Scandinavian sandstone; probably the Cambrian (541–485 Ma) (Fig. 3.5C)	2.65	0.55	1.85	5.85	1.4	3.9	Small hill in the John Paul II Square 52°09′52.1″ N 20°48′30.0″ E
6. Småland granite from southeastern Sweden; 1.75–1.5 Ga (Fig. 3.5D)	1.2	0.8	1.08	3.45	0.54	1.49	Bersohn Square 52°10′06.1″ N 20°48′02.1″ E
7. Gneiss from the Baltic Shield; 1.96–1.75 Ga	1	0.65	1.1	3.15	0.37	1.03	Next to the crossroads of Polish Army and Mira Zimińska-Sygietyńska Streets 52°10′06.1″ N 20°48′24.8″ E
8. Rapakivi granite from the Åland Islands; 1.7–1.54 Ga; Fig. 3.7B	1.25	0.7	0.8	3.4	0.37	1.01	John Paul II Square, in front of the Register Office building 52°09′53.9″ N 20°48′28.1″ E

Boulder volume was calculated using the following formula: 0.523 × length × width × height (Schulz, 1999); boulder weight was estimated on the assumption that 1 m^3 = 2.75 t.

FIGURE 3.5 (A) Småland granite is the largest erratic in Pruszków. This is the only monument of inanimate nature in Pruszków; No. 1 in Table 3.2. (B) Gaize with carbonate cement and preserved internal molds is the third largest erratic in Pruszków; No. 3 in Table 3.2. (C) Sandstone is the fifth largest erratic in Pruszków; No. 5 in Table 3.2. (D) Småland granite in Bersohn's Square, with a commemorative inscription; the sixth largest erratic in Pruszków. No. 6 in Table 3.2. *Source: (A) Photo by Górska-Zabielska, M. (2011), (B) Photo by Górska-Zabielska (2015), (C) Photo by Górska-Zabielska (2011), (D) Photo by Górska-Zabielska, M., 2016. Erratic disappearances, http://science-online.pl/nasze-teksty/nauki-o-ziemi/item/522-erratic-disappearances.*

gravel pit located about 10 km from Pruszków. The corners and edges of the largest erratic (of almost 5 t in weight) are rounded. This feature indicates glacial transport by means of a significant amount of water carrying, among other things, gravel and sand grains. They acted as an abrasive material that at first rounded the edges of the boulder. Such a process could have taken place in englacial tunnels carrying water from a melting ice sheet. The sides of the boulder are smooth. This is the result of wearing down the protruding elements of the boulder by wind-sand-snow streams (abrasion). This process took place in dry and cold environments in the foreland of the retreating ice sheet. A memorial plaque fixed on the boulder is dedicated to "Julian Gomuliński (1894–1961), long standing director of vocational schools, teacher of many generations of specialists for the machine tool industry and the great friend of youth."

Adjacent, a smaller erratic boulder shows a perfect example of crescentic fractures. These features were, most likely, made by the removal of pieces of rock between the ice body and the substratum (Fig. 3.6).

The third largest erratic (Fig. 3.5B; No. 3 in Fig. 3.2 and Table 3.2) is most probably a gaize, a light and porous sedimentary rock, with carbonate cement and a few internal molds/casts/steinkerns preserved. Such large erratics of sedimentary rocks are very rare in the deposition

FIGURE 3.6 (A) Crescentic fractures on the top side of gneiss (one of three erratic boulders in front of the Sport School Complex). (B) Marked crescentic fractures. *Source: From Photo by Górska-Zabielska, M., 2016. Erratic disappearances, http://science-online.pl/nasze-teksty/nauki-o-ziemi/item/522-erratic-disappearances.*

area of the Warthian ice sheet. Therefore, the described erratic is a unique object on the geotourist map of Pruszków. The erratic likely came from the gaize bed that was exposed in the Lower Vistula valley and in the bottom of Gdańsk Bay (Górska-Zabielska, 2008). The dimensions given in Table 3.2 relate to the aboveground part of the boulder.

The fourth largest erratic in Pruszków (No. 4 in Fig. 3.2 and Table 3.2) is granite, probably from southeastern Sweden. It is characterized by a thorough roundness of the edges due to glacial abrasion and its sides are smooth as a result of wind abrasion. The boulder is partly sunk into the ground due to its weight. The dimensions of the part above ground are given in Table 3.2.

Another large erratic that occurs in the town center (Fig. 3.5C; No. 5 in Fig. 3.2 and Table 3.2) is a sandstone most probably of Cambrian age. Primary stratification of loose deposits formed during the deposition of sand grains on the bottom of a water body (sea, lake), which is perfectly visible on the side of the boulder. After deposition, the sand was compacted, cemented, and became sandstone. The boulder, a fragment of sandstone bedrock plucked out by the ice sheet, had to be exposed to atmospheric factors. It was mainly wind erosion, that is, aeolization that resulted in distinct smoothing of one side of the boulder. Some anthropogenic destruction can be seen in the form of screw holes that are probably the traces of a memorial plaque.

The sixth largest erratic in Pruszków (Fig. 3.5D; No. 6 in Fig. 3.2 and Table 3.2) is an indicator erratic of Småland granite, which was transported by the ice sheet from the Småland region in southeastern Sweden (Górska-Zabielska, 2008). One side of the boulder is well-smoothed as a result of glacial abrasion. This process could have occurred when the boulder was carried in the bottom part of the ice sheet and scraped against the substratum over which the ice sheet moved. It is also possible that the boulder sank into the ground and the ice sheet moved over it and polished it. This surface, which is a glacial polish, has been used to place a commemorative inscription ("In remembrance of heroes who were killed in the struggle for national and social liberation during the Nazi occupation. Society of Pruszków, July 22, 1960").

Looking more closely at the surface of the boulder from the other less exposed side, one can see characteristic parallel-arranged fine crests and troughs (Fig. 3.7A). These are the so-called microribs (a form of microrelief) that result from abrasion, which affects the boulder in dry and cold periglacial environment in the foreland of the retreating ice sheet. The destruction

FIGURE 3.7 (A) Parallel-arranged fine crests and troughs (microribs) are the result of the abrasion. (B) Rapakivi granite, an indicator erratic from the Åland Islands, located near the building of Register Office, John Paul II Square. No. 8 in Table 3.2. *Source: Photos by Górska-Zabielska, M., 2016. Erratic disappearances, http://science-online.pl/nasze-teksty/nauki-o-ziemi/item/522-erratic-disappearances.*

also takes place today and its traces can best be seen in the upper part of the boulder in the form of surface exfoliation. The main factors of this process are the changes in air temperature and the circulation of water and solutions in cracks and microvoids between the minerals composing the rock that lead to the disintegration of the rock. The northern side of the boulder is covered with colonies of lichens. The root system of this epilithic flora penetrates the microvoids and has an impact on the development of the present-day exfoliation of the boulder.

The last two boulders, from among eight described, are of moderate size and weigh about 1 t each. The one placed at the crossroads of Polish Army and Mira Zimińska-Sygietyńska (a Polish actor and director of the ensembles, 1901–1997) Streets is the larger of the two (No. 7 in Fig. 3.2 and Table 3.2). It is gneiss, which is a metamorphic rock with typical lineation, that is, the linear arrangement of rock components called blasts (= metamorphic minerals). The boulder surface from the side of the crossroads is glacial polish that until recently was used for the location of a memorial plaque. The traces of its presence (glue layer and screw holes) are still clearly visible.

One of the most beautiful erratics in Pruszków is placed (together with two other boulders, already described) in the central John Paul II Square (Fig. 3.7B; No. 8 in Fig. 3.2 and Table 3.2). It is the smallest of the boulders described here. The boulder has been cut and large (5–15 mm in diameter) pink generally circular crystals of potassium feldspar are clearly visible on the polished surface. These crystals are surrounded by grayish-green rims of plagioclase (sodium-calcium feldspar). Feldspars are accompanied by gray, circular crystals of quartz. Such a structure indicates that the boulder came from the outcrops of Rapakivi granite occurring on the Åland Islands in the middle of the Baltic Sea (Górska-Zabielska, 2008). The boulder is an indicator erratic that is now "upside down" in relation to the position in which it was left after the ice sheet melted and was then subjected to wind abrasion in the ice-sheet foreland. At the bottom part of the boulder (originally the top), one can see the traces of wind abrasion in the form of aeolian microribs.

By contrast, traces of rounding, which are the results of glacial transport and erosion activity of meltwater in englacial tunnels, are visible in the upper part of the boulder. Meltwater carried sand and gravel grains that acted as an abrasive material and rounded the originally sharp edges of the boulder during its movement.

Erratics in the urban space of Pruszków are also present in many other places. Their location, quantity, form, and main features are given in Table 3.3.

3.6 STONES IN AN OPEN URBAN SPACE

Stones, although not necessarily Scandinavian erratics, are also present in gabions (Fig. 3.8) and in the elements decorating the interior of churches. Stones commonly occur as gravestones in the cemeteries located in the two city districts of Tworki and Gąsin as well as in Gordziałkowski and Lipowa Streets (Jewish cemetery). However, it should be noted that in recent times some gravestones have been made of artificial stones that are deceptively similar to their natural counterparts.

Gabions (Italian *gabbione*, which means cage) are large steel cages filled with coarse gravel (Fig. 3.8). Beside their aesthetic and decorative functions, they also have practical applications, as for example they are used to stabilize slopes in Żwirowisko Park (adjacent to the kindergarten in Antek Street; P in Fig. 3.2 and in Table 3.1) that was established in the place of the so-called pits (former "Count's clay pit ponds," which are old mining pits of clays). Gabions also create a unique fence for private property in Podhalańska Street No. 10 (R in Fig. 3.2 and Table 3.1).

The outer walls of the building of the District Authority Office (S in Fig. 3.2 and Table 3.1) are covered with Novabrik elevation brick. It is made of a mixture of broken granite, marble, and mica. Plasticizers, binders, and colorants were added to the mixture to improve the visual and technical characteristics of the brick.

Most of the *matzevahs* in the Jewish cemetery (Fig. 3.9; T in Fig. 3.2 and Table 3.1) in Lipowa Street are made of Kunów sandstone (Lower Jurassic), which was exploited in the vicinity of Kunów near Ostrowiec Świętokrzyski (Walendowski, 2010). This sandstone has been commonly used as the material for architectural details and for, the same reason, can be found in many cemeteries in the Mazovia region.

TABLE 3.3 List of Other Erratics Present in the Urban Space

Form of erratics	Location	Characteristic features
Group of 12 erratics	Northern part of the Falcon Park, located in the city center	1. All three main types of rocks: igneous, metamorphic, and sedimentary are represented 2. Some igneous and sedimentary rocks are rounded (because they were transported by high-energy flows in englacial tunnels) 3. Metamorphic rocks (gneisses) are, by contrast, angular in shape 4. Within the group there is one indictor erratic: Rapakivi granite from the Åland Islands
Three medium-sized erratics	In the base of a fountain in the southern part of Falcon Park	1. Erratics have rounded edges and smooth sides, so they are ball-shaped objects 2. They play a decorative role there
Two erratics—granite and gneiss	At the rear of Bank Millennium at the crossroads of Polish Army and Independence Streets	1. Visible abrasive microrelief on both erratics 2. Gneiss is additionally a wind-polished stone or ventifact
Approximately a dozen boulders (several granites and one sandstone)	In the greenbelt area at the roadside and in the square next to the renewed (in 2016) building in Prus' Street No. 66	Traces of wind abrasion on the surfaces of these erratics indicate that they were exposed to dry and cold climate in the ice-sheet foreland
Several dozen small-sized cobbles and several large erratics	The grotto of Our Lady of Fatima near the main city church	1. Well-rounded boulders 2. They play an aesthetic role

FIGURE 3.8 Gabions: decorative reinforcement (built of rock fragments) of the slopes of the so-called pits, that is, former "Count's clay pit ponds." *Source: Photo by Górska-Zabielska (2015).*

It is worth mentioning that the presence of erratics is not limited only to Pruszków. The number of known erratics changes often because more and more new buildings are constructed around the town and their garage levels reach deep into the subsurface layer of glacial deposits that still contain undiscovered erratics.

3.7 FINAL REMARKS

Socioeconomic changes in Poland in the last 25 years have had a strong influence on Pruszków. The town has always been an industrial and technical center (for ancient metallurgy, railway workshops, American mechanics). For example, in 1977 it provided 21 000 jobs for its residents. Today, Pruszków is being transformed into another satellite community of Warsaw. The local authorities are working hard to make sure that the town attracts new inhabitants. They organize both recurring and occasional artistic and other cultural events. The two museums also offer artistic and cultural activities as a part of their regular programs. All of this activity cannot be ignored.

It seems, however, that some objects that have been located for ages within the town limits are still underestimated. Their educational, scientific, and equally important aesthetic values should be highlighted. This will not involve any start-up money because the objects already exist. However, they should be protected and promoted in order to draw the attention of residents and tourists. We are referring to the following objects of abiotic heritage: old glacial relief, Pliocene clays, till, erratics, stones, and water. They occur in Pruszków in different combinations; their location is shown in the Fig. 3.2. At present, bog iron is the only georesource that is featured in the Museum of Ancient Mazovian Metallurgy.

FIGURE 3.9 ***Matzevahs*** **in the Jewish cemetery in Pruszków are made mainly of Kunów sandstone (Lower Jurassic), which was exploited in the vicinity of Kunów near Ostrowiec Świętokrzyski.** *Source: Photo by Górska-Zabielska, M., 2016. Erratic disappearances, http://science-online.pl/nasze-teksty/nauki-o-ziemi/item/522-erratic-disappearances.*

Erratics are especially noteworthy objects of an inanimate nature. These undervalued and lightly undertreated georesources are valuable evidence of regional geodiversity and the geological past. They enhance the geotouristic value of the natural environment. Some of them are used to commemorate historical events that are important to a town or country. Such a use of boulders confirms timeless, permanent importance of stone obelisks (e.g., Kopczyński and Skoczylas, 2006; Skoczylas and Żyromski, 2007). It seems that erratics deserve more attention on the part of conservation institutions and local authorities that should take care to preserve the natural heritage.

Clever promotion of the objects of abiotic heritage by local tourist societies and/or other organizations popularizing the town certainly could stimulate the development of tourism, including sustainable geotourism in southwestern Mazovia. For example, possible promotions, among others, are: a leaflet, folder, occasional publication, educational path, geotourist trail, information board, lecture on a scientific theme for the general public, or a link on the official website of the town (http://www.pruszkow.pl/poznaj-miasto). Surely, this would lead to job creation in the services to visitors (in hotels, restaurants, tour agencies) and in the production of consumer goods (for inhabitants and visitors) as well as in the expanded service sector (cf. Dowling, 2013). The development of urban tourism is very important, as evidenced by the "Pruszków Stop" survey (http://www.pruszkow.pl/poznaj-miasto/przystanek-pruszkow) conducted in 2007 on behalf of the Commune Pruszków authorities, which showed that up to 75% of the surveyed citizens (group of 450 people aged 25–60 years) are unaware of the tourist attractions of their town.

It is also important to inform the residents of Pruszków about the scientific value of erratics (and other objects of inanimate nature mentioned above) for the reconstruction of the geological past of the region. Urban infrastructure and multifamily buildings with deep-level garages are being built around the town and the earthworks reach deep into the subsurface layer of glacial deposits that still contain many undiscovered Scandinavian erratics (the local press: www.wpr24.pl No. 129 of July 20, 2011). Therefore, it would be worthwhile to take care of newly dug erratics and to place them in prestigious areas of the town. It would certainly contribute to an increase in the geotouristic values of Pruszków and, at the same time, assist in the promotion of Pruszków as an attractive satellite community of Warsaw.

Appropriately exhibited inanimate objects maintain and strengthen the geographical character of the place, its environment, culture, beauty, heritage, and prosperity of its citizens (Reynard, 2008). Their role in the sustainable socioeconomic development of the district and town cannot be overestimated. They contribute to the image creation of the town that adapts the elements of inanimate nature for touristic purposes in accordance with the principles of environmental protection. Finally, it should be noted that local initiatives, increasing awareness among inhabitants, and the promotion of geotouristic values will certainly help draw attention to the need for stronger protection of inanimate resources of the Earth.

Acknowledgments

We would like to thank Maria Wilgat (Maria Curie Skłodowska University, Lublin, Poland) for translating, Hugh Hunt (Kennesaw State University, Kennesaw, Georgia, US) for improving the English version of the text and Małgorzata Gościńska-Kolanko for drawing the figures. The research was conducted within the project No. 612502 "The potential of geotourism in central Poland – critical recognition, protection, and dissemination of geovalues" funded by the Ministry of Science and Higher Education in Poland.

References

Bielawski, P., 2009. Plan Miasta (City plan), Pruszków. Urząd Miasta w Pruszkowie (City Hall in Pruszków) (in Polish).

Czubla, P., Gałązka, D., Górska, M., 2006. Eratyki przewodnie w glinach morenowych Polski (in Polish) (Fennoscandian indicator erratics in the glacial tills of Poland). Przeg. Geol. 54 (4), 245–255.

del Lama, E.A., de La Corte Bacci, D., Martins, L., da Gloria Motta Garcia, M., Kazumi Dehira, L., 2015. Urban geotourism and the old centre of São Paulo City, Brazil. Geoheritage 7 (2), 147–164.

del Monte, M., Fredi, P., Pica, A., Vergari, F., 2013. Geosites within Rome City center (Italy): a mixture of cultural and geomorphological heritage. Geog. Fisica Dinam. Quart. 36, 241–257.

Dowling, R.K., 2013. Global geotourism: an emerging form of sustainable tourism. Czech J. Tour. 2 (2), 59–79.

Gordon, J.E., 2012. Rediscovering a sense of wonder: geoheritage, geotourism and cultural landscape experiences. Geoheritage 4, 65–77.

Górska, M., 2006. Fennoscandian erratics in glacial deposits of the Polish Lowland: methodical aspects. Stud. Quart. 23, 11–15.

Górska-Zabielska, M., 2008. Obszary macierzyste skandynawskich eratyków przewodnich osadów ostatniego zlodowacenia północno-zachodniej Polski i północno-wschodnich Niemiec (in Polish) (Source regions of the Scandinavian indicator erratics in Vistulian glacial deposits from NW Poland and NE Germany). Geologos 14 (2), 177–194.

Górska-Zabielska, M., 2010. Analiza petrograficzna osadów glacjalnych: zarys problematyki (in Polish) (Petrographic study of glacial sediments—an outline of the problem). Landf. Anal. 12 (2), 49–70.

Górska-Zabielska, M., 2016. Erratic disappearances. Available from: http://science-online.pl/nasze-teksty/nauki-o-ziemi/item/522-erratic-disappearances.

Hose, T.A., 1995. Selling the story of Britain's stone. Environ. Inter. 10 (2), 16–17.

Jakubowski, T.H., 2009. Lata prawie bezgrzeszne (in Polish) (Years Almost Sinless). Wydawnictwo Powiatowa i Miejska Biblioteka Publiczna im., H. Sienkiewicza w Pruszkowie.

Kaleta, J., 2010. Pruszków przemysłowy (in Polish) (Industrial Pruszków). Wydawnictwo Powiatowa i Miejska Biblioteka Publiczna im., H. Sienkiewicza w Pruszkowie.

Kondracki, J., 2013. Geografia regionalna polski (in Polish) (Regional Geography of Poland). Wydawnictwo Naukowe PWN, Warsaw, Poland.

Kopczyński, K., Skoczylas, J., 2006. Kamień w religii, kulturze i sztuce (in Polish) (Stone in Religion, Culture, and Art). Wydawnictwo Naukowe Uniwersytetu im Adama Mickiewicza, Poznań, Poland.

Kowalczyk, A., 2000. Geografia turyzmu (in Polish) (Geography of Tourism). Wydawnictwo Naukowe PWN, Warsaw, Poland.

Kowalczyk, A., Nowicki, Z., 2007. Warszawa. In: Nowicki, Z. (Ed.), Wody podziemne miast wojewódzkich Polski (in Polish) (Groundwater of voivodship cities in Poland). Informator Państwowej Służby Hydrogeologicznej, Warsaw, Poland, pp. 221–242.

Kożuchowski, K., 2005. Walory przyrodnicze w turystyce i rekreacji (in Polish) (Natural values in tourism and recreation). Wydawnictwo Kurpisz, Poznań, Poland.

Krzyczkowski, H., 2009. Dzielnica milionerów (in Polish) (District of millionaires). Wydawnictwo Powiatowa i Miejska Biblioteka Publiczna im, H. Sienkiewicza w Pruszkowie, Poland.

Lewin, M., Korzeń, J., 2008. Park Kulturowy Gminy Michałowice jako narzędzie ochrony walorów i środowiska kulturowego gminy Michałowice (in Polish) (Cultural Park of Commune Michałowice as a tool to protect its natural values and cultural environment). Wydawnictwo Gmina Michałowice.

Lijewski, T., Mikułowski, B., Wyrzykowski, J., 2002. Geografia turystyki Polski (in Polish) (Tourism geography of Poland). PWE (Polskie Wydawnictwo Ekonomiczne), Warsaw, Poland.

Lisiecki, L.E., Raymo, M.E., 2005. A Pliocene-Pleistocene stack of 57 globally distributed benthic δ18O records. doi: 10.1029/2004PA001071.

Lollino, G., Giordan, D., Marunteanu, C., Christaras, B., Yoshinori, I., Margottini, C. (Eds.), 2015. Engineering Geology for Society and Territory: Vol. 8, Preservation of Cultural Heritage. Springer, New York, NY, United States.

Marks, L., Karabanov, A., Nitychoruk, J., Bahdasarau, M., Krzywicki, T., Majecka, A., Pochocka-Szwarc, K., Rychel, J., Woronko, B., Zbucki, Ł., Hradunova, A., Hrychanik, M., Mamchyk, S., Rylova, T., Nowacki, Ł., Pielach, M., 2016. Revised limit of the Saalian ice sheet in central Europe. Quat. Int., 1–16. doi: 10.1016/j.quaint.2016.07.043.

Mazurek, S., 2011. Zapomniana ruda darniowa (in Polish) (Forgotten bog ore). Nowy Kamieniarz 87 (2), 70–74.

Migoń FP., 2012. Geoturystyka (in Polish) (Geotourism). Wydawnictwo Naukowe PWN, Warsaw, Poland.

Mojski, J.E., 2005. Ziemie polskie w czwartorzędzie. Zarys morfogenezy (in Polish) (Polish Lands in the Quaternary. Outline of Morphogenesis). Wydawnictwo Państwowy Instytut Geologiczny, Warsaw, Poland.

Newsome, D., Dowling, R., 2006. The scope and nature of geotourism. In: Newsome, D., Dowling, R. (Eds.), Geotourism. Sustainability, Impacts and Management. Elsevier/Heineman Publishers, Oxford, United Kingdom, pp. 3–25.

Newsome, D., Dowling, R., 2010. Geotourism: The Tourism of Geology and Landscape. Goodfellow Publisher, Oxford, United Kingdom.

Palacio-Prieto, J.L., 2015. Geoheritage within cities: urban geosites in Mexico City. Geoheritage 7 (4), 365–373.

Pica, A., Vergari, F., Fredi, P., del Monte, M., 2015. The Aeterna Urbs Geomorphological Heritage (Rome, Italy). Geoheritage. doi: 10.1007/s12371-015-0150-3.

Pica, A., Grangier, L., Reynard, E., Kaiser, Ch., del Monte, M., 2016. GeoguideRome, urban geotourism offer powered by mobile application technology. Abstracts with Programs—EGU General Assembly 2016, April 17–22, 2016 in Vienna, Austria, p. 941.

Pruszków Stop et al., 2007. A study commissioned by the Commune Pruszkow by PBS DGA Company o.o. Available from: http://www.pruszkow.pl/poznaj-miasto/przystanek-pruszkow.

Ratajczak, T., Rzepa, G., 2011. Polskie rudy darniowe (in Polish) (Polish bog ore). Wydawnictwo Akademii Górniczo-Hutniczej, Kraków, Poland.

Reynard, E., 2008. Scientific research and tourist promotion of geomorphological heritage. Geog. Fisi. Dinam. Quat. 31, 225–230.

Reynard, E., Kaiser, Ch., Martin, S., Regolini, G., 2015. An application for geosciences communication by smartphones and 2ts. In: Lollino, G. (Ed.), Engineering Geology for Society and Territory, Vol. 8, Preservation of Cultural Heritage. Springer, New York, NY, United States, pp. 265–272.

Rodrigues, M.L., Machado, C.R., Freire, E., 2011. Geotourism routes in urban areas: a preliminary approach to the Lisbon geoheritage survey. Geo J. Tour. Geosites 8 (2), 281–294.

Rubinowski, Z., Wójcik, Z., 1978. Odsłonięcia geologiczne Kielc i okolic oraz problemy ich ochrony i zagospodarowania (in Polish) (Geological outcrops of Kielce and the surrounding area and the problems of their protection and development). Prace Muz. Ziemi 20, 95–121.

Rutkowski, M., 2001. Żelazne łąki (in Polish) (Iron meadows). Wiedza i Życie 5. Available from: http://archiwum.wiz.pl/2001/01050500.asp.

Schulz, W., 1999. Sedimentäre Findlinge im norddeutschen Vereisungsgebiet (in German). Archiv. Geschieb. 2 (8), 523–560.

Skoczylas, J., Żyromski, M., 2007. Symbolika kamienia jako element procesu legitymizacji władzy w cywilizacji europejskiej (in Polish),(The symbolism of stone as an element of legitimacy in European civilization). Wydawnictwo Naukowe Uniwersytetu im. Adama Mickiewicza, Poznań, Poland.

Skwara, M., 2002. Pruszków. Nasze miasto (in Polish) (Pruszków. Our town). Wydawnictwo Powiatowa i Miejska Biblioteka Publiczna im. H. Sienkiewicza w Pruszkowie.

Słomka, T., Kicińska-Świderska, A., 2004. Geoturystyka: podstawowe pojęcia (in Polish) (Geotourism: basic concepts). Geoturystyka 1 (1), 5–7.

Tomczak, E., 2007. Starożytne centrum metalurgiczne koło Warszawy. Zagadnienia dyskusyjne (in Polish) (Ancient metallurgical center near Warsaw. Discussion topics). Archeol. Polski 52 (1–2), 177–186.

Walendowski, H., 2010. Piaskowce kunowskie i dolskie (in Polish) (Sandstones from Kunów and Dolsk, Holy Cross Mountains). Nowy Kamieniarz 45, 82.

Woyda, S., 2002. Mazowieckie centrum metalurgiczne z młodszego okresu przedrzymskiego i okresu wpływów rzymskich (in Polish) (Mazovian metallurgical center of the earlier pre Roman and Roman periods). In: Orzechowski, S. (Ed.), Hutnictwo Świętokrzyskie Oraz Inne Centra I Ośrodki Starożytnej Metalurgii Żelaza Na Ziemiach Polskich (Metallurgy in the Holy Cross Mountains and other centers of ancient iron metallurgy in Polish territory). Wydawnictwo Świętokrzyskie Stowarzyszenie Dziedzictwa Przemysłowego, Kielce, Poland, pp. 151–154.

Woyda, S., 2002. Mazowieckie Centrum Metalurgiczne z czasów Imperium Rzymskiego (in Polish) (Mazovian Center of Metallurgy from the Roman Empire Period). In: Horban, I., Chmurowa, Z., Zegadło, G. (Eds.), Hutnictwo Świętokrzyskie Oraz Inne Centra I Ośrodki Starożytnej Metalurgii Żelaza Na Ziemiach Polskich (Metallurgy in the Holy Cross Mountains and other centers of ancient iron metallurgy in Polish territory). Przegląd Pruszkowski, Pruszków, 1-2–5-9.

CHAPTER

4

Anthropogeomorphological Metamorphosis of an Urban Area in the Postglacial Landscape: A Case Study of Poznań City

Zbigniew Zwoliński, Iwona Hildebrandt-Radke, Małgorzata Mazurek, Mirosław Makohonienko

Institute of Geoecology and Geoinformation, Adam Mickiewicz University in Poznań, Poznań, Poland

OUTLINE

4.1 Introduction 55

4.2 Geological and Sedimentological Setting 60

4.3 Geomorphological Setting 64

4.4 Anthropogenic Changes in Morphological Landscapes 66

4.5 Urban Geosites 71

4.6 Conclusions 73

References 74

4.1 INTRODUCTION

In its surface geological structure, the Central European Plain is mainly composed of Scandinavian Pleistocene glaciation deposits. These deposits predominantly consist of various types of glacial tills, fluvioglacial sands and gravels, and sometimes lacustrine clays. In lithological terms, the Quaternary sediments of the Central European Plain are among the most genetically and facially diverse (Lindner, 1988, Marks et al., 2016). They are additionally

Urban Geomorphology. http://dx.doi.org/10.1016/B978-0-12-811951-8.00004-7

diversified with a variety of superimposed Holocene deposits. The thickness of Quaternary sediments varies from a few meters to several hundred meters. In Poland, the thickness of these sediments reaches up to 335 m in the Szeszupa basin (Zwoliński et al., 2008).

Just like the Central European Plain, the Polish Lowland is composed of Quaternary sediments. The Pleistocene sediments comprise moraine plateaus, frontal moraine hills, outwash plains, and crevasse forms. Within these forms, Holocene sections of river valleys have developed. This very general lithological and morphological outline represents nearly two-thirds of Poland's area (PGI, 2006). Human activity on Polish land, especially during the last millennium, has significantly transformed the lithology and morphology of the terrain and the local river network, adjusting the occupied land for settlement and economic purposes. Nowadays, the degree of transformation in urban areas is so great, and the accessibility of these areas for research is so limited by development, that it is sometimes impossible to fully recognize the scale of relief transformation. Hence, in many cases, it is difficult to distinguish which sediments have been deposited by the sedimentation processes and which are anthropogenic. Similarly, in morphology, it is sometimes difficult to distinguish which landforms are natural and which are anthropogenic in origin. The aim of this chapter is to examine geological and morphological characteristics of the areas created by chronologically and spatially overlapping geomorphic and morphogenetic processes, mainly during the last glaciation, and —against such background— to present the major landform transformation of the city of Poznań from the Middle Ages to the present day, coinciding with the definition of urban geomorphology by Thornbush (2015).

As for many cities of the Central European Plain and the Polish Lowlands, the characteristic feature of Poznań is its location on the Warta River. Poznań is located relatively close to the geometrical center of Europe, in the central-western part of Poland, halfway between Warsaw and Berlin. The city's topography uses the natural arrangement of landforms connected with the Warta valley's relief and surrounding formations, mainly consisting of plateau and outwash forms dissected by the tributaries of the Warta. Using Kondracki's (2011) physiographic system of division for Poland, Poznań is located in the Wielkopolska Lake District. At the hierarchical level of the mesoregions, Poznań occupies the middle section of the Poznań Warta Gap running south-north, the eastern part of the Poznań Lake District, the western part of the Gniezno Lake District, and the northwestern part of the Września Plain. The Poznań Warta Gap divides the city into left- and right-bank parts (Fig. 4.1). Left-bank Poznań is connected to the Poznań Lake District and the right-bank to the Gniezno Lake District and Września Plain.

The area of Poznań is covered by a relatively dense network of surface waters (0.75 km km^{-2}), but the most important hydrographic axis of the city is the Warta River, which flows across the city in a regulated and navigable 15-km long stretch. The length of the Warta is 808 km, while the surface area of the drainage basin is 54 529 km^2 (KZGW, 2017). Within Poznań, the Warta occupies three significant left-bank tributaries (from south toward north): the Junikowski (12 km in length), Bogdanka (also 12 km in length), and Różany Streams (6 km in length). It also hosts three significant right tributaries (again, from south toward north): the Głuszynka River (34 km in length), Cybina River (41 km in length), and Główna River (37 km in length) (Fig. 4.2). Together with other environmental and artificial factors, this hydrographic system is responsible for frequent high-water levels in Poznań, which has contributed to constant changes in the river network in the city. The Warta River used to be a multichannel river and,

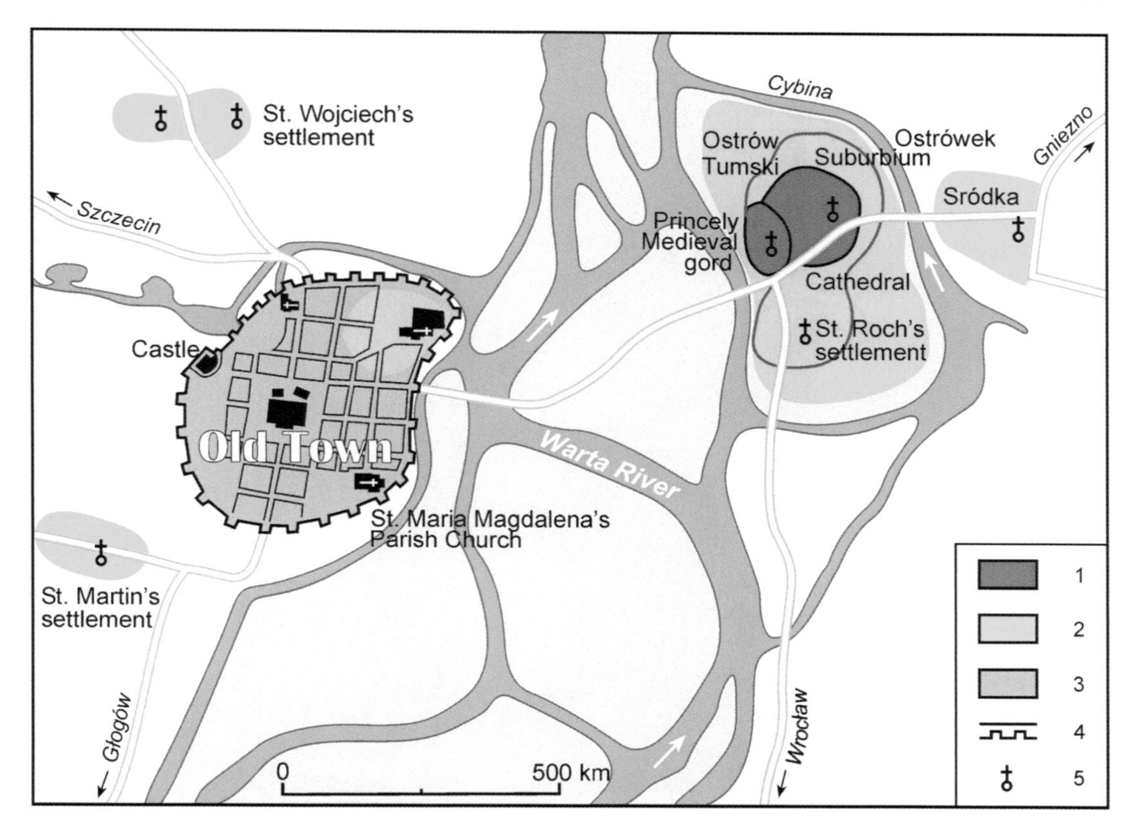

FIGURE 4.1 Location of Poznań in the Medieval Age and names used in text. (1) Gord and suburbium in 10th century; (2) market, craft, and church settlement from 11–13th centuries; (3) location of the city in 1253 AD; (4) defensive walls; and (5) churches and chapels.

according to Topolski (1988), the Warta River flowed in three channels during the 18th century: eastern, central, and western (Fig. 4.3). Today, the Warta River in Poznań comprises one river channel that is partly engineered (Fig. 4.2).

In the opinion of Pasławski (1956), in the valley of the Warta River, the following conditions were favorable in the Middle Ages: the island location of the town; the multichannel river that facilitated the defense of the castle; the possibility of utilizing water for energy purposes; and terraced slopes of the valley. The centuries-old settlement on the islands in the Warta valley was possible due to the continual build-up of surfaces due to flooding during tens of floods as well as through the build-up of anthropogenic deposits (Fig. 4.4). According to Topolski (1988) in Ostrów Tumski, the area was built up by 4–5.5 m over 300 years and Kaniecki (2013) reconstructed these elevations over the centuries, indicating that not only the build-up of the islands took place, but also lowering of the terrain (Fig. 4.5). In the Middle Ages, after flooding on the Warta River in 1253, the town was moved to higher river terraces (Kóčka-Krenz, 2015a) and later to the moraine plateau on the left bank of the valley. As previously, factors conducive to the development of the city (i.e., the island location of the

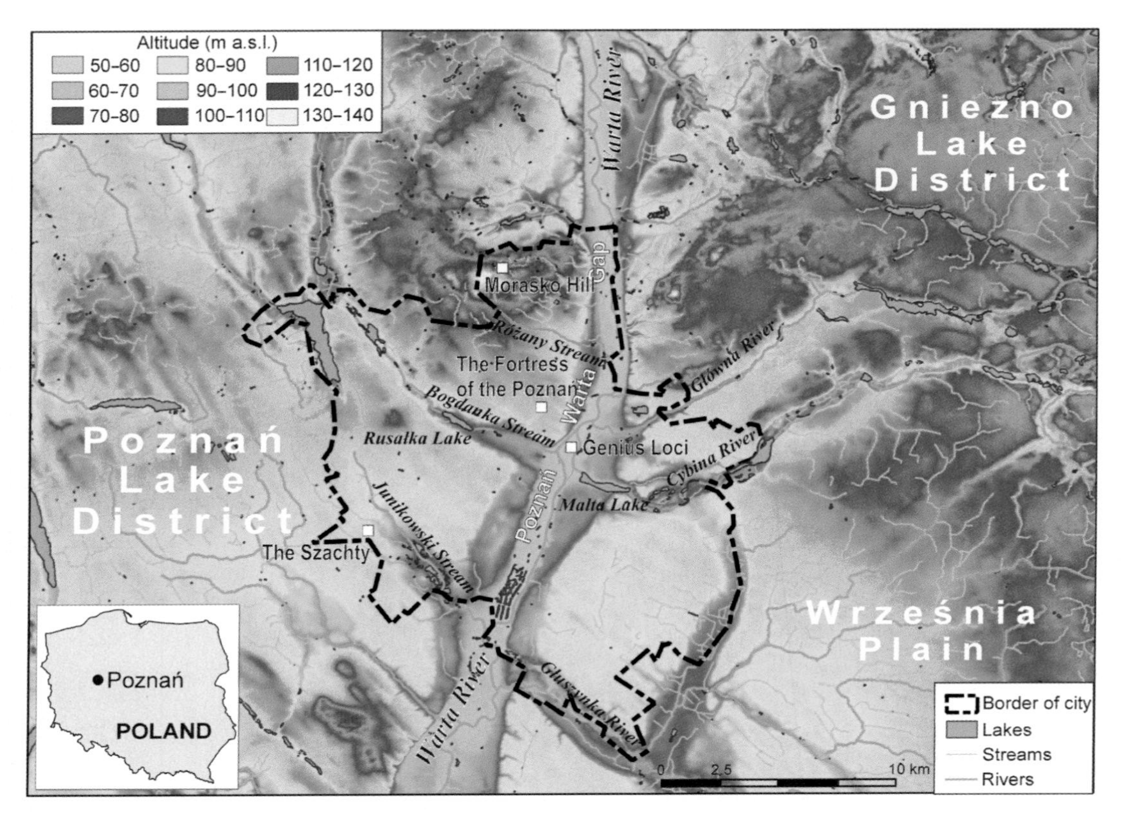

FIGURE 4.2 Digital elevation model (DEM) and main elements of hydrographic network for Poznań vicinity and names used in text. Urban geosites are marked with a *white square*. DEM elaborated on the basis of Vmap Level 2 (J. Jasiewicz).

town, the river multichannel in the 13th century) became constrained by urban sprawl (i.e., including due to the demographic development of Poznań residents).

The first geomorphological studies of the Poznań area have included Pawłowski's (1929) paper, which refers to the gap section of the Warta valley. Bartkowski (1957) continued researching this problem, delimiting seven levels of terraces in the Warta River valley. The geomorphological background of Poznań, together with the areas around Poznań, are shown in the maps by Bartkowski and Krygowski (1959); Karczewski, (2007); Krygowski (1953); and Tomaszewski (1960) as well as geomorphological sketches for detailed geological map sheets of 1:50 000 scale (Chachaj, 1996; Chmal, 1996, 1997; Cincio, 1996) (Fig. 4.6). The wide geomorphological context of Poznań can be found in the works of Bartkowski (1973); Biedrowski (1968); Hildebrandt-Radke (2016); Kozarski (1995); Krygowski (1961); and Żynda (1996). Within recent geomorphological and geological studies, the work of Milecka et al. (2010); Troć (2005); and Troć and Milecka (2008) also contain key reports and analyses. The monumental publications of Kaniecki (1993, 2004), focusing on the changes of the river network of Poznań in the Warta valley, often refer to the changes in the valley

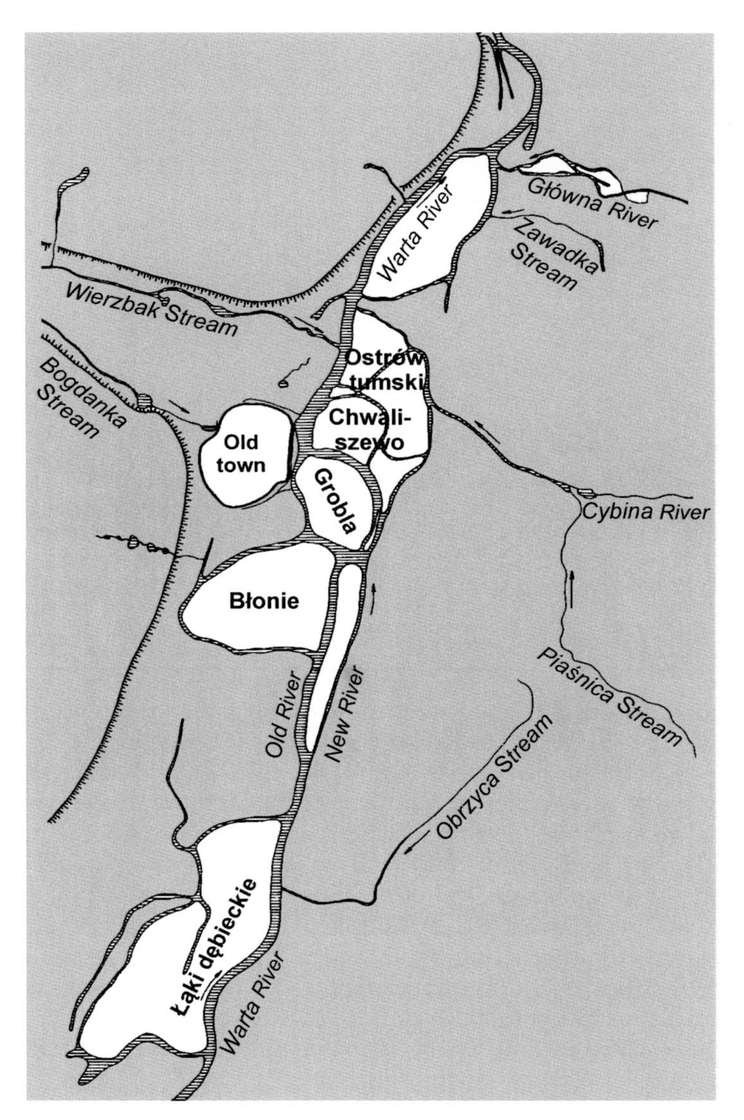

FIGURE 4.3 Hydrographic network in Poznań in the 18th century. *Source: Modified from Topolski, J., 1988. Dzieje Poznania, T. 1–2. PWN, Warszawa.*

bottom and higher levels of terraces and sometimes to plateaus (Kaniecki, 2013). Recently, an interesting investigation by Szałata et al. (2016) was published, showing the distribution of settlements from the Stone Age to the late Middle Ages in the context of contemporary elements of the geographic environment, including some characteristics of relief based on Shuttle Radar Topography Mission (SRTM) data and the geomorphological map of Krygowski (Karczewski, 2007).

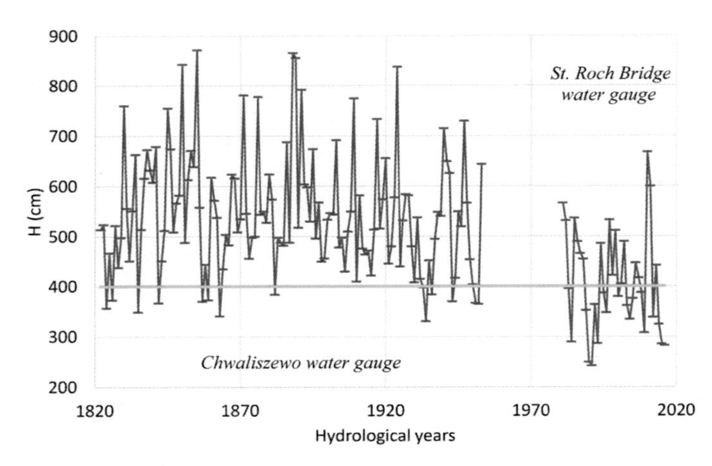

FIGURE 4.4 Maximum water levels (H) in consecutive years 1822–1953 and 1981–2016: altogether, 168 hydrological years for the Warta River in Poznań. Note inundation of floodplain begins circa 400 cm. It means that overbank flows of the Warta River were observed 135 times. *Source: Chwaliszewo water gauge, data taken from Pasławski, Z., 1956. Wybitne wezbrania Warty pod Poznaniem i prawdopodobieństwo wystąpienia najwyższych rocznych stanów wody. Przegląd Geograficzny I(IX/1). St. Roch Bridge water gauge, data taken from the Institute of Meteorology and Water Management (2017).*

4.2 GEOLOGICAL AND SEDIMENTOLOGICAL SETTING

The area of Poznań lies on the border of two geological-tectonic units: the Fore-Sudetic monocline and the Szczecin-Łódź-Miechów syncline (Pożaryski, 1974). Older geological structures are built from the Permian series folded during the Variscan orogeny. The next series contains younger Permian and Mesozoic rocks inclined to the north and northeast. Numerous faults are associated with Paleogene and Neogene tectonic horsts and grabens (Grocholski, 1991). One of these tectonic grabens is used by the Poznań Warta gap, which passes south-north through the central part of the city (Pawłowski, 1929). The varied pre-Cenozoic relief was covered with nearly 250 m of an almost horizontal layer of Neogene clay, mud, and sand, with alternating lignite in the form of 3–8 layers with thickness up to 25 m (Chmal, 1997; Ciuk, 1978). Of importance to the Poznań area are the Pliocene lacustrine clays that are up to 110 m thick (Dyjor, 1970; Kunkel, 1975) and often exposed in glaciotectonic structures (Krygowski, 1961), as for instance on the Warta River valley slopes (Chmal, 1997).

Pleistocene and Holocene sediments cover Pliocene sediments in the region and the presence of tills from all Scandinavian glaciers have been found. However, the San glaciation is poorly represented by glacial sediments (circa 700–500 ka), whose occurrence was found only in subglacial channels cutting down into Neogene and Paleogene deposits. These are gray clayey tills with a high degree of firmness and, from subsequent glaciations, separated by sand deposits that accumulated during the Mazovian interglacial, filling the Wielkopolska Fossil valley (Troć, 2005). The deposits of the Odra glaciation (circa 200 ka) are much better represented. There are two levels of glacial tills that are separated by a series of glaciofluvial or lacustrine clay deposits: the bottom series is characterized by a greater share of local provenience rocks and Scandinavian rocks dominate the upper series (Chmal, 1997). The flora

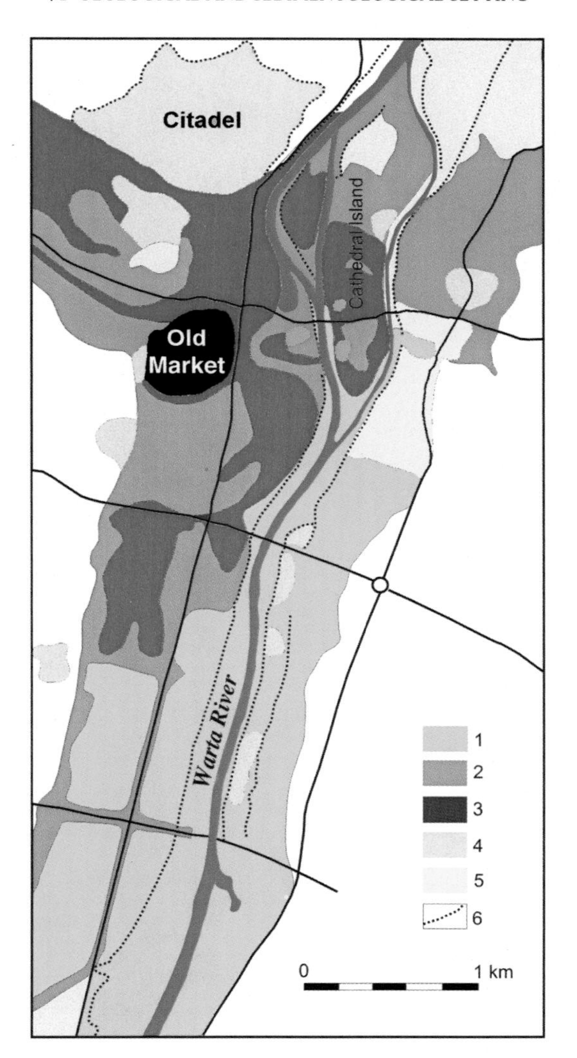

FIGURE 4.5 Hypsometric changes of terrain surface because of leveling out works along the Warta River valley. (1) Raising by 0–2 m; (2) raising by 2–5 m; (3) raising above 5 m; (4) lowering the terrain surface; (5) lowering, then raising the terrain surface; and (6) steep slopes. *Source: Modified form Kaniecki, A., 2013. Wpływ antropopresji na przemiany środowiskowe w dolinie Warty w Poznaniu. Landf. Anal. 24, 23–34.*

and fauna of the Eemian interglacial were found in the deposits of the Odra glaciation and interglacial deposits are formed as fine dusty sands with organic remains or as a peat layer (Chmal, 1997).

Another stage of the geological development of the Poznań area, which proved most crucial, was the Weichselian (Vistulian) glaciation (circa 117–14 ka). According to Krygowski and Żurawski (1968), the ice sheet entered the area through the two main elements of Poznań's

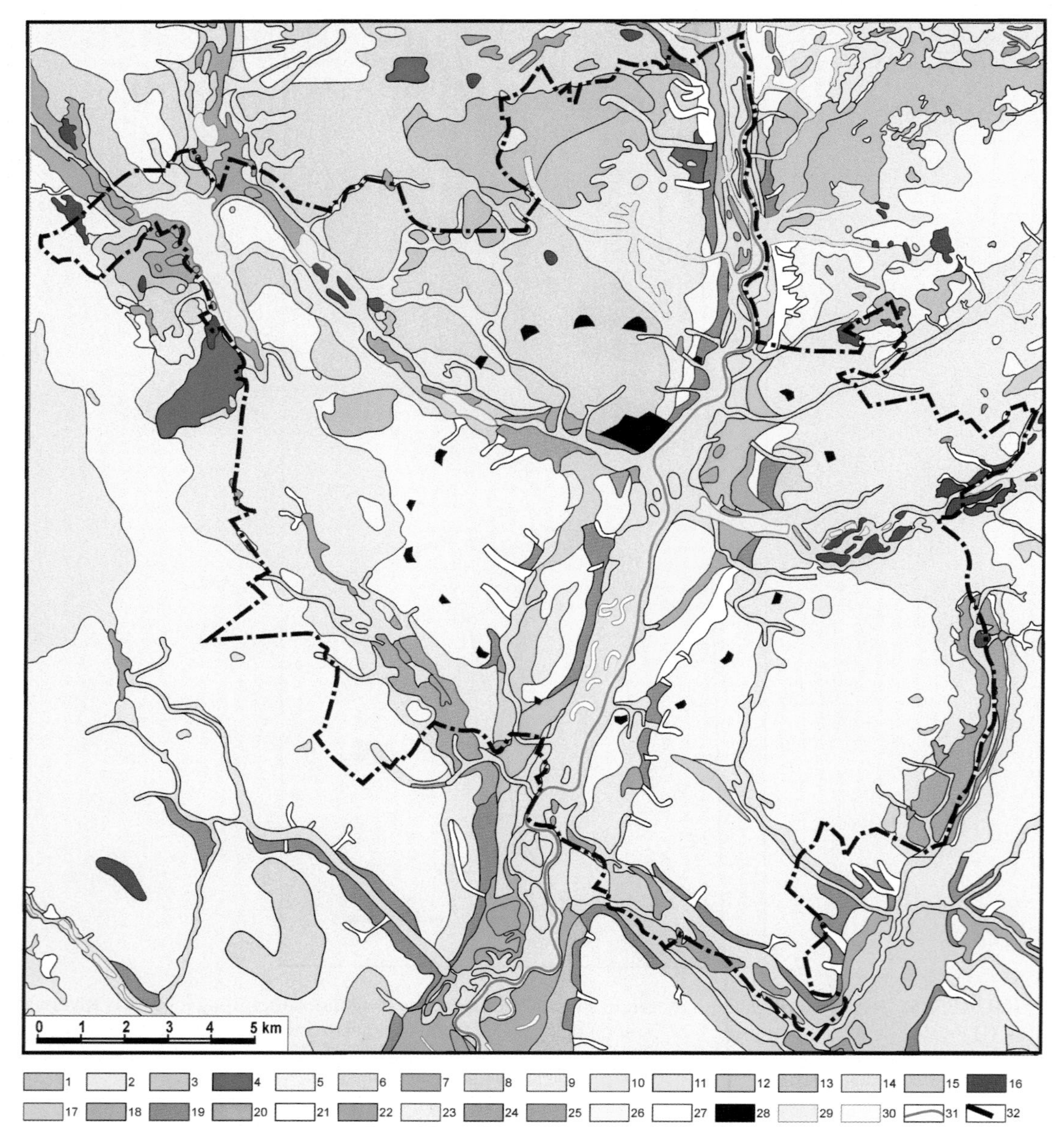

FIGURE 4.6 Geomorphological map of the Poznań area. (1) Morainic hills of accumulation origin; (2) morainic hills of glaciotectonic origin; (3) accumulative morainic hills; (4) terminal morainic hills; (5) flat morainic upland; (6) undulating morainic upland; (7) highest outwash level; (8) higher outwash level; (9) lowest outwash level; (10) first erosion-accumulation terrace; (11) second erosion-accumulation terrace; (12) accumulation terrace; (13) ice-marginal valley terrace; (14) bottom of valley; (15) eskers; (16) kames and kame terraces; (17) denudation plains; (18) alluvial remnants; (19) slopes; (20) alluvial fans; (21) slope wash plains; (22) aeolian sands; (23) lacustrine plains; (24) glaciolacustrine plains; (25) peat plains; (26) erosion plains of meltwaters; (27) valley; (28) forts; (29) lakes; (30) dry

relief, that is, the basins of the Warta, of the Bogdanka, and the Junikowski Streams as well as the morainic hills of the Dziewicza and Morasko. The ice sheet advance of the Leszno phase in the region has been recorded as a layer of glaciofluvial sands and gravels, and a glacial till layer with a thickness up to 12 m (Chmal, 1997). According to Kozarski (1995), their age can be estimated at uncal. 19 000 BP. The Poznań phase is made up of sands and gravels as well as glacial tills of frontal moraines in the northern part of the city (Suchy Las, Morasko, and Czerwonak). They occur in hills of relative height of 5–15 m, resulting from glaciotectonic processes, as confirmed by the presence of Pliocene clays near or on the surface (Chmal, 1997). During the stationing of the ice sheet north of Poznań, a three-level outwash cover was developed in the marginal zone made of 2-m thick fluvioglacial deposits (Biedrowski, 1968). The age of the formation of the marginal zone of Poznań was determined by Kozarski (1995) at uncal. 18 400 BP. Also, sedimentation of the clay deposits, which locally formed in the northwestern part of the city (Chmal, 1997), can also be associated with the melting period of the Poznań phase. The Pomeranian phase (uncal. 16 200 BP; Kozarski, 1995) around Poznań is associated with the formation of the sediments of the Warta River meadow terraces in the gap section of the valley. At that time, the Warta River was a bifurcation river that directed its waters mainly southward. Bifurcation ceased at the beginning of the Oldest Dryas, about uncal. 13 000 years BP and since then erosion of the meadow terrace has been observed (Kozarski, 1995). The characteristic feature of the meadow terrace is its structure of medium-grained sands and gravels (Bartkowski, 1957). In the Late Glacial (Bölling to Younger Dryas), deep erosion of the bottom of the valley ceased and the accumulation of fine and medium sands formed another terrace.

The upper terraces in the Warta River valley were formed in the early Holocene, while the lowest river terrace system was formed during the late Holocene. Later, there was a stage of deep erosion of the upper terraces, even 9–10 m down. Strong erosion was also observed in the valleys of the Warta tributaries, namely the Bogdanka and Cybina. The maximum erosion phase, according to Chmal (1997), took place in the subBoreal period. After a period of erosion, a distinct phase of accumulation of sands and muds containing organic material was observed in the Warta River valley. These sediments formed another floodplain level. The thickness of alluvia reaches in some places up to 9 m (Troć, Milecka 2008). Under the optimum climatic and hydrological conditions of the subAtlantic period, peat accumulation took place in the

◀ and wet oxbow lakes; (31) rivers; and (32) border of city. *Source: Modified from Hildebrandt-Radke, I., 2016. Środowisko geograficzne Poznania. In: Kara, M., Makohonienko, M., Michałowski, A. (Eds.), Przemiany osadnictwa i środowiska przyrodniczego Poznania i okolic od schyłku starożytności do lokacji miasta. Bogucki Wydawnictwo Naukowe, Poznań, 23–46, based on previous geomorphological maps according to Bartkowski, T., 1957. Rozwój polodowcowej sieci hydrograficznej w Wielkopolsce Środkowej. Zeszyty Naukowe UAM w Poznaniu. Geografia 8(1), 3–79; Krygowski. B., 1961. Geografia fizyczna Niziny Wielkopolskiej. Cz. I. Geomorfologia. Poznańskie Towarzystwo Przyjaciół Nauk, Wydział Matematyczno-Przyrodniczy, Poznań, 204 p; Tomaszewski, E., 1960. Mapa geomorfologiczna Polski w skali 1:50 000. Galon, R. (Ed.). Arkusz Poznań i Kostrzyn. Instytut Geografii PAN; and geomorphological sketches attached to the Detailed Geological Map of Poland by Bartczak, E., 1993. Szczegółowa Mapa Geologiczna Polski w skali 1:50 000, arkusz Kórnik z objaśnieniami. PIG, Warszawa; Chachaj, J., 1996. Szczegółowa Mapa Geologiczna Polski w skali 1:50,000, arkusz—Mosina z objaśnieniami. PIG, Warszawa; Chmal, R., 1992. Szczegółowa Mapa Geologiczna Polski w skali 1:50 000, arkusz Stęszew z objaśnieniami. PIG, Warszawa; Chmal, R., 1996. Szczegółowa Mapa Geologiczna Polski arkusz 471—Poznań (N-33–130-D). PIG, Warszawa; Chmal, R., 1997. Objaśnienia do Szczegółowej Mapy Geologicznej Polski, arkusz 471—Poznań (N-33–130-D). PIG, Warszawa; Cincio, Z., 1996. Szczegółowa Mapa Geologiczna Polski w skali 1:50,000, arkusz Swarzędz z objaśnieniami. PIG, Warszawa; Gogołek, W., 1993. Szczegółowa Mapa Geologiczna Polski w skali 1:50 000, arkusz Buk z objaśnieniami, PIG, Warszawa; and Sydow, S., 1996. Szczegółowa Mapa Geologiczna Polski w skali 1:50 000, arkusz Murowana Goślina z objaśnieniami. PIG, Warszawa.*

oxbow lakes of the Warta River drainage basin and in the valleys of Warta's tributaries—the Bogdanka and Cybina as well as the Junikowski Streams; their thickness is 7–10 m (Troć, Milecka 2008, Milecka et al., 2010). All alluvial sediments connected to the Warta River valley and its tributaries represent the channel, overbank, and oxbow facies. Channel sediments are formed of sands of different granulation. They can be identified at the bottom of the channel and in the meander sandbars, where they accumulate because of lateral sedimentation. The overbank formation developed by inundation of the river terraces with Warta River waters and is composed of fine-grained sediments: silty till, silty sand, silt, and clay (Zwoliński, 1992). The oxbow facies are associated with the accumulation of peat, gyttja, and muds in the valleys of the Warta River and its tributaries (Milecka et al., 2010; Troć, 2005).

4.3 GEOMORPHOLOGICAL SETTING

Despite the lowland character of the Wielkopolska (Great Poland) region, where Poznań has been developing (on the regional scale, on the Polish Lowland within the North European Plain), its terrain shows exceptional diversification (Figs 4.2–4.6). The largest part of Poznań is taken by the moraine plateau, which is the effect of the last Scandinavian ice sheet—the aforementioned Poznań phase of the Weichselian (Vistulian) glaciation (Kozarski, 1995)—on which forms of frontal and areal deglaciation have been imposed. The flat moraine plateau reaches an average elevation of 80–100 m ASL in the southern part of the city. In the northern part of the city, the more varied relief is associated with the area of undulating moraine plateau (altitude 90–100 m ASL, with slopes 8–16°) and the hills and ramparts of the frontal accumulative and glaciotectonic push moraines (Krygowski, 1961). The relative heights of these formations with respect to the upland is circa 50 m, while to the bottom of the Warta River it is circa 100 m. The dominant elevation is Morasko Hill (154 m ASL). In contrast, the southern and western parts of the city stretch out on the flat moraine plateau (altitude 80–85 m ASL, slopes up to 0.5°; Żynda, 1996).

The surface area of all the moraine plateaus within the modern boundaries of the city is over 30%. The next most extensive geomorphological unit delimited in Poznań are outwash plains, which take up circa 30% (Fig. 4.6). The seven main outwash areas developed in the marginal zone of the Poznań phase of the Weichselian (Vistulian) glaciation are proglacial forms with almost flat surfaces characterized by gentle slopes in the southern directions: SE, S, and SW (Biedrowski, 1968; Krygowski, 1961). They are found in various places in the city on the flat bottom moraine plateau, with the largest patches found in the eastern and western parts of the city.

The convex forms in the morphology of the terrain, but occurring sporadically, are crevasse forms, such as eskers, kames, and kame terraces. Eskers occur in the form of narrow, long and winding hills formed from sands and gravels deposited by glacial water in subglacial channels. Two NW–SE oriented eskers are in the southeastern part of the city. Kame forms around Poznań are more common, mainly in the eastern and northwestern parts of the city. Typical kames are small hills formed from sediments deposited by water flowing in crevasses and hollows of dead ice. In glacial channels and ice-filled depressions, kame terraces were formed from the material deposited by water flowing in the spaces between the slopes of the river valleys of the Cybina and Bogdanka and ice filling the depressions. Single kame hills in Poznań are found in the north and northwest parts of the city.

However, the most important and largest form of the meridional course is the Warta River's gap section. Within Poznań's city limits, the south-north part of the Warta River valley reaches a length of 15 km, while the width varies from 3 km at the southern end of the city to 1.5 km in the northern end (Fig. 4.2), downcutting through the moraine plateau at an average of 20–40 m. Although Pawłowski (1929) described four terraces in the Warta valley: the floodplain, the lower terrace (2–6 m), the central terrace (7–12 m), and the upper terrace (15/17–21 m), Bartkowski (1957) developed a Warta River terraces system concept that, in his opinion, included seven terraces in the Poznań Warta gap (also Kaniecki, 2004):

1. Floodplain (altitude of 53 m ASL, relative height up to 3 m above average water level) covering the bottom of an 800-m wide valley on both sides of the river; it is built of fluvial muds, sand, and gravel; the morphology of this area has been greatly elevated by construction embankments; the oldest districts of Poznań are located on this terrace: Ostrów Tumski and Śródka.
2. Terrace II (meadow terrace, altitude of 55–57 m ASL, relative height of 3–7 m above average water level), which is preserved in a small section south of the Bogdanka Stream valley; the left-bank Medieval Poznań, translocated from Ostrów Tumski (location rights of 1253), developed in this area.
3. Terrace III (bifurcation, braided terrace, altitude of 58–59 m ASL, relative height of 8–9 m above average water level), which is fragmentary on the right bank of the Warta River, north of the main river valley.
4. Terrace IV (erosion terrace dissected in glacial tills, altitude of 60–64 m ASL, relative height of 10–14 m above average water level); in the second half of the 20th century, new districts of the city (Rataje) were built there.
5. Terrace V (see discussion below list).
6. Terrace VI (altitude of 67–70 m ASL, relative height 17–20 m above average water level), is preserved in a small portion on the right bank in the southern part of the city.
7. Terrace VII (high, outwash terrace, altitude of 71–73 m ASL, and relative height of 21–23 m above the average water level) (Bartkowski, 1957: 19–20 m), whose width is about 1–2 km); a drop of 7‰ on the left bank and 5‰ on the right; from the lower levels it was separated by sharp edges, but at the turn of the 19 and 20th centuries the terrace was severely softened during earthworks carried out during the expansion of the city; within it a part of downtown is built (e.g., the area between Liberty Square and the Main Railway Station).

The sequence of the terrace levels lacks a Terrace V because of fluvial erosion and the supposed height above the bottom of the valley about 15–16 m. Its traces were found only north of Poznań. On the lowest floodplain in the bottom of the Warta River, on the islands in the riverbed of the anastomosing river, the site of Poznań, considered to be the beginning of the Polish state (Kóčka-Krenz, 2015a), was located at the end of the 10th century. On the second left bank meadow terrace, the location of Medieval Poznań (Kóčka-Krenz, 2015b) commenced in 1253.

The moraine plateau is dissected by numerous concave forms in the shape of subglacial valleys and erosional lateral river valleys, whose slope range is 1–4°, depths reach 20–30 m, and widths are up to 500 m. These forms show the NW–SE orientation on the left bank of the Warta River (valleys of the Junikowski, Bogdanka, and Różany Streams) and consequently on the right bank of the river show SE–NW orientation (Głuszynka River valley) as well as NE–SW orientation (Cybina and Główna valleys). In some of them, in the channels of which the Cybina and Bogdanka Streams, the bottom is uneven, with numerous riffles and pools. On the other hand, others show more leveled bottom (valleys of the Junikowski Stream and Główna River). Moreover, endorheic basins of different genesis and sizes are numerous. The basins can reach a depth of 1–5 m.

The area of Poznań is an example of the presence of endorheic basins of meteorite origin (Fig. 4.7). They are unique in Poland and one of only a few in Europe. Evidence for such forms can be seen at the slope of Morasko Hill in the northern part of Poznań, which described in greater detail throughout subsequent sections in this chapter.

4.4 ANTHROPOGENIC CHANGES IN MORPHOLOGICAL LANDSCAPES

In urban areas, anthropogenic forms are a very common and frequent, but important type of terrain. They are particularly significant in the areas inhabited from prehistoric and early historical periods. Prehistoric times did not leave clear signs of relief transformation in Poznań. However, in the early Middle Ages, in Poznań, an important center of the Piast dynasty in the area of western Slavs was established—the gord of Ostrów Tumski. Settlement also developed on the patches of the floodplain of the Warta River (Bartkowski, Krygowski, 1959). The low position of the settlement among the Warta and Cybina channels in the wetlands and backwaters served for defense and facilitated the transport of building materials, but also contributed to the significant transformation of the relief of this area. Ostrów Tumski, one of the islands within the floodplain, was developed from the turn of the 8 and 9th centuries and intensively since the middle of the 10th century (Krąpiec, 2013, Kóčka-Krenz, 2015). Since then, the growth of cultural depositional layers of a thickness 5–7 m was noted by Kaniecki (1993, 2013). Looking for protection against the potential inundation of the Warta River, and in connection with progressive settlement, many river terraces were raised and leveled. Due to the described processes, several changes in the Warta River system and its tributaries were made (Kaniecki, 2013).

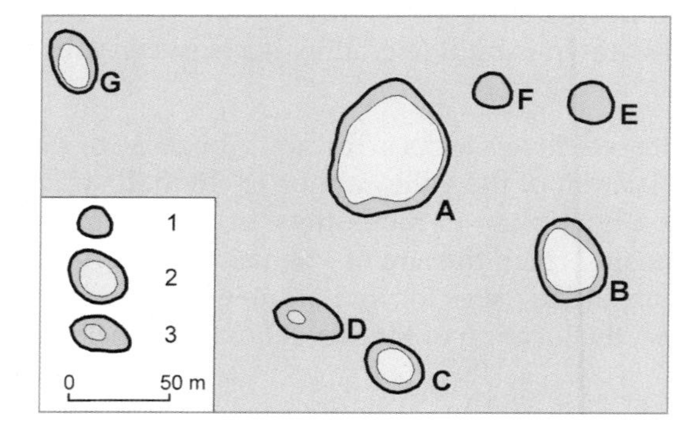

FIGURE 4.7 Spatial distribution of main impact craters on Morasko Hill. (1) Dry craters; (2) craters permanently or occasionally filled with water; and (3) swamp.

The second area of the oldest and largest terrain transformation is the area of the modern downtown. This was due to the relocation of the settlement from Ostrów Tumski and the establishment of the town in 1253 (Kóčka-Krenz, 2015a), around which the suburbs were formed. In the 15th century, a prominent urban center (Fig. 4.1) was formed, consisting of four independent units: the Old Bishop's town of Śródka together with Ostrów Tumski; the new royal city of Poznań (Market), with defensive walls, together with suburbs on the left riverbank; as well as church towns of Chwaliszewo and Ostrówek (Zagórski, 2008). In the following 16–18th centuries, the city continued its spatial and demographic development. In its further development, several stages marked by breakthrough events can be identified (Zagórski, 2008, revised; Figs 4.8 and 4.9):

1. At the end of the 18th century, the Medieval city walls began to be destroyed and in turn the suburbs on the left bank of the Warta River were developed: St. Martin, St. Wojciech, and Nowa Grobla and in the right-bank area: Chwaliszewo, Śródka, and St. Roch.
2. From 1828, the construction of the Prussian fortress of Poznań took place—consisting of 18 forts, including the citadel, located in a ring of 9.5 km diameter and a circumference of 30 km of walls and moats.
3. In 1925, satellite towns were merged with Poznań: Główna, Komandoria, Winiary, Naramowice, Rataje, Starołęka Mała, and Dębiec and in 1933: Golęcin, Podolany, parts of Wola, Sołacz, and Sytków; in the years of the Nazi occupation of 1940–1942, 20 more towns were added to Poznań.
4. After World War II, the area of the city has been enlarged several times: since 1966 toward southeast by the development of a large residential area Rataje, from 1968 northward through the development of the Winogrady district and further (since 1976), when the residential area of Piątkowo was built; form 70 of 20th century can be observed real urban sprawl.

When referring to all anthropogenic processes, it is important to note the predominance of terrain-building processes in the form of, for example, road or rail embankments and leveling, but in the history of urban development there are cases of hill-leveling and constructing excavations. In the history of settlement of the area of Poznań, it was characteristic to move from the lowest areas within the floodplain of the Warta River valley, through higher and higher levels of terraces, up to the level of moraine plateaus and outwash areas. The highest located areas were settled and, thus, transformed on a larger scale only in the 19 and 20th centuries (Kaniecki, 2013) during the Anthropocene.

The largest areas among the anthropogenic forms occupy the leveled plains of residential areas. Beginning with the construction of a gord at Ostrów Tumski, with embankments and moat, the church of the Blessed Virgin Mary, and the cathedral, significant relief transformation took place in this area (Kaniecki, 2004, 2013). They consisted of successive raising of the lowest areas that were threatened by Warta River inundations. The process also included the interior of the gord. The area was often raised by several tens of centimeters, repeating if necessary this human-made action (Kaniecki, 2013). Archaeological studies conducted at Ostrów Tumski indicate the leveling of the basement of both the cathedral and the Duke's Palace, with wooden frame construction up to 1.5 m high, filled with sand and stone pavement (Dębski and Sikorski, 2005; Józefowiczowa, 1963).

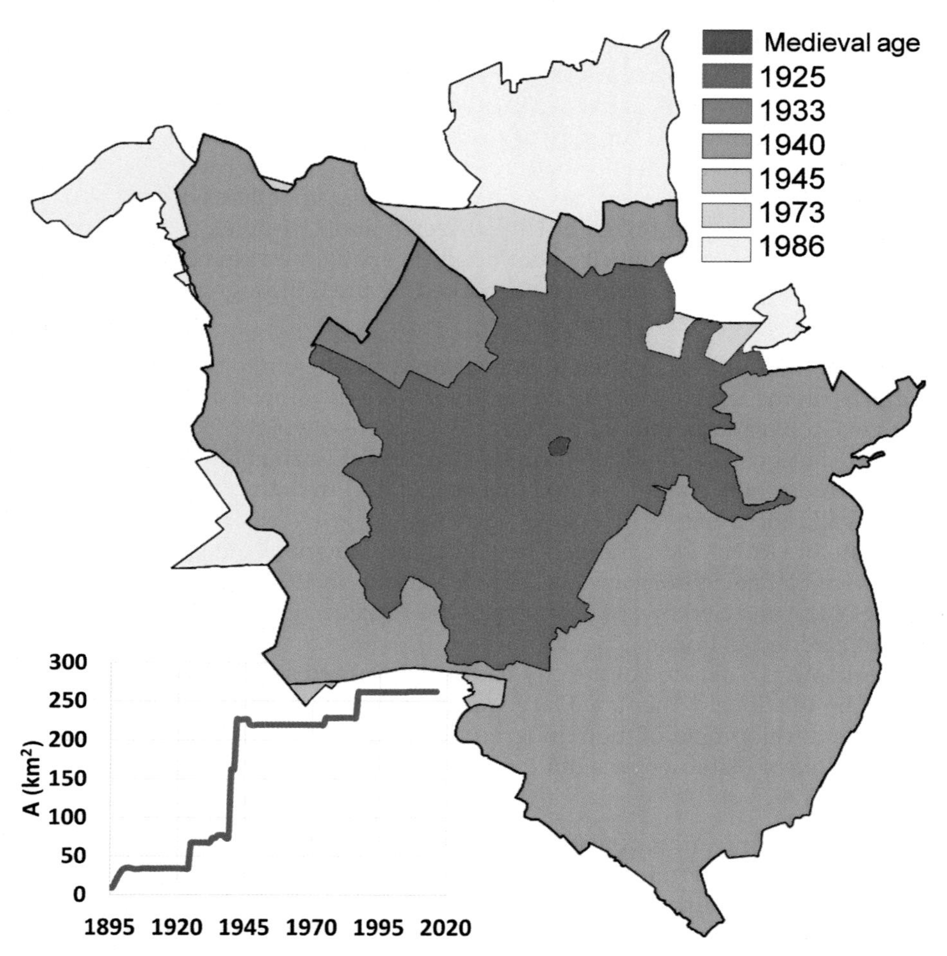

FIGURE 4.8 Urban sprawl of Poznań city. *Source: Zagórski, Z. (Ed.), 2008. Nazewnictwo geograficzne Poznania. Wydawnictwo Naukowe UAM, Poznań; Kruszka, K. (Ed.), 2008. Statystyczna karta historii Poznania. Urząd Statystyczny w Poznaniu.*

Throughout the 12th century, the inhabitants of Ostrów Tumski struggled with floods, responding to them with successive ground layers of overburden. Only at the end of the 12th century had the hydrological situation stabilized. Kaniecki (2013) estimated that the area of Ostrów Tumski was raised 3–5 m over the period of 300 years. The island's location between the main channel of the Warta River and its side channels—the Old River and Cybinka Stream—was a constraint for the spatial development of the Medieval settlement. Therefore, Duke Przemysł I decided to relocate the settlement to the left bank of the Warta River (Kóčka-Krenz, 2015a). Since then, Poznań has been developing mainly toward the west, climbing to the next and next levels of meadow terraces. Since the 15th century, the development referred to the area along transportation embankments across the Warta River valley and between the Chwaliszewo and Nowa Grobla islands within the floodplain. In the 16th century and in the

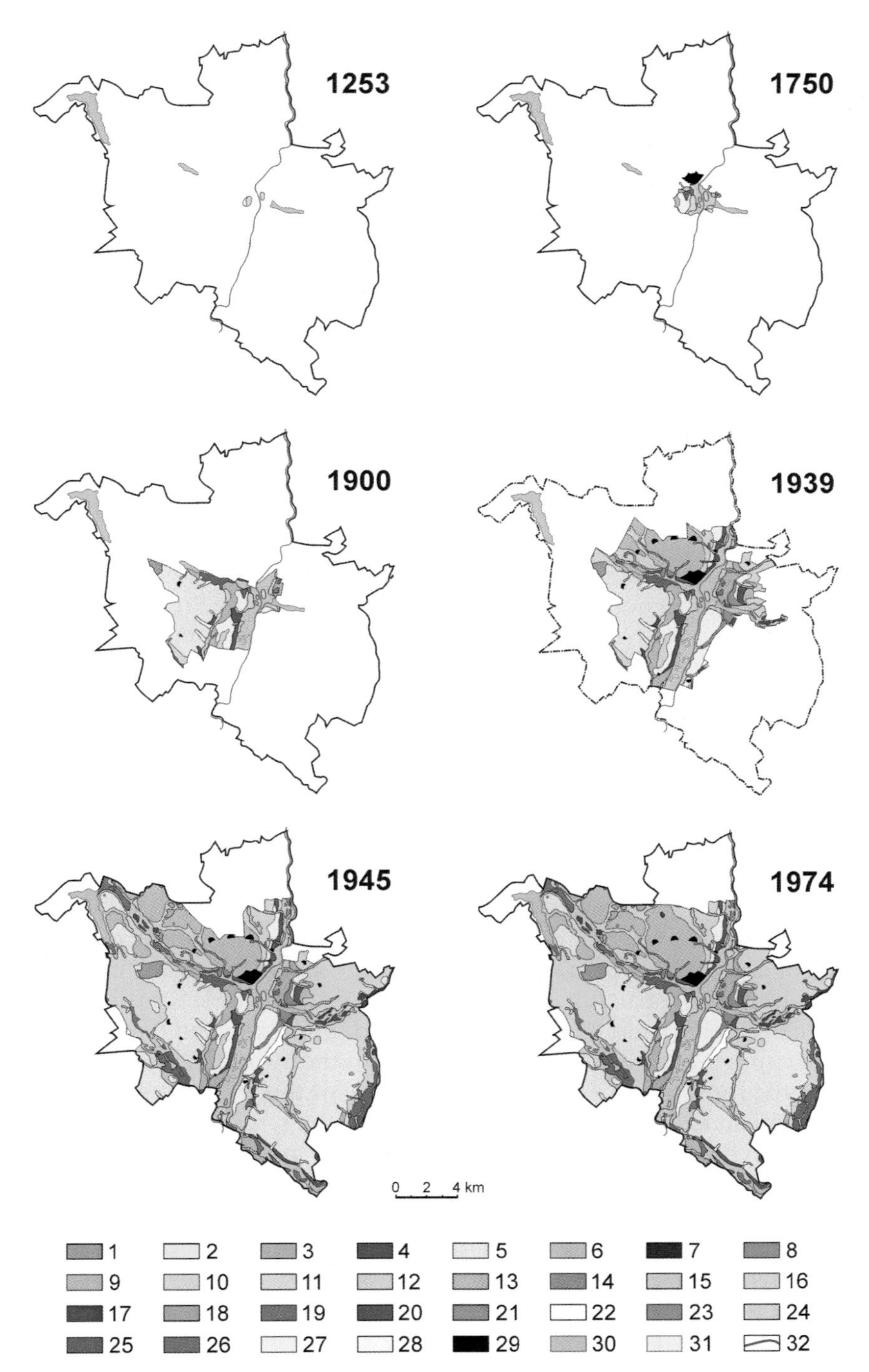

FIGURE 4.9 **Urban sprawl of Poznań city on the background of geomorphological map (Fig. 4.6).**

first half of the 17th century, the development of the gentler slopes of the valley began and later also the edges of the upper terraces and moraine plateau. The basin areas in the city wetlands, especially in the periods of humidification of the climate, were filled up and leveled. According to Kaniecki (2013), around 1536, the level of Terrace II of 55–57 m was raised to an elevation of 58 m ASL, that is, to the level of Terrace III—the same height to which the settlement areas of Ostrów Tumski were also raised. The material used for this purpose was a mixture of native ground, mainly sand and less often till, as well as additives related to human activities, such as debris and rubbish.

The 17th century generally did not favor the development of the city. From the time of the Swedish invasion of Poland (1655–1660) comes information about the construction of Swedish earthworks in front of the city walls. In the first half of the 17th century, these structures were eroded by floods, dug up by the inhabitants of Poznań, and over time leveled.

Since the Prussian partition (1772–1918) there has been an intensive development of road and rail networks. Road and railway escarpments and ditches appeared in the landscape of the city. But this process accelerated in the post-World War II period. Nowadays, communication anthropogenic forms are common, including examples of spectacular communication incisions for southern highway Warsaw–Berlin or the Poznań Fast Tramway, which resembles an overground metro.

Another significant period of transformation of the terrain in Poznań during the Prussian partition relates to fortification works aimed at the creation of the Prussian fortress of Poznań. In 1828–1895, the town was surrounded by a ring of walls, forts, and moats (Biesiadka et al., 2006). Over time, these fortifications became an important barrier to the spatial development of the city, so that, within its boundaries (looking for land for development), streams, meadows, and old moats were leveled (Kaniecki, 2013). It was only at the beginning of the 20th century that parts of fortifications and embankments could be demolished (Kaniecki, 2004). However, even today in the Poznań landscape, there is a citadel, the remains (sometimes ruins) of 17 forts, and several sluices. The material from the demolition of the fortification was used to fill up surrounding wetlands and, thus, to raise urban areas.

Because of the great floods at the end of the 19th century, the construction of the flood embankments in the downtown area, Chwaliszewo and Ostrów Tumski on the left and right riverbanks, respectively, started. Further floods in the 20th century contributed to the continuation of embankment expansion. In addition, significant transformation of the terrain was possible by constructing two artificial water reservoirs: Rusałka (from damming of the Bogdanka Stream) and Malta (through the damming of the Cybina River),which is today used for sports and recreational purposes. In the 19 and 20th centuries, a new type of anthropogenic form appeared in the topography of the city: these are disused excavation clay, till, and sand pits (Graf, 1995). At present, the largest concentration of this type of forms occurs in the valley of the Junikowski Stream in the southern part of Poznań. All excavations are filled with underground and rainfall waters and are gradually overgrowing with vegetation.

Among anthropogenic forms in agricultural areas in Poznań, especially on the outskirts of the city, are agricultural terraces, high field borders, and incisions of unpaved roads. One of spectacular anthropogenic forms created just outside of Poznań is a landfill of municipal waste in Suchy Las, which creates a completely new and distinctive landscape form. Around Poznań, there are nine of this type of form.

4.5 URBAN GEOSITES

Among the protected objects and areas of Poznań are geosites. Geosites in urbanized areas are relatively rarely introduced as forms of abiotic nature conservation. However, even though they usually occupy small areas of terrain and their location depends on the size of the city, its population, as well as its spatial structure, urban geosites have recently been noticed and more attention is being paid to their significance for a variety of reasons (Reynard et al., 2017). Still, few exist in town centers, with most geosites being found on the outskirts of cities, where there are more open spaces. Even so, urban public spaces contribute to the visibility and protection of the city's geodiversity in the form of geosites (Zwoliński et al., 2017). The Polish Central Register of Geosites of the State Geological Institute contains six geosites in Poznań, but only one geosite has distinct anthropogenic traces.

The geosite Morasko Hill originated in the earlier phases of the Quaternary (Stankowski, 2011). It is a glaciotectonically pushed culmination section of frontal moraines, which were formed during the Poznań phase of the Weichselian (Vistulian) glaciation (Bartkowski and Krygowski, 1959; Karczewski, 1961, 1976; Kozarski, 1986; Krygowski, 1961). Morasko Hill, at an altitude of 154 m ASL, is the highest late-glacial elevation in the landscape of Poznań and central Wielkopolska. It has unique scenic values: relative heights in the lowland landscape of Wielkopolska are rather low, but between the elevation of Morasko Hill and the nearby Poznań Warta gap, they reach over 100 m. The landscape of Morasko Hill is significantly diverse—there are some late-glacial forms, such as frontal moraine hills, moraine plateau, outwash plains, erosion gullies, and dead ice meltout basins as well as boulders found in forests and fields, which have been used since the Middle Ages for local construction. Its present-day morphology is the result of processes connected to the activity of the last ice sheet, followed by Holocene morphogenetic processes. Morasko Hill, due to its exposure in the surrounding landscape, may have played an early cultural role in the local community, judging from the numerous traces of prehistoric settlement occurring in its vicinity (Makohonienko et al., 2016). In the 19th century, there was a wood triangulation station on the culmination of Morasko Hill and this could be reconstructed under the planned Morasko Geopark and should favor the perception of landscape values. Near the geosite Morasko Hill is the geosite Morasko Meteorite, which is located on the northern outskirts of Morasko Hill. This site represents the well-preserved remnants of the meteorite impact that occurred in the middle Holocene, circa 5000 years ago (Stankowski, 2001, 2008; Szczuciński et al., 2016; Tobolski, 1976). The Morasko impact represents the largest documented iron meteorite shower in central Europe and it is unique in the world due to the presence of impact traces in soft glaciogenic sediments (Muszyński et al., 2014). The site includes seven impact craters, the largest of which reaches a diameter of 100 m and the deepest one has a depth of up to 11.5 m (Stankowski, 2009). The described craters are circular and bowl-shaped; they display a symmetric distribution around the largest crater (Włodarski et al., 2017; Fig. 4.8). Some of the craters are continuously or periodically filled with water. Since 1976, the site has been protected as the impact site in vicinity of hornbeam-oak forest.

Alongside these registered geosites, there is a new proposal to establish three other geosites that are connected to human activity in Poznań. The first two are in the center of Poznań: Genius Loci and the Fortress of Poznań. The third geosite, the Szachty, is located at th e southern edges of the city.

The Genius Loci geosite includes relics of a fragment of a defensive rampart surrounding a Medieval gord on Ostrów Tumski Island on the Warta River in Poznań. At present, the face of the city wall is exposed under the modern glass architectural form. The length of protected rampart is 500 m (Antowska-Gorączniak, 2013, Olek et al., 2016). In the exposed part, sediments of both the foundations of the gord and the anthropogenic sediments that were superimposed in the later stages of city development can be traced.

The exposure of the cross-section gives the possibility of observing these deposits in horizontal (spatial) as well as vertical (temporal) layout. Geomorphologically, it is a floodplain. Hypsometric differences show that the gord was situated in areas prone to frequent floods and periodical stagnation of surface waters, as evidenced by layers of alluvia containing malacofauna (Kurzawska, 2013). Alluvial ground sediments occur at a height of 52 m ASL, with an average surface area of approximately 58 m ASL (Kaniecki, 2013). It can be concluded that about 7 m of anthropogenic sediments are deposited in the rampart section (Fig. 4.10). On the other hand, the rampart itself is constructed of an extra sand layer that levels the area

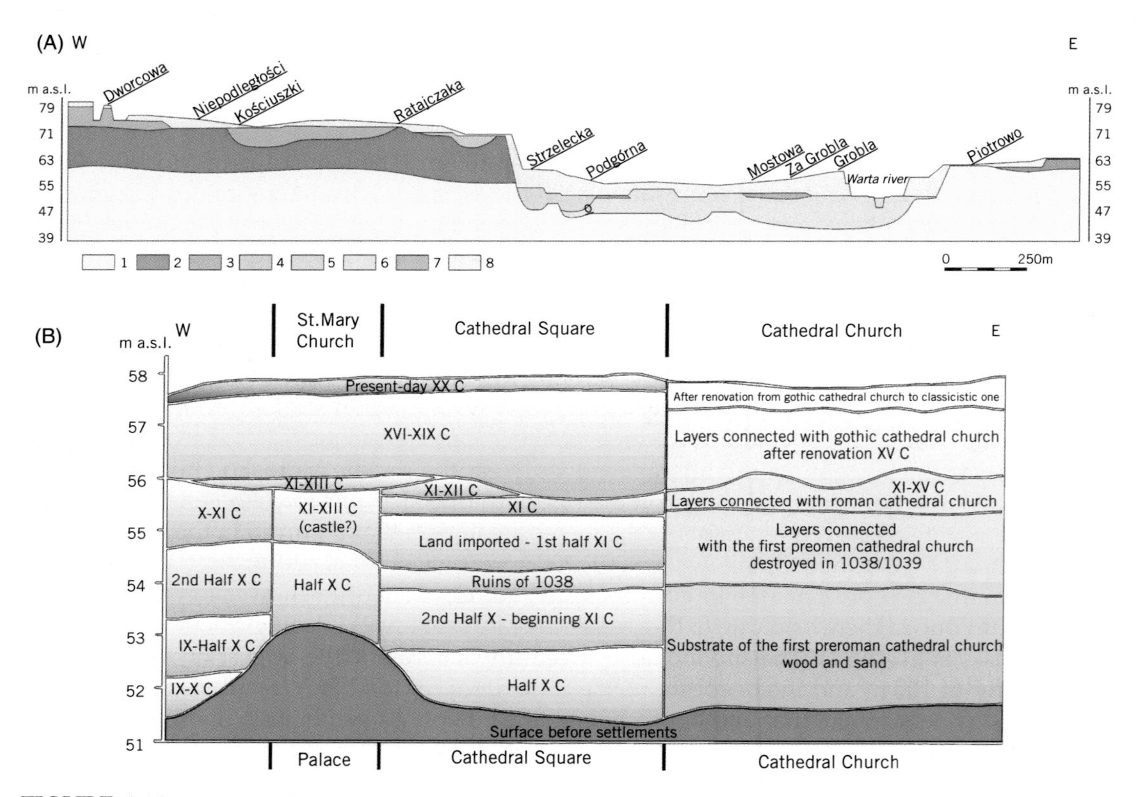

FIGURE 4.10 Geological cross-sections through Poznań. (A) Cross-section for Poznań by Troć (2005), (1) Neogene clays; (2) tills of the Poznań phase; (3) fluvioglacial sands and gravels of the Poznań phase; (4) lacustrine clay and muds of the Poznań phase; (5) Holocene alluvial deposits, Terrace II; (6) Holocene alluvial deposits, Terrace I; (7) Holocene muds; and (8) anthropogenic deposits. (B) Anthropogenic deposits on Ostrów Tumski by Kaniecki (2013). *Source: Troć M., 2005. Warunki geośrodowiskowe na obszarze Starego Miasta w Poznaniu. MS; Kaniecki, A., 2013. Wpływ antropopresji na przemiany środowiskowe w dolinie Warty w Poznaniu. Landf. Anal. 24, 23-34.*

and is hardened with fascines (branches, bark, wood roof planks, small stones, and braids). A fragment of the 16th-century wall was recorded in the ceiling layers of the rampart. Two layers have been preserved: the compensating layer formed because of its construction and the postdemolition layer consisting of brick rubble and lime mortar. In various places of archaeological excavation, the 16th-century layers are covered by anthropogenic layers from the 18 to 19th centuries of numerous historical materials. The last segment recorded during the excavation research was the humus layer.

The second proposed geosite; The Fortress of Poznań includes 22 military forts, mainly from the second half of the 19th century (Kaniecki, 2013). Together with the enormous citadel, they constitute the then-most modern Prussian fortress, an example of military architecture with widespread ground reworked around forts, especially near the citadel.

The other form worth paying attention to in the future is the geosite: The Szachty (shafts). This geosite comprises 40 disused pits turned into artificial waterbodies, which at present are associated aquatic ecosystems (Stępniewska, Abramowicz, 2016). They are connected to the excavation of till on a flat moraine plateau at the beginning of the 20th century (Graf, 1995). Now, it is of ecological utility and a recreation area for citizens. These last two geosites are described in a subchapter, Anthropogenic changes in morphological landscape, and are presented in more detail by Zwoliński et al. (2017).

4.6 CONCLUSION

Poznań, located in the postglacial area, belongs to cities in Poland with Medieval roots. A common feature of many significant Medieval urban centers was the location by the river, which could provide both defensive and communication functions. In their development, one can distinguish gord centers—the seat of princely authorities (Poznań was founded on the Warta River floodplain in the 10th century) and then a fortified town with walls and moats (Poznań was established on the left-bank moraine plateau on April 23rd, 1253) (Jurek, 2005). The Prince's gord in Poznań had its beginnings on the Warta River floodplain and its right-bank tributary Cybina River, whereas the town was located on the second and third left-bank meadow terraces (Kaniecki, 2004, 2013). This stage of Poznań's development was limited to the Warta River valley and, in particular to the terrace system. From this period, the stages of building up the second terrace to the level of the third terrace are recorded in valley deposits. Due to frequent floods from at least the Middle Ages until the middle of the 20th century, the largest thickness of anthropogenic sediments within Poznań is related to the area of the Warta River valley (Kaniecki, 2015). In Poznań's historical part of Ostrów Tumski, depending on local conditions, these sediments range from 5 to 7 m in thickness to about 3 m thick on the Old Market Square (Kaniecki, 2004, 2013). Along with the development of the town, there were also anthropogenic changes in the pattern of the Warta River channels and its tributaries. The multichannel Warta River finally became a forced straight and partly meandering river flowing in one channel in the 19 and 20th centuries (Kaniecki, 2013). Thus, until the end of the 18th century, the town developed within the Warta valley.

The settlement did not spread past the Medieval walls until the end of the 18th century, leaving the river valley and entering neighboring areas of the moraine plateaus on the left and right banks of the Warta River. This caused the inhabitants of Poznań to turn away or at least move away from the Warta River. In the first half of the 19th century, the city was surrounded by a ring of Prussian fortifications, which began to be dismantled in the early 20th century (Wilkaniec, Urbański, 2010). However, to this day, these fortifications are partially developed public spaces and partially secured ruins. In the second half of the 20th century, the Warta River channel was stabilized, the riverbed was partly straightened and shortened, and in the area of the city center its riverbed gained concrete embankments. In the 1980s, the town reached its current administrative boundaries, including many neighboring satellite towns, thus, creating the center of the Poznań metropolis. Currently, the dominant direction of Poznań's development is toward the north, where there are precious areas of abiotic and biotic nature, including meteor craters on terminal moraines covered by compact afforestation. The expansion of Poznań toward the north was caused, among other reasons, by the foundation of the Morasko Campus of the Adam Mickiewicz University in the vicinity of Morasko Hill.

In the 21st century, the inhabitants of Poznań are turning back towards the Warta River valley. Recreational, sports, and tourist areas are created on the section of the regulated river channel. Such a solution is possible because of the ever-smaller maximum discharges of the river in the last 40 years and the regulation of water flows as a result of the construction of the Jeziorsko reservoir in the upper reaches of the Warta River in 1986 (Kaczmarek, 2010).

Poznań is undoubtedly a typical central European lowland town, located on the river, with a history of urban development (*de facto* urban sprawl) associated with the hydrological history of the river. In turn, the hydrological history of the river affected the geomorphological evolution of the river valley and its system of terraces. Similar to many cities in central Europe, Poznań's proximity to the Warta River channel and valley, together with its hydrogeomorphological controls and evolution, has clearly influenced the city's location and development. Further, as demonstrated in this case study, river valleys in urban areas with rich natural history and anthropogenic metamorphoses are predisposed places for establishing geosites that have valuable geomorphological and geoarchaeological advantages (Zwoliński et al., 2017).

References

Antowska-Gorączniak, O., 2013. Badania na stanowisku przy ul. Posadzego 5 w 2009/2010 r. – charakterystyka nawarstwień. In: Kóčka-Krenz, H. (ed.), Poznań we wczesnym średniowieczu, 8. Wydawnictwo Nauka i Innowacje, Poznań, 19–60.

Bartczak, E., 1993. Szczegółowa Mapa Geologiczna Polski w skali 1:50 000, arkusz Kórnik z objaśnieniami. PIG, Warszawa.

Bartkowski, T., 1957. Rozwój polodowcowej sieci hydrograficznej w Wielkopolsce Środkowej. Zeszyty Naukowe UAM w Poznaniu. Geografia 8 (1), 3–79.

Bartkowski, T., 1973. Fizjografia Poznania. In: Błaszczyk, W. (Ed.), Początki i Rozwój Starego Miasta w Świetle Badań Archeologicznych i Urbanistyczno-Architektonicznychpp. 19–36, Mat. ogólnopol. symp., 18–19 października 1973, Poznań.

Bartkowski, T., Krygowski, B., 1959. Próba kartograficznego ujęcia geomorfologii najbliższej okolicy Poznania. Zeszyty Naukowe UAM. Geografia 2 (21), 87–94.

Biesiadka, J., Gawlak, A., Kucharski, Sz., Wojciechowski, M., 2006. Twierdza Poznań: o fortyfikacjach miasta Poznania w XIX i XX wieku. Wydawnictwo Rawelin, Poznań.

Chachaj, J., 1996. Szczegółowa Mapa Geologiczna Polski w skali 1:50 000, arkusz—Mosina z objaśnieniami. PIG, Warszawa. Biedrowski, Z., 1968. Sandry Okolic Poznania. Studium geomorfologiczno-sedymentologiczne. UAM, Poznań, MS Ph.D. Thesis.

Chmal, R., 1992. Szczegółowa Mapa Geologiczna Polski w skali 1:50 000, arkusz Stęszew z objaśnieniami. PIG, Warszawa.

Chmal, R., 1996. Szczegółowa Mapa Geologiczna Polski arkusz 471—Poznań (N-33–130-D). PIG, Warszawa.

Chmal, R., 1997. Objaśnienia do Szczegółowej Mapy Geologicznej Polski, arkusz 471—Poznań (N-33–130-D). PIG, Warszawa.

Cincio, Z., 1996. Szczegółowa Mapa Geologiczna Polski w skali 1:50,000, arkusz Swarzędz z objaśnieniami. PIG, Warszawa.

Ciuk, E., 1978. Geologiczne podstawy dla nowego zagłębia węgla brunatnego w strefie rowu tektonicznego Poznań–Czempiń–Gostyń. Przegląd Geologiczny 26, 588–594.

Dębski, A., Sikorski, A., 2005. Ostrów Tumski 10 – charakterystyka warstw i materiałów źródłowych w wykopie XXIV. In: Kóčka-Krenz, H. (Ed.), Poznań we wczesnym średniowieczu, vol. 4, pp. 23–58.

Dyjor, S., 1970. Seria poznańska w Polsce Zachodniej. Kwartalnik Geologiczny 14 (4), 818–833.

Gogołek, W., 1993. Szczegółowa Mapa Geologiczna Polski w skali 1:50 000, arkusz Buk z objaśnieniami. PIG, Warszawa.

Graf R., 1995: Zmiany stosunków wodnych w dorzeczu Strumienia Junikowskiego związane z kopalnictwem. In: Kaniecki, A. (Ed.), Dorzecze Strumienia Junikowskiego, Stan obecny i perspektywy, Wydawnictwo Sorus, Poznań, 14–21.

Grocholski, W., 1991. Budowa geologiczna przedkenozoicznego podłoża Wielkopolski. Przewodnik 62 Zjazdu PTG, Poznań, 7–18.

Hildebrandt-Radke, I., 2016. Środowisko geograficzne Poznania. In: Kara, M., Makohonienko, M., Michałowski, A. (Eds.), Przemiany osadnictwa i środowiska przyrodniczego poznania i okolic od schyłku starożytności do lokacji miasta. Bogucki Wydawnictwo Naukowe, Poznań, pp. 23–46.

Józefowiczowa, K., 1963. Z Badań nad Architekturą Przedromańską i Romańską w Poznaniu. IHKM, PAN,, Ossolineum, Wrocław.

Jurek, T., 2005. Przebieg lokacji Poznania. In: Civitas Posnaniensis: Studia z dziejów średniowiecznego Poznania, Poznańskie Towarzystwo Przyjaciół Nauk, Poznań, 173–191.

Kaczmarek, H., 2010. Analiza zdjęć lotniczych oraz wyników pomiarów geodezyjnych w badaniach dynamiki strefy brzegowej sztucznych zbiorników wodnych - zbiornik Jeziorsko, rzeka Warta. Landform Analysis 13, 19–26.

Kaniecki, A., 1993. Poznań: dzieje miasta wodą pisane. Przemiany rzeźby i sieci wodnej, część 1, Wyd. Aquarius, Poznań.

Kaniecki, A., 2004. Poznań. Dzieje Miasta Wodą Pisane. Cz. I–III. Wydawnictwo PTPN, Poznań.

Kaniecki, A., 2013. Wpływ antropopresji na przemiany środowiskowe w dolinie Warty w Poznaniu. Landf. Anal. 24, 23–34. doi: 10.12657/landfana.024.003.

Kaniecki A., 2015. Rola i znaczenie Warty w dawnym Poznaniu. Badania Fizjograficzne VI, ser. A, Geografia Fizyczna 66, 47–69.

Karczewski, A., 1961. Morasko Hill: example of a terminal push moraine of the Poznań stage. Vth Congress INQUA, Guide–Book App. 21–22.

Karczewski, A., 1976. Morphology and lithology of closed depression area located on the northern slope of Morasko Hill near Poznań. Hurnik, H. (Ed.), Meteorite Morasko and the region of its fall, vol. 2, Adam Mickiewicz University Press, Poznań, Ser. Astronomia, pp. 7–20.

Karczewski, A., Mazurek, M., Stach, A., Zwoliński, Zb., Dmowska, A., 2007. Mapa geomorfologiczna Niziny Wielkopolsko-Kujawskiej pod redakcja B.Krygowskiego, 1:300 000. Instytut Paleogeografii i Geoskologii, Uniwersytet im. Adama Mickiewicza, Poznań.

Kóčka-Krenz, H., 2015a. Poznań—od grodu do miasta. Archeologia Histotrica Polona 23, 121–138. doi: 10.12775/AHP.2015.005.

Kóčka-Krenz, H., 2015b. Proces formowania się państwa Piastów. Folia Praehistorica Posnaniensia, 205–218. doi: 10.14746/fpp.2015.20.12.

Kondracki, J., 2011. Geografia regionalna Polski. PWN, Warszawa.

Kozarski, S., 1986. Skale czasu a rytm zdarzeń geomorfologicznych vistulianu na Niżu Polskim. Czasopismo Geograficzne 52 (2), 247–270.

Kozarski, S., 1995. Deglacjacja północno-zachodniej Polski: warunki środowiska i transformacji geosystemu (~20 ka–10 ka BP). Dokumentacja Geograficzna, 1.

Krąpiec M., 2013. Dendrochronologiczne datowanie wału grodu poznańskiego na podstawie drewna wyeksplorowanego podczas badań prowadzonych przy ulicy ks. Posadzego 5 w 2009 r. In: Kóčka-Krenz, H. (ed.), Pozna we wczesnym średniowieczu, 8. Wydawnictwo Nauka i Innowacje, Poznań, 285–292.

Kruszka, K. (Ed.), 2008. Statystyczna karta historii Poznania. Urząd Statystyczny w Poznaniu.

Krygowski B., 1953. Mapa geomorfologiczna Polski w skali 1:300,000.

Krygowski, B., 1961. Geografia Fizyczna Niziny Wielkopolskiej. Cz. I. Geomorfologia, Poznańskie Towarzystwo Przyjaciół Nauk. Wydział Matematyczno-Przyrodniczy, Poznań, 204 p.

Krygowski, B., Żurawski, M., 1968. Objaśnienia do Przeglądowej mapy hydrogeologicznej Polski 1:300 000. Arkusz Poznań. Wydawnictwa Geologiczne, Warszawa, 1–151.

Kunkel, A., 1975. Osady iłowe neogenu młodszego Wielkopolski środkowej w świetle bibułowej chromatografii rozdzielczej. Poznańskie Towarzystwo Przyjaciół Nauk, vol. 14, Prace Komisji Geograficzno-Geologicznej, Warszawa-Poznań.

Kurzawska, A., 2013. Analiza malakologiczna. Kóčka-Krenz, H. (Ed.), Poznań we wczesnym średniowieczu, vol. 8, Wydawnictwo Nauka i Innowacje, Poznań, pp. 331–336.

KZGW (Krajowy Zarząd Gospodarki Wodnej), 2017. Podział Hydrograficzny Polski, skala 1:50 000.

Lindner L., 1988. Stratigraphy and extents of Pleistocene continental glaciations in Europe. Acta Geologica Polonica 38 (1-4), 63–83.

Makohonienko, M., Kara, M., Hildebrandt-Radke, I., Jasiewicz, J., Antczak-Górka, B., Michałowski, A., 2016. Rozwój wczesnomiejskich założeń środkowej Wielkopolski—środowisko i gospodrka miasta Poznania rekonstruowana na podstawie archiwów kopalnych. Zarys problematyki. In: Kara, M., Makohonienko, M., Michałowski, A. (Eds.), Przemiany osadnictwa i środowiska przyrodniczego Poznania i okolic od schyłku starożytności do lokacji miasta.Bogucki Wydawnictwo Naukowe, Poznań, pp. 13–21.

Marks, L., Dzierżek, J., Janiszewski, R., Kaczorowski, J., Lindner, L., Majecka, A., Makos, M., Szymanek, M., Tołoczko-Pasek, A., Woronko, B., 2016. Quaternary stratigraphy and palaeogeography of Poland. Acta Geologia Polonica 66(3), 403–427.

Milecka, K., Nyćkowiak, M., Troć, M., 2010. Wiek osadów międzyglinowych na lewym brzegu Warty w Poznaniu w świetle badań palinologicznych. Badania Fizjograficzne A61, 105–118.

Muszyński, A., Stankowski, W., Szczuciński, W., 2014. Field excursion to the "Morasko Meteorite" Reserve. Mars—Connecting Planetary Scientists in Europe, Poznań, 6 June 2014, 1–11.

Olek, W., Majka, J., Stempin, A., Sikora, M., Zborowska, M., 2016. Hydroscopic properties of PEG treated archaeological wood from the rampart of the 10th century stronghold as exposed in the Archaeological Reserve Genius Loci in Poznań (Poland). Journal of Cultural Heritage 18, March-April, 299–305.

Pożaryski, W., 1974. Podział obszaru Polski na jednostki tektoniczne. In: Pożaryski, W. (Ed.), Budowa geologiczna Polski. Tektonika, Cz. 1. Niż Polski: 24–34.

Pawłowski, S., 1929. Rozważania nad morfologią doliny Warty pod Poznaniem. Badania Geograficzne nad Polską Północno-Zachodnią 4 (5), 91–106.

Pasławski, Z., 1956. Wybitne wezbrania Warty pod Poznaniem i prawdopodobieństwo wystąpienia najwyższych rocznych stanów wody. Przegląd Geograficzny I, (IX/1).

PGI [Polish Geological Institute], 2006. Geological map of Poland, scale 1:500 000, Warszawa.

Reynard, E., Pica, A., Coratza, P., 2017. Urban geomorphological heritage: an overview. Quaestiones Geographicae 36 (3), 7–20.

Stankowski, W., 2001. The geology and morphology of the natural reserve "Meteoryt Morasko". Planet Space Sci. 49, 749–753.

Stankowski, W., 2008. Morasko meteorite a curiosity of the Poznań region: Time and Results of the Fall. Adam. Mickiewicz University Press, Poznań, Ser. Geologia, 9, 91 p.

Stankowski, W., 2009. Meteoryt Morasko. Wydawnictwo Naukowe UAM, Poznań.

Stankowski, W., 2011. Rezerwat Meteoryt Morasko: morfogeneza kosmiczna zagłębień terenu. Landf. Anal. 16, 149–154.

Stępniewska, M., Abramowicz, D., 2016. Social on and the use of ecosystem services on municipal post-mining lands. An example of Szachty in Poznań. Ekonomia i Środowisko 4(59), 252–262.

Sydow, S., 1996. Szczegółowa Mapa Geologiczna Polski w skali 1:50 000, arkusz Murowana Goślina z objaśnieniami. PIG, Warszawa.

Szałata, A., Makohonienko, M., Jasiewicz, J., 2016. Analiza przestrzenna rozmieszczenia dawnego osadnictwa na obszarze miasta Poznania w świetle źródeł archeologicznych. In: Kara, M., Makohonienko, M., Michałowski, A. (Eds.), Przemiany osadnictwa i środowiska przyrodniczego poznania i okolic od schyłku starożytności do lokacji miasta. Bogucki Wydawnictwo Naukowe, Poznań, pp. 167–204.

Szczuciński, W., Pleskot, K., Makohonienko, M., Tjallingii, R., Apolinarska, K., Cerbin, S., Goslar, T., Nowaczyk, N., Rzodkiewicz, M., Słowiński, M., Woszczyk, M., Brauer, A., 2016. Environmental effects of small meteorite impact in unconsolidated sediments—case of iron meteorite shower in Morasko, Poland. 79th Annual Meeting of the Meteoritical Society, Berlin: paper, vol. 6433, p. 1.

Thornbush, M., 2015. Geography, urban geomorphology and sustainability. Area 47 (4), 350–353. doi: 10.1111/area.12218.

Tobolski, K., 1976. Palynological investigations of the bottom sediments in closed depressions Meteorite Morasko and the region of its fall. Hurnik, H. (Ed.), Meteorite Morasko and the Region of its Fall, 2, Adam Mickiewicz University Press, Poznań, pp. 21–26, Ser. Astronomia.

Tomaszewski, E., 1960. Mapa geomorfologiczna Polski w skali 1:50 000. In: Galon, R. (Ed.) Arkusz Poznań i Kostrzyn. Instytut Geografii PAN.

Topolski, J., 1988. Dzieje Poznania, T. 1–2, PWN, Warszawa.

Troć M., 2005. Warunki geośrodowiskowe na obszarze Starego Miasta w Poznaniu. MS.

Troć, M., Milecka, K., 2008. Wiek osadów aluwialnych doliny Warty oraz doliny Cybiny i Bogdanki w rejonie śródmieścia w Poznaniu. Badania Fizjograficzne nad Polską Zachodnią ser. A 59, 145–160.

Wilkaniec, A., Urbański, P., 2010. Twierdza Poznań w krajobrazie na przestrzeni XIX i XX wieku – od krajobrazu rolniczego po zurbanizowany. Acta Scientiarum Polonorum, Administratio Locorum 9(2), 147–158.

Włodarski, W., Papis, J., Szczuciński, W., 2017. Morphology of the Morasko crater field (western Poland): influences of pre-impact topography, meteoroid impact processes, and post-impact alterations. Geomorphology 295, 586–597.

Zagórski, Z. (Ed.), 2008. Nazewnictwo geograficzne Poznania. Wydawnictwo Naukowe UAM, Poznań.

Zwoliński, Zb., Kostrzewski, A., Stach, A., 2008. Tło geograficzne współczesnej ewolucji rzeźby młodoglacjalnej. In: Starkel, L., Kotarba, A., Kostrzewski, A., Krzemień, K. (Eds.), Współczesne Przemiany Rzeźby. SGP, IGiGP UJ, IGiPZ PAN, Kraków.

Zwoliński, Zb., 1992. Sedimentology and geomorphology of overbank flows on meandering river floodplains. Geomorphology 4(6), 367–379.

Zwoliński, Zb., Hildebrandt-Radke, I., Mazurek, M., Makohonienko, M., 2017. Existing and proposed urban geosites values resulting from geodiversity of Poznań City. Quaestiones Geographicae 36 (3), 125–149. doi: 10.1515/quageo-2017–0031.

Żynda, S., 1996. Rzeźba terenu, geomorfologia. In: May, J., Stelmasiak, S., Ludwiczak, I., Niezborała, M. (Eds.), Środowisko Naturalne Miasta Poznania, Część 1. Urząd Miejski w Poznaniu. Wydział Ochrony Środowiska, Poznań, pp. 15–22.

SECTION II

ANTHROPOGEOMORPHOLOGY

5 *Urban Stream Geomorphology and Salmon Repatriation in Lower Vernon Creek, British Columbia (Canada)* 81

6 *Landform Change Due to Airport Building* 101

CHAPTER

5

Urban Stream Geomorphology and Salmon Repatriation in Lower Vernon Creek, British Columbia (Canada)

Alexander MacDuff, Bernard O. Bauer

University of British Columbia, Kelowna, BC, Canada

OUTLINE

5.1 Introduction 81

5.2 Methods 84

5.2.1 Reach and Subreach Characterization by Stream Walk 84

5.2.2 Topographic Surveys 84

5.2.3 In-Channel Flow Hydraulics 85

5.2.4 Bed and Bank Characterization 85

5.2.5 Hydraulic Modeling 86

5.3 Results 87

5.3.1 Reach Differentiation based on Visual Descriptors 87

5.3.2 Substrate Size 89

5.3.3 Flow Hydraulics 90

5.3.4 Spawning Habitat Assessment 93

5.4 Discussion 96

5.5 Summary and Conclusions 97

References 98

5.1 INTRODUCTION

The government of British Columbia, Canada undertook a series of engineering projects in the mid-1950s that led to the channelization and transformation of the Okanagan River and many of its tributaries for purposes of flood control and navigation improvement (Okanagan Nation Alliance, 2004). Channel straightening reduced the total length of the Okanagan River by one-half, with significant reductions in the quantity and quality of natural aquatic and riparian habitat. The construction of the McIntyre Dam, near Oliver, BC (Fig. 5.1), as well as two other dams, at Okanagan Falls and Penticton, and 17 vertical drop structures between Osoyoos Lake and Skaha Lake were required to manage discharge, reduce flow velocity, and prevent

Urban Geomorphology. http://dx.doi.org/10.1016/B978-0-12-811951-8.00005-9

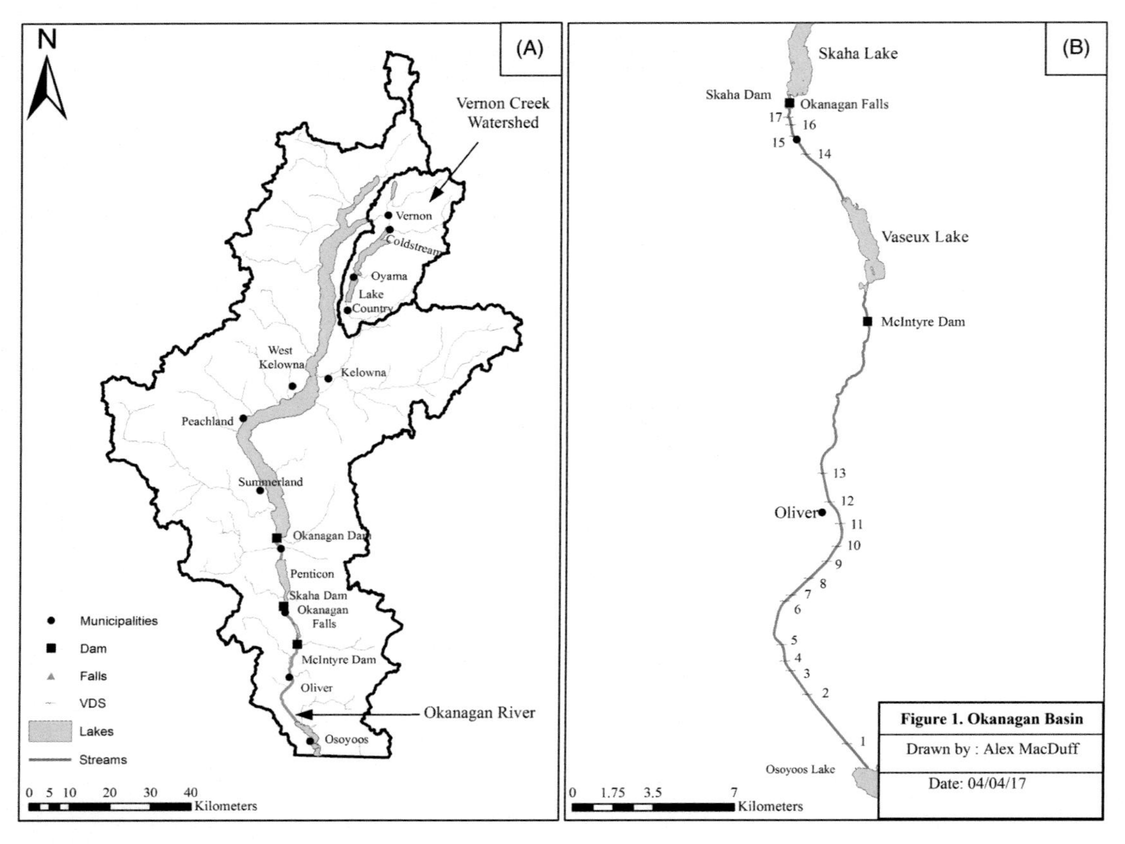

FIGURE 5.1 Location of (A) the Vernon Creek watershed within the Okanagan basin and (B) dams and vertical drop structures along the Okanagan River, north of the Canada-US border.

channel bottom scour. These flow-control structures effectively precluded fish passage by anadromous salmon into Okanagan main-stem lakes and their tributaries (Okanagan Nation Alliance, 2004).

One such tributary at the northern end of Okanagan Lake, Lower Vernon Creek, is an 11-km stream that emanates from Kalamalka Lake and flows northwest through the city of Vernon (Fig. 5.1). At Polson Park, in the heart of the city, the creek turns southwest and flows toward the north end of Okanagan Lake. Less than 2 km of the stream, mostly near the downstream end, can be considered natural, whereas the upstream 9 km have been subject to intensive engineering modifications that were intended to protect properties and infrastructure from erosion and flooding in the urban area.

Nevertheless, Lower Vernon Creek was identified (and continues to serve) as a major spawning ground for Kokanee salmon that reside in Okanagan Lake. Kokanee salmon are a land-locked, nonanadromous variant of Sockeye salmon that bears the same species name (*Oncorhynchus nerka*) and is used in this study as a surrogate model for Sockeye salmon.

It has been estimated that approximately 90% of the spawning habitat for Kokanee has been lost in the Okanagan region since the 1950s (Shepard and Ptolemy, 1999), with only about 1% remaining in Lower Vernon Creek.

Very little is known about anadromous salmon populations (e.g., Sockeye, Coho, Chinook) that would have been in Lower Vernon Creek historically because recorded fish counts only date back to 1971 (Sisiutl Resources, 1986). These fish counts focused on Kokanee, Trout, and other indigenous and invasive species because anadromous salmon were no longer in the system this far north of the Canada-US border. Historical maps and photographs show that large portions of Lower Vernon Creek had been channelized by the early 1920s, including the reaches that flow through Polson Park and the Vernon Golf and Country Club. According to newspaper articles from that period (available in the Vernon Museum), the original dam at the Kalamalka Lake outlet (completed in 1903) likely blocked anadromous salmon runs into the Upper Vernon Creek watershed.

Beginning in the 1980s, concern for declining salmon stocks in the Columbia River system led to heightened public awareness and greater interest in stream naturalization efforts that might mitigate the environmental damage done to aquatic systems by channelization and the emplacement of flow-control structures. Of particular note is the Okanagan River Restoration Initiative, spearheaded by the Okanagan Nation Alliance in collaboration with several government agencies in Canada and the US (Okanagan Nation Alliance, 2004). The long-term project objectives were to: (1) renovate McIntyre Dam and several flow-control structures in order to enable fish passage; (2) set back the banks and remeander the river at a site near Oliver; (3) create three spawning beds in the Penticton Channel, just below the outlet of Okanagan Lake; and (4) reintroduce Sockeye salmon into Skaha Lake (and other mainstem lakes). After historically small runs in the 1990s, with only 1500–2000 Sockeye salmon returning to the lower Okanagan River system, the numbers have since rebounded to the largest recorded runs, with 163 000 Sockeye counted in 2012 (Hume, 2014). Sockeye were reintroduced into Skaha Lake in 2014 and the Penticton Channel spawning beds have been active for the past three seasons. This success has translated into discussions on expanding restoration efforts to the Upper Okanagan basin by facilitating fish passage past the dam at the southern end of Okanagan Lake, near Penticton. A new project that will remeander a short reach of Mission Creek (near Kelowna) is close to completion (Davidson, 2016).

In light of the anticipated repatriation of Sockeye salmon into Okanagan Lake and its tributaries, this study was intended to assess the habitat potential of Lower Vernon Creek, which is the only fish-passable linkage between Okanagan and Kalamalka Lakes. Lower Vernon Creek, therefore, serves as a critical control on whether anadromous salmon will have access to other lakes (e.g., Wood, Ellison) and creeks higher in the Okanagan system. Should Lower Vernon Creek present a fish-passage barrier to migrating salmon or be found unsuitable for spawning and rearing, then a significant proportion of potential habitat would be unavailable to Sockeye repatriated to Okanagan Lake.

Despite efforts in the 1990s to enhance the habitat value of the upper reaches of Lower Vernon Creek, the Kokanee population has continued to decline. The most recent enumeration in 2003 counted only 69 spawners (Webster, 2004). If this decline is not symptomatic of a region-wide trend for all Okanagan Lake stock, then it suggests that the habitat value of Lower Vernon Creek continues to be influenced by urbanization in ways that are detrimental to salmonids in general, anadromous or otherwise.

5.2 METHODS

A combination of approaches was employed to accomplish the primary objective of assessing the habitat potential of Lower Vernon Creek in the context of heavy urbanization. An initial stream walk was undertaken in order to define the various reaches using qualitative descriptors and photo-documentation. Engineer's level surveys supplemented with digital geographic positioning (DGP) systems yielded cross-sectional and longitudinal profiles that defined creek geometry. Flow hydraulic parameterization at nine of the cross-sections was undertaken using an electromagnetic current meter and wading rod. Bed material and channel roughness were assessed using Wolman pebble counts (Kondolf, 2000), supplemented with grab samples and sieve analysis. Fish counts and spawning activity were documented with additional stream walks during the typical migration and spawning season (late August through early November). Hydraulic modeling was performed using HEC-RAS (US Army Corps of Engineers, 2015) to assess any potential habitat viability across a range of typical hydrologic conditions.

5.2.1 Reach and Subreach Characterization by Stream Walk

Prior to the stream walks, Lower Vernon Creek was divided into five primary reaches (Fig. 5.2) following the suggestions in Sisiutl Resources (1986) and mapped using Google Earth©. These reaches are numbered 1–5, in the upstream direction, with reach five further divided into five subreaches (A–E, in the upstream direction) because of the complexity inherent to that reach.

A reconnaissance stream walk was conducted on July 9, 2015 and every reach and subreach was described using the Google Earth© map as a reference. Information recorded during the stream walk included descriptions of channel morphology, the presence of fish, and crude measurements of flow hydraulics. A second stream walk was performed on October 4, 2015, with the objective of updating the fieldnotes to fall-time conditions, and using digital photography to document the typical state of each reach and subreach. In addition, the presence or absence of Kokanee salmon was documented (Fig. 5.3) because this was the peak period of spawning.

5.2.2 Topographic Surveys

On the basis of the reconnaissance stream walk, 13 long-term monitoring sites were established. The majority of sites were in Reach 4 (three sites) and Reach 5 (six sites) because those reaches experienced the greatest impact from urbanization and also had "the greatest (natural) potential for restoration" (Sisiutl Resources, 1986, p. 2). Based on gentle slope gradients and the fine substrate in lower reaches, it seemed unlikely that these would provide significant spawning habitat. At each site, the cross-sectional topography of the channel and banks was surveyed and marked with temporary benchmarks. A Topcon RTK-DGP system was used to georeference these benchmarks and a Leica engineer's level, survey rod, and tape measure were used to document the cross-section points relative to the benchmarks.

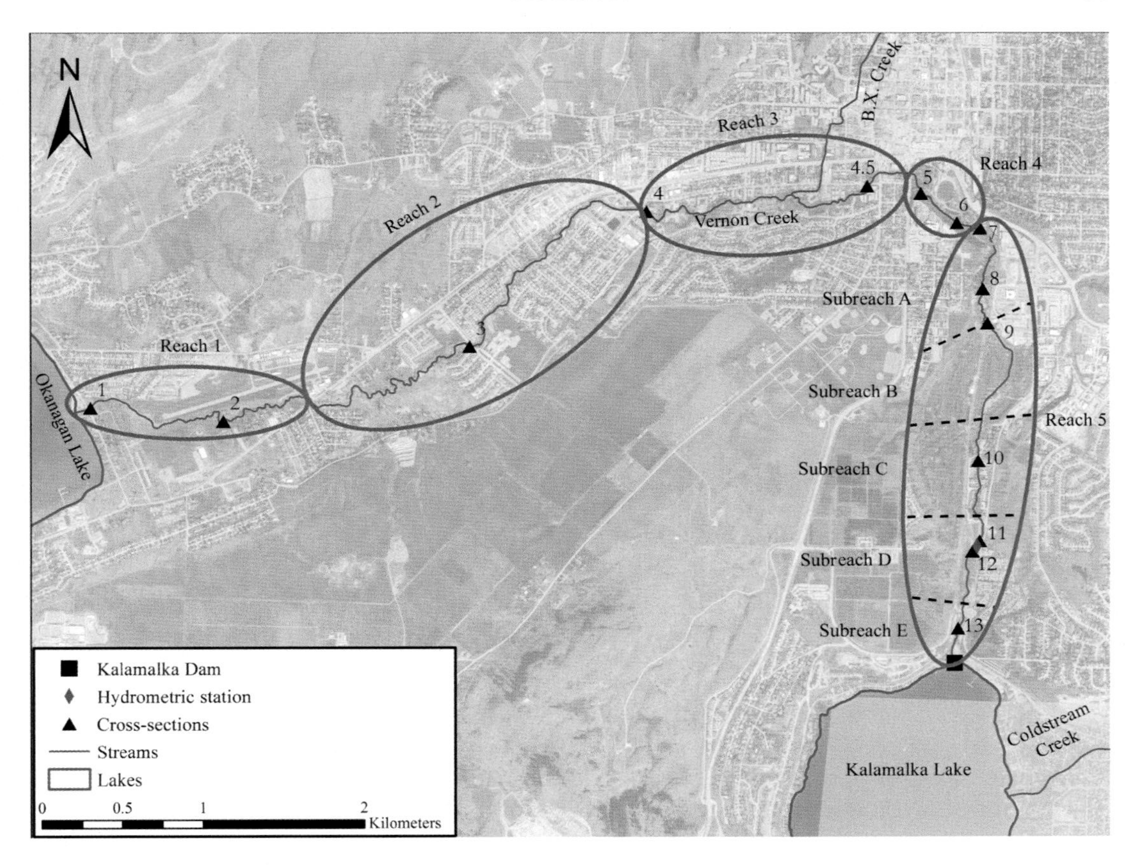

FIGURE 5.2 **Reach division of Lower Vernon Creek showing reach boundaries and cross-sections.**

5.2.3 In-Channel Flow Hydraulics

Measurements of mean flow depth and velocity were made at each cross-section using an electromagnetic current meter (Marsh McBirney Inc., Flow-Mate Model 2000) mounted on a standard wading rod. Because of shallow flow depth in most of the creek, only one measurement at 0.6 times the depth (from the water surface) was possible. A 30-second averaging interval was used. Data for each of the cross-sections were collected twice in early and mid-July 2015. In addition, a third set of hydraulic parameters was acquired in Reach 3 on October 12, 2015. The objective was to parameterize the conditions (velocity, depth) of active Kokanee spawning redds, primarily in glides. Samples were taken immediately downstream of a total of 27 redds, rather than upstream or above the redds in order to minimize sampling disturbances.

5.2.4 Bed and Bank Characterization

Due to the general coarseness of the substrate along most of the channel, the character of the bed-surface material was characterized using the Wolman method (Kondolf, 2000).

FIGURE 5.3 **Pair of spawning Kokanee, each about 200–300 mm long, observed at a glide downstream of the BX Creek confluence in Reach 3. Over 30 Kokanee utilized this specific glide for spawning (October 4, 2016).**

At each cross-section, a minimum of 100 pebbles were sampled at equal intervals across the bottom and a grain-size template was used to estimate the size of the *B*-axis passing through the opening. The data from the pebble counts were analyzed using the GRADISTAT software program (Blott and Pye, 2001). Sand-bedded sections in the lower reaches were excluded from this analysis, as salmonids are unable to utilize substrate dominated by sand and finer grained particles for spawning habitat. Finally, the composition of the banks, whether glaciolacustrine silts, fluvial sands and gravels, cement walls, riprap, or vegetated surfaces, was assessed visually and recorded in the fieldnotes during the stream walks.

5.2.5 Hydraulic Modeling

A one-dimensional hydraulic flow model (HEC-RAS 4.1) developed by the US Army Corps of Engineers (2015) was used to simulate the expected flow conditions at each of the cross-sections under a range of discharge regimes. The cross-section profiles from the surveys were used to populate the model and an initial calibration was conducted using the surveyed water levels and discharge values during the times of the survey. Channel roughness (Manning's n) was adjusted sequentially in the model so as to have the measured and modeled water levels align as closely as possible. Once the model was calibrated, a series of simulations were performed using a range of discharge values that were thought to represent average flow conditions during the spawning season. Discharge data were downloaded from the Water Survey of Canada for the permanent hydrometric station by the Alpine Centre shopping mall (Station: 08NM065).

5.3 RESULTS

5.3.1 Reach Differentiation based on Visual Descriptors

Although there can be considerable intrareach variability in the five reaches identified in this study, Figs 5.4 and 5.5 show a series of photographs intended to portray interreach differences, from downstream to upstream. Reach 1 (Fig. 5.4A) is approximately 1.9 km long, extending from the creek outlet at Okanagan Lake upstream to the intersection with Okanagan Landing Road. The channel is bordered mostly be wetlands, open fields, and parks and, therefore, retains a somewhat natural look. However, this low gradient, meandering reach has relatively slow flow velocities and is prone to sediment deposition. Substrate materials (sand, silt, and clay) are too fine and the water velocities too slow for this reach to be considered to be valuable spawning habitat for salmonids. In contrast, Reach 2 (Fig. 5.4B) has been described as "having an ideal slope for Kokanee spawning and is amply endowed with good spawning gravel" (Sisiutl Resources, 1986, p. 2). The majority of the channel, although confined by riprap along residential areas, has a seminatural meandering form and significant overstory of mature, riparian trees. The many riffle-pool sequences provide flow complexity, and the glides and riffles have high-quality substrate for spawning. On the October 4, 2015 stream walk, 288 Kokanee spawners were counted in Reach 2.

Reach 3 (Fig. 5.4C) extends from the crossing at 43rd Street upstream to the western boundary of Polson Park. The majority of the channel is constrained by private property on either

FIGURE 5.4 (A) Reach 1, XS2, looking downstream (August 10, 2015). (B) Reach 2, above Willow Garden, looking downstream (October 4, 2015). (C) Reach 3, near the BX Creek confluence, looking upstream (October 4, 2015). (D) Reach 4, XS6, Polson Park, looking upstream (November 12, 2015).

FIGURE 5.5 Reach 5, (A) XS8, looking downstream (October 4, 2015). (B) The Vernon Golf and Country Club, looking downstream (October 4, 2015). (C) Subreach C (October 4, 2015). (D) XS12 by Alpine Center mall, looking upstream (October 4, 2015).

side, with extensive lengths of artificial bank and some channel straightening. It is not as freely meandering as Reaches 1 and 2, but also not as straight as upstream reaches. Although there are many riffle-pool-glide sequences that provide good spawning gravels and moderate flows, these sequences are spaced out between long stretches of creek with heavily armored banks and coarse cobble substrate. Siltation by fines appears not to be an issue. During the stream walk of October 4, there were 348 Kokanee spawners counted, making it the most productive reach in the entire system, despite the relatively short length. However, the presence of Kokanee was sporadic, with bunches of 15–50 individuals at glides and very few between.

Reach 4 (Fig. 5.4D) spans Polson Park and contains three cross-sections (XS5–XS7). This very short reach is completely channelized and was described as the reach in most need of extensive improvement (Sisiutl Resources, 1986). The channel was straightened in order to mitigate flooding concerns and in 1990 concrete blocks were installed to armor the banks in order to prevent erosion and bed siltation (Sisiutl Resources, 1986). The flow conditions have been altered radically from the natural conditions, to yield a rectangular conveyance channel with little hydraulic complexity. The bed consists mostly of large cobbles and remnants of concrete blocks. At XS5 (downstream end of Reach 4), there is a single back eddy and pool that has formed in the wake of a large willow root and this is the farthest upstream location where Kokanee were found.

Reach 5 (Fig. 5.5) was subdivided into five subreaches (A–E). Subreach 5A (Fig. 5.5A), which extends from Polson Park to the Vernon Golf and Country Club, is notable because the stream is mostly unconfined and freely meandering through a natural floodplain with ample riparian vegetation. Several pool-riffle sequences are evident, but much of the reach has a layer of fine sediment covering the bed, likely due to persistent mudflow contributions from a steep bluff that abuts the stream as well as backwater effects from a beaver dam. A scenic boardwalk winds along the stream and residents use it extensively.

Subreach 5B is the section of stream that flows through the Vernon Golf and Country Club (Fig. 5.5B). The channel has been straightened and both banks have been armored with large boulders. Because of the steep gradient through this subreach, flow velocities are high and the substrate is dominated by cobbles and boulders. There is very little flow complexity and poor habitat quality for fish. Much of the stream is bordered by grassy fairways and there is virtually no natural riparian vegetation or shade. Subreaches 5C–E have a mix of natural and human-modified channel (Figs 5.5C and 5.5D). In some locations, there are shallow riffle-glide series, with moderate flows and plentiful spawning gravels located adjacent to deeper hiding pools with large woody debris; whereas, in other locations the banks are armored and the bottom substrate is covered with coarse, cobbles, and angular cement-block remnants. Subreach 5D (Fig. 5.5D) was subject to "enhancement" in 1990 in an effort to improve the Kokanee fishery (Sisiutl Resources, 1986). Nevertheless, no salmonids were counted in this subreach or any of the subreaches of Reach 5.

5.3.2 Substrate Size

Table 5.1 provides grain size statistics for cross-sections based on Wolman pebble counts. The substrate at XS1–2 and 8–9 was too fine to conduct pebble counts and much of the bed was covered in sand and silt at those locations. Cross-sections 1 and 2 were in Reach 1, where there are backwater influences from Okanagan Lake that lead to reductions in flow velocity and, hence, deposition. Cross-sections 8 and 9 are in Subreach 5A, where there was a beaver dam as well as silt contributions from the side slopes along the channel margin. Cross-section 7 was

TABLE 5.1 Grain Size Statistics for each Cross-Section based on Wolman Pebble Counts using GRADISTAT Sediment Analysis Program

XS	D_{50}	Arithmetic mean	Sorting index (σ)	<2 mm	2–8 mm	8–64 mm	>64 mm
	(mm)	(mm)	(mm)	(%)	(%)	(%)	(%)
3	7.4	8.7	5.5	4.0	33.6	62.4	0
4	25.9	30.0	23.5	0	10.7	71.3	18.1
4.5	6.1	8.1	5.9	1.7	44.1	54.2	0
5	29.2	38.0	36.5	0	21.6	59.5	19.0
6	29.7	39.7	30.0	0	0	85.6	14.4
10	22.6	45.5	44.6	0	4.4	68.1	27.4
11	16.4	17.4	11.5	4.4	17.7	77	0.9
12	10.6	13.9	9.1	0.9	13.8	85.3	0
13	4.1	22.2	30.6	39.8	16	21.2	23

Source: Blott, S., Pye, K., 2001. GRADISTAT: a grain size distribution and statistics package for the analysis of unconsolidated sediments. Earth Surf. Process. Landforms 26(11), 1237–1248.

covered in large pieces of concrete and asphalt, so a substrate assessment seemed inappropriate. Of the remaining cross-sections for which substrate statistics were assessed (Table 5.1), XS5 and 6 had the largest grain sizes and greatest fraction of substrate materials greater than 8 mm. These two cross-sections appear in Reach 4, which is the most heavily engineered and channelized portion of Lower Vernon Creek through Polson Park (Fig. 5.4D). Similarly, XS4 (Reach 3) and 10 (Subreach 5C) are in portions of the creek that have been heavily modified by humans. In contrast, fine bed materials were found at the lower (XS3) and upper (XS13) ends of Lower Vernon Creek. An additional cross-section was added in Reach 3 (XS4.5) because it was thought desirable to compare sites with little-to-no observed presence of fish (e.g., XS5) to sites immediately downstream that were heavily utilized by spawning Kokanee (XS4.5).

5.3.3 Flow Hydraulics

Field measurements of flow depth and velocity at each cross-section were used to calibrate the HEC-RAS model for Lower Vernon Creek from the outlet of Kalamalka Lake to the north end of Okanagan Lake. It proved difficult to calibrate certain cross-sections in part because of the large distances separating a few of them, but also because of backwater effects from beaver dams and a debris jam downstream of the cross-sections. This was mitigated by creating a pair of virtual (duplicate) cross-sections for each reach with calibration issues. These virtual cross-sections were identical in shape to the measured cross-section in the reach, except that the elevation of each of the survey points was adjusted according to the local slope of the channel in order to accommodate for distance downstream and upstream at which the virtual cross-sections were inserted. Although the addition of such virtual profiles to the HEC-RAS model does add an abstract dimension, it serves to improve the calibration of the measured cross-sections. No inferences are made about these virtual cross-sections with regard to fish habitat and all subsequent calculations are based only on the measured cross-sections.

Historical data on flow discharge hydrographs for Little Vernon Creek (Station: 08NM065) since 1979 were acquired from the Water Survey of Canada (Environment Canada, 2015) and a flood frequency analysis was conducted in order to determine long-term discharge statistics. The mean monthly discharges in September, October, and November as well as the 25 and 75th percentiles for October were used in the HEC-RAS simulations. These discharge values were selected as representative of typical flow conditions during the salmonid spawning period. Only the results from October are presented because the bulk of Sockeye spawning in the region occurs during this month. Additionally, the average discharges for other months are very similar.

Figs. 5.6 and 5.7 show the modeling results for mean flow velocity and mean flow depth, respectively, for each of the 13 cross-sections in the five reaches of Lower Vernon Creek. The recommended guidelines for maximum and minimum flow velocity and for minimum flow depth to enable successful Sockeye salmon spawning, as reported by the BC Ministry of Environment (Slaney and Zaldokas, 1997), are placed on the graphs as reference markers. From these results, it would appear that most of Lower Vernon Creek satisfies the velocity and depth requirements of spawning salmon, except at the creek outlet (Reach 1: XS1) and immediately upstream of Polson Park (Reach 5: XS8–9).

Long (2007) suggested that the spawning behavior of salmonids likely reflects a combination of flow variables and, therefore, depth and velocity should not be considered

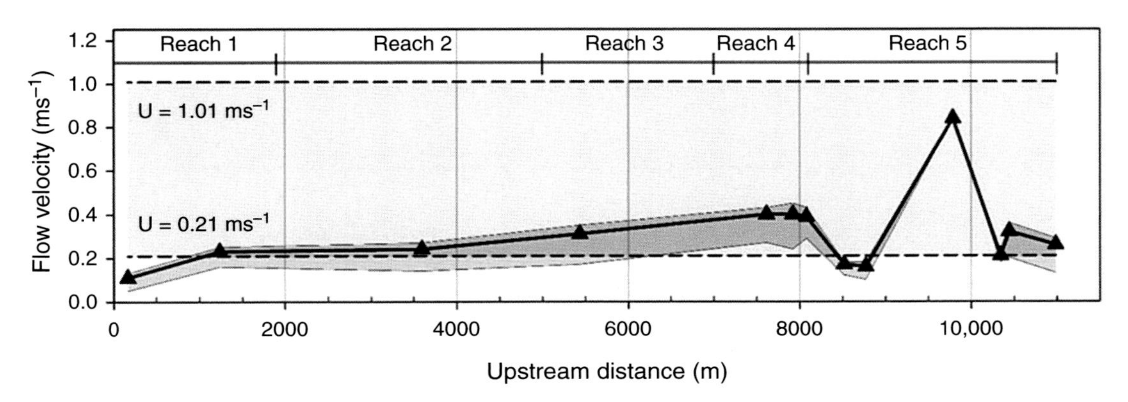

FIGURE 5.6 Longitudinal distribution of modeled flow velocity in Lower Vernon Creek relative to minimum and maximum requirements (*light gray shade*) for Sockeye salmon as recommended by the BC Ministry of Environment (Slaney and Zaldokas, 1997). Diamonds show locations of cross-sections (XS1 at lower end and XS13 at upper end), with the *solid line* indicating mean discharge conditions for October. The *dark gray-shaded* area above and below the *solid line* indicates the range of flow conditions given by the 25 and 75th percentiles for October discharge.

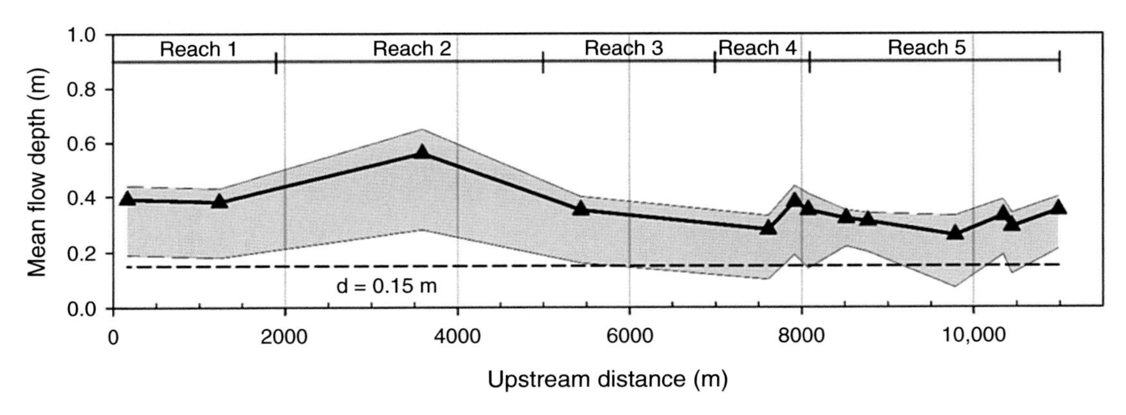

FIGURE 5.7 Longitudinal distribution of modeled flow depth in Lower Vernon Creek relative to minimum requirements for Sockeye salmon as recommended by the BC Ministry of Environment (Slaney and Zaldokas, 1997). Diamonds show locations of cross-sections, with the *solid line* indicating mean discharge conditions for October. *Dark gray-shaded* area above and below the *solid line* indicates the range of flow conditions given by the 25 and 75th percentiles for October discharge.

independently as constraining hydraulic parameters. The Froude number was suggested as a useful nondimensional parameter because it combines flow velocity and depth, as follows:

$$Fr = U / \sqrt{gd} \tag{5.1}$$

where U is mean flow velocity, d is mean flow depth, and g is gravitational acceleration. Values of $Fr > 1$ indicate "supercritical" or streaming flow conditions and values of $Fr < 1$ indicate "subcritical" or tranquil flow conditions (Long, 2007). Fig. 5.8 shows the longitudinal distribution of Fr at each cross-section based on results from the HEC-RAS model simulations for October

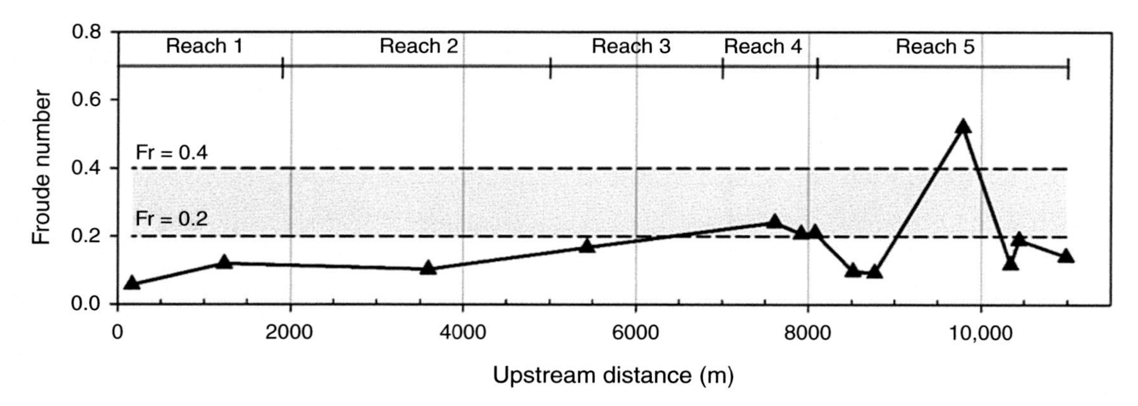

FIGURE 5.8 Longitudinal distribution of modeled *Fr* in the main channel of Lower Vernon Creek. Diamonds show locations of cross-sections, with the *solid line* indicating mean discharge conditions for October. *Gray shade* shows preferred range for Okanagan River Sockeye (Long, 2007).

discharge conditions. In addition, the maximum and minimum ranges for Okanagan River Sockeye salmon, as per Long (2007), are indicated for reference. Most of the cross-section results fall near or slightly below the lower limit ($Fr = 0.2$), whereas one cross-section (XS10 in the middle of Reach 5) falls well above the upper limit ($Fr = 0.4$). If *Fr* is indeed a critical indicator of spawning suitability, then Lower Vernon Creek would not seem to provide significant habitat potential.

On October 12, 2015, during the middle of the Kokanee spawning season, 27 active redds were identified in Reach 3 and depth-velocity measurements were taken in close proximity to, but not directly over, each redd so as not to cause disturbance. Flow velocity was measured at 0.05 m above the bed. The Kokanee data are represented as full circles in Fig. 5.9. The star symbols refer to data collected from the Adams River in the vicinity of redds occupied mostly

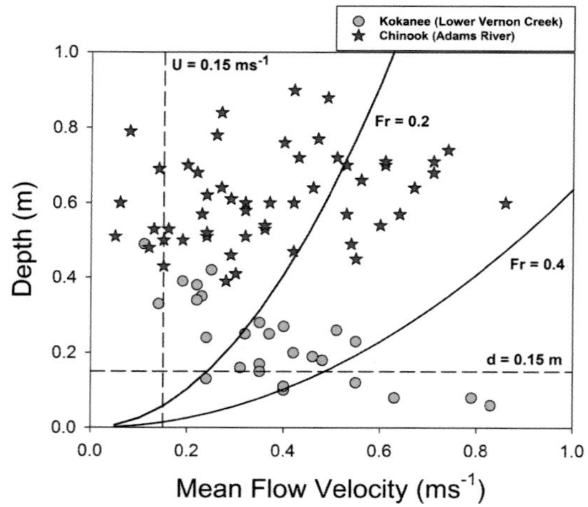

FIGURE 5.9 Depth-velocity data for 27 Kokanee redds in Reach 3 of Lower Vernon Creek (collected October 2015 by Alexander MacDuff) and 52 Chinook and Sockeye redds in the Adams River (collected October 2016 by Samuel Grenier). *Dashed lines* show lower limits for depth and velocity conditions for Sockeye salmon (Slaney and Zaldokas, 1997), whereas *solid curves* show *Fr* limits for Okanagan River Sockeye (Long, 2007).

by Chinook salmon, with some Sockeye salmon. Depth-velocity measurements were taken at 0.1 m above the bed. Reference lines are included for Sockeye salmon depth and velocity limits, as recommended by the Ministry of Environment (Slaney and Zaldokas, 1997), as well as the ranges for *Fr* recommended by Long (2007) for Okanagan River Sockeye salmon. Most of the data fall within the depth and velocity limits, but *Fr* appears to provide a much too constraining condition for these actively spawning fish. Despite the large spread in the data, there is considerable overlap between flow conditions for the Kokanee and Chinook redds. Each species occupies a range of near-bed flow velocities ranging between 0.1 and 0.7 ms^{-1}. The separation evident in the flow depth is explained partly by the fact that the Adams River is wider and deeper than Lower Vernon Creek and because Chinook are much larger than Kokanee and require deeper water in which to swim and spawn. Sockeye salmon are of intermediate size and can occupy shallower water than Chinook.

5.3.4 Spawning Habitat Assessment

The study was motivated by the likelihood of future reintroduction of Sockeye salmon to their historic range, which includes Okanagan Lake and its tributaries, and by the need to assess the impact of human modification to the parafluvial environment through the urbanized reaches of Lower Vernon Creek. Any attempt at applying best-management practices leading to repatriation of indigenous salmon to these waters must be adequately informed by scientific evidence. In particular, salmon spawning habitat suitability must consider substrate conditions and hydraulic-flow conditions.

5.3.4.1 Substrate Suitability

Table 5.1 provides measured substrate size characteristics for nine cross-sections where sediment-size analysis was judged useful, whereas Table 5.2 presents preferred size requirements for spawning Sockeye salmon. Most studies have focused on the median (D_{50}) or mean grain size of a representative substrate sample in subreaches where salmon are known to spawn. In addition, there is often concern for the proportion of fine-grained material (i.e., sand and silt) because of the potential to smother eggs or prevent adequate circulation of oxygenated water through the bed of redds. Kondolf (2000), for example, reported that for all

TABLE 5.2 Preferred Substrate Size for Sockeye Salmon during Spawning

Location (source)	D_{50} (mm)	Geometric mean (mm)	Range (mm)	<8 mm (%)	>64 mm (%)
Okanagan River, BC Kondolf and Wolman (1993)	25	18			
Little Wenatchee, WA Kondolf and Wolman (1993)	17.8	13.4			
Quartz Creek Kondolf and Wolman (1993)	19	18.4			
BC Average Slaney and Zaldokas (1997)			13–102		
This study				30	20

salmon redds sampled, the proportion of fines (defined as <8 mm) is <28%. For this study, the proportion of the sample with sizes >64 mm was also considered because a large proportion of cobble material on the bed would be detrimental to habitat productivity given the inability of medium-sized salmon, such as Sockeye, to excavate viable redds.

A comparison of Tables 5.1 and 5.2 suggests that only five of the 14 sites sampled meet the criteria for substrate size that define preferred spawning requirements of Sockeye salmon (XS4–5, XS10–12). Cross-sections 1–3, 8–9, and 13 were excluded from further consideration because the mean grain size was too small (e.g., XS1–2, XS13) or the proportion of fines was too large (e.g., XS3, XS13). Cross-sections 8 and 9 had veneers of fine materials over coarser substrate due to landslide contributions and beaver ponding. In contrast, XS6 and 7 (in Reach 4, through Polson Park) were deemed unsuitable for spawning because the substrate was tightly packed, cemented, and armored to the extent that particles were very difficult to extract during manual sampling. Given this, it seems inconceivable that fish would be capable of excavating redds, despite the mean grain sizes being in the preferred range.

Of the five cross-sections that were found to satisfy the preferred substrate characteristics for spawning Sockeye salmon, two of them (XS4–5) were in Reaches 3 and 4 and three of them (XS10–12) were upstream of Polson Park. XS4–5 and XS10 had relatively coarse means and a large fraction (>15%) of cobbles. Despite not being "well endowed with excellent spawning gravels" (as per Sisiutl Resources, 1986, p. 2), they satisfied the basic requirements for Sockeye spawning. In addition, XS4 is located in Reach 3, where most of the spawning Kokanee were found during the stream walks. Cross-sections 11 and 12 would appear to provide an ideal substrate for spawning salmon, despite there being no evidence of Kokanee in the reaches upstream of Polson Park. Subreach 5D upstream of the Alpine Center was heavily reengineered to create a sequence of steps and pools with the addition of large logs anchored to banks with steel cable (Fig. 5.5D). It is not clear from the construction reports whether gravel was added to the system in this subreach, but it is known that gravel was added elsewhere, such as in the channel through Polson Park (Young, 1991).

5.3.4.2 Hydraulic Suitability

Figs. 5.6–5.8 show the results of HEC-RAS steady-flow simulations for average October conditions along Lower Vernon Creek and these results are compared to recommended flow criteria for spawning salmon (Table 5.3). Based on flow depth and velocity as independent criteria, most cross-sections appear to provide suitable conditions for spawning. Indeed, flow depth appears not to be a limiting factor, except in a few cases and only if low-flow situations persist (i.e., lower 25% percentile of historic October flows). Minimum flow velocity is a more restrictive constraint, but even then only three cross-sections (XS1, XS8–9) fail to meet the minimum velocity requirements for Sockeye salmon spawning. Again, the low-flow states

TABLE 5.3 Preferred Flow Depths, Velocities, and *Fr* for Spawning Sockeye Salmon

Study (source)	Minimum depth (m)	Minimum velocity (m s^{-1})	Maximum velocity (m s^{-1})	Lower Froude number	Upper Froude number
BC Average Slaney and Zaldokas (1997)	0.15	0.21	1.01		
Okanagan River, BC Long (2007)	0.18	0.37	0.88	0.2	0.4

may be of concern because many cross-sections would fall below the minimum threshold of 0.21 ms^{-1}. However, as Kokanee seemed averse to spawning at many of the cross-sections that provided adequate hydraulic conditions, it was hypothesized that a combination of depth and velocity as expressed in *Fr* (as per Long, 2007) might provide additional insight. Fig. 5.8 shows that very few cross-sections satisfy the criteria proposed by Long (2007). Interestingly, only XS5–7 met *Fr* criteria and these cross-sections were all in Reach 4 through Polson Park, which is the most heavily engineered reach. Cross-section 10 was the only location with *Fr* values well above the target maximum of 0.4. Therefore, it seems that the *Fr* criterion is too restrictive as a viable metric for assessing hydraulic suitability for spawning salmon, consistent with the results from active redds shown in Fig. 5.9.

5.3.4.3 Integrated Assessment

The potential of Lower Vernon Creek to serve as spawning habitat for Sockeye salmon depends, in part, on how strictly the hydraulic criteria are adhered to as limiting factors. For example, if *Fr* values recommended by Long (2007) are combined with the substrate size limits presented in Table 5.2, then only XS5 satisfies the combined conditions for spawning suitability. However, given that Fig. 5.9 seems to suggest that *Fr* criteria are too restrictive, and if only the depth and velocity criteria in Table 5.3 are adhered to, then XS4–5 and 10–12 provide viable spawning conditions. Table 5.4 gives an overview of the integrated assessment of spawning habitat potential for all cross-sections, which also includes information on where spawning Kokanee were found during the stream walks.

In the lower reaches of Lower Vernon Creek, below the heavily engineered section through Polson Park (Fig. 5.4D), the most suitable spawning conditions are found at XS3–5. The cross-sections closest to Okanagan Lake (i.e., XS1–2) are generally not suitable because the grain

TABLE 5.4 Rating of Cross-Sections for Sockeye Spawning Suitability based on Hydraulic (Depth, Velocity) and Substrate Size Criteria.

XS	Hydraulics	Substrate	Presence of Kokanee	Overall suitability
1	No	No	No	No
2	Possible	No	No	No
3	Possible	Possible	Yes	Possible
4	Yes	Yes	Yes	Yes
5	Yes	Yes	Yes	Yes
6	Yes	No	No	No
7	Yes	No	No	No
8	No	No	No	No
9	No	No	No	No
10	Yes	Yes	No	Possible
11	Possible	Yes	No	Possible
12	Yes	Yes	No	Possible
13	No	No	No	No

sizes are too fine and flow velocities too slow because of backwater effects. The concrete-sided section through Polson Park (XS6–7) appears to have suitable flow hydraulics, but the substrate is not usable because of the tight packing and cementation of the streambed. Immediately upstream of Polson Park, XS8 and XS9 are unsuitable because the substrate is too fine and flow velocity too slow, reflecting the sluggish nature of Reach 5A. However, the remaining upstream sections (XS10–12), with the exception of XS13 right below the outlet of Kalamalka Lake, provide suitable hydraulic and substrate conditions even though no Kokanee were found there.

5.4 DISCUSSION

Given the likelihood that Sockeye salmon will be repatriated to Okanagan Lake (and its tributaries) in the near future, it is timely to contemplate how Lower Vernon Creek could be rehabilitated to improve potential spawning habitat. A key observation with respect to Table 5.4 is that there already exists viable habitat upstream of Polson Park, in Reach 5 (i.e., XS10–12), which currently supports stocks of Rainbow Trout and Common Carp, but not Kokanee salmon. MacDuff (2016) considered a range of explanations as to why Kokanee might not be found in the upstream reaches based on a variety of factors, including upstream trends in temperature, pH, and conductivity. For each of these three water-quality parameters, there was a marked transition between XS5 and XS9, with an increase in temperature and pH and a decrease in conductivity. Nevertheless, none of these transitions were deemed to be limiting for salmonids, given the range of acceptable values reported in the literature. MacDuff (2016) concluded that the heavily engineered section through Polson Park presented an insurmountable barrier to Kokanee passage largely because the 1-km stretch of channel provides no refuge or resting habitat because the banks are either riprapped or walled with concrete. Larger, stronger species of fish, such as Rainbow Trout, seem not to be deterred by continual migration through such a long reach of fast-flowing, shallow water and this may also be true for Sockeye salmon. But Kokanee are a smaller variant of Sockeye and, therefore, are not able to endure (or are unwilling to undertake) the migration. In instances where Kokanee attempt to move past XS5, they invariable turn back. Removing the hardened bank materials and allowing the creek to adopt a more naturally meandering downstream pattern through Polson Park would enhance the hydraulic complexity and biodiversity of the system. This may be sufficient to allow for upstream migration of Sockeye and Kokanee salmon to the base of the control structure at Kalamalka Lake.

Over the longer term, the issue of sediment supply and gravel recruitment in the stream system looms large. Historically, Lower Vernon Creek experienced annual floods during the spring freshet, when Kalamalka Lake was at high stage. Such large hydrologic events had the potential to redistribute sediment in important ways by ripping up bottom armor, remobilizing bed material, and eroding channel banks and thereby maintaining aquatic and riparian ecosystem complexity and replenishing spawning beds. It is interesting to observe the large amount of embedded cobble material that exists in much of Lower Vernon Creek today. The primary cause of this somewhat unnatural state, it is argued, is an anthropogenic one driven by two factors: (1) regulated discharge releases from Kalamalka Lake and (2) channelization and bank stabilization.

Under the current discharge regime, the flows are perennially constrained within an unnaturally muted flow hydrograph with the conspicuous absence of large annual floods.

Thus, there is no longer the capacity to rip up bed materials and mobilize large particles that occur naturally in the glaciofluvial deposits that characterize the sedimentology of the floodplain and valley sides. This is the only source of sediment to resupply the creek because the upstream boundary of Lower Vernon Creek is the outlet of Kalamalka Lake, which provides only fine-grained silt and mud to it. In other words, Lower Vernon Creek is not connected geomorphically to its headwaters, where there is ample sediment production from colluvial processes. Kalamalka Lake (as well as Wood and Ellison Lakes) serve to interrupt the sediment conveyor belt that typically exists along a classic longitudinal profile of a river. Furthermore, the straightening and entrenching of the channel through the city of Vernon (to protect against overbank flooding events that inundate the urban environment) led to an increase in channel slope through its urban reaches. The net effect was to increase average flow velocity (i.e., the energy grade line), thereby causing erosion of the channel banks and bed. To prevent further erosion, the banks were stabilized with concrete and riprap, thereby cutting off local sediment supplies to the creek. Over the long term, these modifications caused the majority of fine- and medium-grained bed materials to be winnowed and transported downstream, effectively leaving a coarse cobble lag on the streambed (i.e., static armor). These conjectures are supported by the strongly cemented and gravel-free conditions through Polson Park (Reach 4), despite an attempt 25 years ago to artificially enrichen the channel by introducing large quantities of gravel-sized sediment (Young, 1991). However, very little of this gravel material can be found in the channelized reach today.

Gravel material of a size suitable for spawning salmonids was found at a few locations, mostly in Reaches 2 and 3, and these may be a combination of naturally sourced gravel or artificially replenished gravel. These potentially productive areas were more sinuous than those upstream and had exposed gravels in some portion of the banks. The suitability of these sites between XS3 and XS5 was confirmed when over 600 spawning Kokanee were observed during the October stream walk. However, the suitable areas did not span the entire length of these reaches and there were extensive stretches that were engineered where no fish were found.

5.5 SUMMARY AND CONCLUSIONS

A series of qualitative stream walks along Lower Vernon Creek identified several reaches that were utilized by spawning Kokanee salmon or occupied by other fish species (e.g., Rainbow Trout), despite heavy urbanization and channelization along virtually the entire length from Kalamalka Lake to Okanagan Lake. A model of the expected hydraulic conditions during the October spawning season showed that only certain sections of Reaches 2–3 and 5C–D would be suitable for Sockeye salmon should they be repatriated into Okanagan Lake and its tributaries. However, an analysis of the channel substrate showed that very little of the stream (~1.6 km) had the loose gravels required for Sockeye spawning, which is a major concern. Unlike what might be expected for a natural valley-bottom creek with continuous sediment supply from its headwaters, more than half of Lower Vernon Creek had bed material that was too large or too embedded for redd excavation by Kokanee and Sockeye salmon.

The observed relationships between channel form, flow hydraulics, sediment supply, and substrate size suggest two critical points with regard to stream rehabilitation. First, where meandering is evident, the slope of the channel is sufficient to transport gravels, but not too excessive to induce intensive erosion of the streambed and banks, given the range of discharges that presently characterize the hydrologic regime. Second, if the most productive salmon spawning areas are being supplied by only a few gravel outcrops that were observed sporadically on the channel margins during the stream walks, then it implies that overall stream capacity is relatively limited with regard to sustaining a healthy salmonid population. Without an adequate sediment supply to which the creek has periodic access, there will be an excess of energy, especially in the channelized reaches, that is capable of winnowing the bed and leaving behind an armored, immobile surface of cobbles that are too large for salmonid spawning. Thus, the historical effort to prevent flooding and property erosion in the city of Vernon via channelization and hardening of the banks has served to increase the available stream power, while also eliminating gravel recruitment from the banks and floodplain materials.

Building on these study results, the most significant enhancements that could be made to Lower Vernon Creek by way of a rehabilitation initiative are: (1) to naturalize the channel banks by removing concrete and riprap and restore some of the meandering form of a naturally dynamic and productive stream; (2) identify areas of gravel deposits along the stream margin and, if possible, allow for relatively controlled erosion at a few of these outcrops, so that there is a natural resupply of essential gravel and coarse sand; and (3) place large woody structures in the channel to create pool structures, habitat complexity, and refugia for fish rearing. The goal of these actions would be to increase the density of riffle-pool-glide structures and to provide a long-term supply of spawning gravels. This is especially true of Reach 4, which flows through Polson Park, because this section presents a barrier to the upstream migration of Kokanee salmon and potentially to Sockeye salmon. Improvements to Reach 4 are likely: (1) the most cost-effective because the property is owned by the city of Vernon and there would be no displacement of private homes and (2) most beneficial, despite its short length, because several locations upstream already assessed as suitable for Sockeye salmon would be made accessible.

References

Blott, S., Pye, K., 2001. GRADISTAT: a grain size distribution and statistics package for the analysis of unconsolidated sediments. Earth Surf. Proc. Landforms 26 (11), 1237–1248.

Davidson, K. 2016. Okanagan Nation Alliance reintroduce salmon to Okanagan Lake. *Global News*. Available from: https://globalnews.ca/news/2756106/okanagan-nation-alliance-reintroduce-salmon-to-okanagan-lake/.

Environment Canada, n.d. Environment and Climate Change Canada Historical Hydrometric Data—Station:08NM065. https://wateroffice.ec.gc.ca/mainmenu/historical_data_index_e.html.

Hume, M., 2014. Okanagan Sockeye Restoration Successful With Decade-Long Effort. The Globe and Mail, September 21. https://www.theglobeandmail.com/news/british-columbia/okanagan-sockeye-restoration-successful-with-decade-long-effort/article20714952/.

Kondolf, G., 2000. Assessing salmonid spawning gravel quality. Trans. Am. Fish. Soc. 129 (1), 262–281. doi: 10.1577/1548-8659(2000)1292.0.co;2.

Kondolf, G., Wolman, M., 1993. The sizes of salmonid spawning gravels. Water Resour. Res. 29 (7), 2275–2285. doi: 10.1029/93wr00402.

Long, K., 2007. The Effects of Redd Selection and Redd Geometry on the Survival of Incubating Okanagan Sockeye Eggs (MSc thesis), Biology. University of New Brunswick., pp. 9–31.

MacDuff, A., 2016. An Assessment of Sockeye Salmon Habitat Potential in Lower Vernon Creek, British Columbia, (BSc Honours thesis), Earth and Environmental Sciences. University of British Columbia.

Okanagan Nation Alliance, 2004. Okanagan Sockeye Reintroduction. http://www.turtleisland.org/news/oksalmon.pdf.

Shepard, B., Ptolemy, R., 1999. Flows for Fish requirements for Okanagan Lake tributaries. http://a100.gov.bc.ca/pub/acat/public/viewReport.do?reportId(15913.

Sisiutl Resources, 1986. Vernon Creek Fisheries Enhancement Project. http://a100.gov.bc.ca/pub/acat/public/viewReport.do?reportId=13362.

Slaney, P., Zaldokas, D., 1997. Fish habitat rehabilitation procedures. Watershed Restoration Technical Circular No. 9, Ministry of Environment, Lands and Parks, Vancouver, BC, Canada. http://www.env.gov.bc.ca/wld/documents/wrp/wrtc_9.pdf.

US Army Corps of Engineers, 2015. HEC-RAS 4.1 [Software]. http://www.hec.usace.army.mil/software/hec-ras/downloads.aspx.

Webster, J., Chara Consulting, 2004. 2003 Kokanee Stream Spawner Enumeration – Okanagan Drainage. http://a100.gov.bc.ca/pub/acat/public/viewReport.do?reportId=1509.

Young, D., 1991. Habitat Conservaton Fund, 991. Vernon Creek Improvement 1990–991—Final Report. http://a100.gov.bc.ca/pub/acat/public/viewReport.do?reportId=3364.

CHAPTER

6

Landform Change Due to Airport Building

Edyta Pijet-Migoń, Piotr Migoń***

*Institute of Tourism, Wrocław School of Banking, Wrocław, Poland; **Institute of Geography and Regional Development, University of Wrocław, Wrocław, Poland

OUTLINE

6.1 Introduction 101

6.2 Types of Geomorphic Change 103

6.2.1 Land Reclamation 104

6.2.2 Coastline Alteration 105

6.2.3 Land Leveling 106

6.2.4 Artificial and Partly Artificial Islands 108

6.2.5 Small-scale Alteration of Relief 109

6.3 Conclusions 110

References 111

6.1 INTRODUCTION

Among the most evident global societal trends experienced since at least the mid-20th century is the massive increase in human mobility. People travel for a multitude of reasons, mainly for business and leisure, and to countless destinations worldwide. As recently as 1970, the total number of passengers using air transportation was a little more than 310 million. This figure tripled by 1990 (1.025 billion), tripled again by 2013 (3.048 billion), and in 2016 as many as 3.7 billion people used commercial aircrafts for travel (http://www.iata.org/). This unprecedented growth would not be possible without parallel technological advances in the aviation industry. Aircrafts are now bigger and faster than they used to be and can accommodate more people, flying longer distances in shorter time. The frequency of air operations has increased too and flights are less and less weather-dependent thanks to modern navigation

Urban Geomorphology. http://dx.doi.org/10.1016/B978-0-12-811951-8.00006-0

systems. Airports are gateways to air travel and contemporary cities, especially megacities, cannot really function without airports, which have become both a necessity and a sign of metropolitan status. However, alongside this growth in traffic and technological progress, the demand for space for modern airports is increasing. Runways capable of suiting wide-body intercontinental aircrafts need to be at least 2500 m long and big airports require at least two runways. More and more space is needed for taxing routes and the apron. At the same time, space occupied by airport ground infrastructure increases, not only because airports serve more and more flights and travelers, but also due to the development of related services, such as cargo terminals, maintenance units, fuel stations, and others. Further competitors for space are connecting railways and motorways, parking lots, hotels, conference facilities, and shopping centers. Some airports, along with their business surroundings, have grown so much that they form a new type of urban space termed "aerotropolis" (Kasarda, 2001; Kasarda and Lindsay, 2011). Therefore, it is no wonder that the biggest airports of the world occupy 20–30 km^2 (http://www.worldatlas.com/articles/the-world-s-10-largest-airports-by-size.html) and even medium-size airports need a few km^2 to perform their roles effectively (Fig. 6.1). In addition, there are special terrain requirements to ensure safety of aviation. The crucial one is that the airport area must be essentially flat within the limits of the airport itself and lacking substantial elevations along the approach and take-off paths. Such conditions are not met everywhere and, hence, airport building becomes inevitably associated with considerable ground engineering, which, in turn, implies landform change. In specific examples, competition for space is so severe that no suitable ground for airport building or enlargement is available. The only solution is to create a new land surface and some of the largest artificial islands in the world

FIGURE 6.1 **The area occupied by high-capacity modern airports, such as the Brussels Airport, may reach tens of squared kilometers and requires extensive terrain leveling.**

are those built to host modern airports. Thus, aviation has its geomorphological dimension and the purpose of this chapter is to provide a review of various anthropogenic alterations of relief implemented during airport building or extension.

The subject of surface change due to the development of airport infrastructure, despite the colossal earthworks involved, has been scarcely discussed in the literature. Douglas and Lawson (2003) were probably the first authors to address it explicitly, focusing on several case studies (Hong Kong, Osaka Kansai, Singapore, Incheon-Seoul). Szabó et al. (2010) extensively covered various aspects of anthropogenic geomorphology, but only one specific case of land reclamation for the new Hong Kong Airport was presented in more detail by Dávid et al. (2010). Very little or no mention of the subject can be found in more general summaries of human impact on the geomorphic environment (e.g., Cooke and Doornkamp, 1990; Goudie and Viles, 2016). Thus, web-based resources provide most information and this chapter is considerably based on them as well as on the authors' own observations. Google Earth, for example, is an excellent resource to appreciate the magnitude of surface change associated with building modern airports.

6.2 TYPES OF GEOMORPHIC CHANGE

Landform change associated with airport construction has many facets and shows a certain hierarchy (Table 6.1). Some major relief alterations are ubiquitous, whereas others vary depending on the geomorphic setting of the airport. Land leveling is a prerequisite everywhere and may involve both removal of minor topographic highs as well as filling depressions. Even if the magnitude of vertical change may seem small in lowland terrains, being spread over several squared kilometers, it implies substantial relocation of earthen materials. In more hilly to mountainous terrains, the complete removal of minor bedrock hills may be necessary, not only in the vicinity of urban centers, but also in more remote areas, such as next to the landing strip in Mulu National Park in Malaysia. In coastal settings, the demand for large space to accommodate an airport in otherwise heavily built-up and industrialized areas resulted in three major types of intervention: land reclamation at the expense of shallow sea/lake or coastal wetlands; coastline alteration due to building of artificial spits; and the most extreme building of entirely artificial islands, occasionally several kilometers offshore. In the past, when air transport using hydroplanes was more popular, coastal dredging was also accomplished.

TABLE 6.1 Landform Change due to Airport Building

Major	Minor	Second-order effects
Land grading	Earth embankments	Disruption of long-shore transport of sediment
Land reclamation	Observation hills	Ground subsidence
Coastline alterations	Causeways	Enhanced weathering of exposed rock slopes
Hill removal	Breakwaters and seawalls	Rill erosion
Artificial and partly artificial islands	Drainage ditches	
	Tunnels	
	Channel relocation	
	Sea-bottom dredging	

Minor changes superimposed on the major ones are multiple and include construction of both positive and negative relief features. The former include earth embankments to lessen noise pollution, artificial hills serving as observations points, causeways connecting artificial islands with the mainland, and so forth, whereas the latter are represented by drainage ditches, ponds, and tunnels. Artificial drainage channels may replace natural channels if drainage diversion is necessary.

Finally, there are second-order changes understood as alterations of natural processes subsequent to intentional landform change due to airport building. Among them are perturbations in littoral drift, resulting in enhanced erosion and deposition, accelerated ground subsidence due to load of the fill, exposure to weathering and mass movements of rock slopes cut during airport construction, and rill erosion on unconsolidated dumped ground.

6.2.1 Land Reclamation

Low coastal areas are favored locations for airports for three main reasons. First, topographic obstacles are usually minimal, facilitating safety of aviation operations. Second, there was often very limited antecedent urban development in these localities, as coastal wetlands and marshes created too many geotechnical engineering problems for city expansion and shallow bays did not encourage their use in marine transportation due to navigation problems. Third, future airport extension appears more feasible than in inland locations with established land use. Thus, land reclamation in coastal areas is carried out for two purposes: to build a brand new airport, but also to extend the existing runway, or to add an additional one.

One of the early examples of building an airport on former coastal marshes and tidal flats comes from San Francisco. In the 1930s, approximately 1.4 km^2 encircled by a 3-km long seawall were reclaimed, allowing for the building of two new runways (https://www.flysfo.com/about-sfo/history-sfo) (Fig. 6.2). Similarly, the former airport serving Rio de Janeiro, Santos Dumont (now a secondary airport for domestic connections), occupies a piece of land inside Guanabara Bay that has been entirely reclaimed. Land gain was about 0.7 km^2 (circa 1.4 $\times$ 0.6 km).

A much more recent, and much more impressive example in terms of the area reclaimed, is Incheon International Airport, which serves Seoul's metropolitan area in South Korea (in addition to the existing Gimpo Airport, closer to the capital). Its construction began in 1992, was brought to an end in 2000, and the airport was officially opened in March 2001. It is located partly on existing coastal flatlands, but mainly lies on the former tidal flats between the Islands of Yeongjong and Youngyu circa 53 km to the west of Seoul. As much as 56 km^2 of muddy tidal flats, originally submerged to 1 m depth, have been reclaimed to make space for the airport that includes two runways, each 4-km long (Fig. 6.3). Construction of the main terminal building involved removal of 25 million tons of earthen material (Douglas and Lawson, 2003) and the total cost of building the airport is estimated at US$6.3 billion (Bowen and Cidell, 2011). Another mega airport in Southeast Asia, whose construction required drainage and reclamation of swampy ground, is Changi Airport in Singapore. Earthworks began in 1991 and the total gain of >20 km^2 was accomplished in several phases (Bo et al., 2005). Of similar size is the area occupied by Hamad International Airport in Doha, Qatar, built over the period 2006–2016, where approximately half of it is on human-made ground within the former marine embayment.

FIGURE 6.2 An example of early (1930s) reclamation works to provide space for an enlarging airport: plans to extend the San Francisco airport. *Source: https://livingnewdeal.org/projects/san-francisco-airport-san-francisco-ca/ig.*

6.2.2 Coastline Alteration

Apart from land reclamation, which is coastline alteration itself, more localized changes are causally associated with the extension of runways. This type of upgrade at many airports is dictated by the use of wide-body jet aircrafts, such as Boeing 747, Airbus A340, and more recently Airbus A380, which need particularly long take-off and landing strips. Replacement of turboprops by smaller jets typically also required runway extension, for example to accommodate medium-size jet aircraft, such as B737, which is the most popular jet aircraft ever produced. For many airports, especially those located on islands that are popular tourist destinations, extension into the sea was the only option due to constraints imposed by topography and land use. Hence, significant land reclamation and coastline alteration were inevitable. Examples include Gibraltar Airport, Male in the Maldives, and Denpasar in Bali, Indonesia.

In Gibraltar, the only place to build the airport was a narrow neck connecting the peninsula of high relief and the mainland that is, however, the territory of Spain. To allow landings of medium-size jet aircrafts, runway extension into the bay was carried out and approximately half of the 1828-m long runway belt is on human-made land. Far more dramatic was the

FIGURE 6.3 Incheon International Airport in Seoul occupies most of the reclaimed tidal flats between two already existing islands. *Broken line* indicates outlines of the islands before reclamation. *Source:* © *Google, CNES/Airbus.*

extension of the runway in Male. The original one was 900 m long and fitted the atoll fringe, whereas the contemporary one is 3200 m long. Enlargement of the apron is currently at the planning stage and, if decided, will result in even more land reclamation inside the atoll. In Denpasar, a runway originally 1200 m long was enlarged to 2700 m as early as the 1960s. To do so, an artificial spit was built, whose construction affected long-shore drift and resulted in perturbations in the pattern of sand movement in the nearshore zone.

The airport in Wellington, New Zealand, was built on a tombolo that emerged from the sea during a land uplift episode associated with the 1855 earthquake. Its flat surface provided a convenient location for the first airport built in the 1930s, but its further development in the 1950s to cope with increasing traffic and larger aircrafts required significant earthworks, including removal of a low hill in the northeast part of the tombolo and reclamation of approximately 0.5 km^2 of land in Lyall Bay (Fig. 6.4). The current runway is about 2 km long and plans are being discussed to extend it further, providing an opportunity for more intercontinental flights. The only way to accomplish the extension would be massive reclamation works in Lyall Bay and the building of a strong breakwater to protect the runway against high waves in Cook Strait.

6.2.3 Land Leveling

Modern airports occupy huge spaces, even measured in tens of squared kilometers (Table 6.2), and such extensive flat terrains may be scarce. Therefore, leveling and filling minor topographic depressions is a necessity and even seemingly perfectly level ground requires significant earthworks, accompanied by ground dewatering and drainage diversion.

FIGURE 6.4 Location of Wellington Airport on a tombolo raised during the 1855 earthquake. The far end of the runway is on artificial ground that reduced the area of Lyall Bay.

TABLE 6.2 Top 10 World Airports by Size

Airport name	City, country	Area occupied (km^2)
King Fahd International	Riyadh, Saudi Arabia	780
Denver International	Denver, CO, USA	137.3
Dallas/Fort Worth International	Dallas, TX, USA	78.0
Shanghai Pudong International	Shanghai, China	33.5
Charles de Gaulle	Paris, France	32.0
Barajas	Madrid, Spain	30.5
Bangkok International	Bangkok, Thailand	29.8
O'Hare	Chicago, IL, USA	26.1
Cairo International	Cairo, Egypt	25.5
Beijing Capital International	Beijing, China	23.3

Source: http://www.worldatlas.com/articles/the-world-s-10-largest-airports-by-size.html.

Occasionally, the partial or complete removal of terrain elevations is involved in the enlargement of airports or extension of runways. This kind of operation was carried out in the notoriously dangerous Toncontín Airport in Tegucigalpa, Honduras. Hilly morphology significantly constrained the space for the airport and the runway was less than 1900 m long, which was not enough to ensure safe landings of medium-size jet aircraft. After an aviation accident on May 30, 2008, the decision to upgrade (extend) the runway was made and the only way to achieve this goal was to flatten a natural hill located just beyond the runway threshold. Some 180 thousand m^3 of material was bulldozed and removed (Lopez, 2012). Similar earthworks had to be carried out at Beigan Island (Taiwan), where a hillside projected too much into the approach area. To comply with safety regulations, a slope section approximately 330 m long, 150 m wide, and 60 m high was carved out (Fig. 6.5).

FIGURE 6.5 Cut-off hillside at Beigan Island, Taiwan, to increase safety during take-off and landing operations. This example shows that even small airports may be associated with significant landform change.

6.2.4 Artificial and Partly Artificial Islands

Perhaps the most famous airport located entirely on an artificial island is Kansai International Airport in Osaka Bay, Japan (Fig. 6.6). It is also the oldest one of its kind. Before the decision to build a new airport was made, the agglomeration of Osaka was served by Itami Airport, which has been in operation since 1931. However, with its capacity nearly exhausted, no airport enlargement was thought possible within the densely built urban space and experience with strong opposition against the construction of the new Narita Airport for Tokyo on agricultural land made such an option unfeasible too. In addition, a location was sought, where airport operations could have been conducted in 24 h, without causing noise pollution for nearby residents (Mesri and Funk, 2015).

Building operations of the first phase started in 1987 and the official opening of the airport took place on September 4, 1994 (Puzrin et al., 2010). The human-made island that hosts the airport is located in Osaka Bay circa 5 km from the shore and connected to the mainland by a causeway. The original depth to the sea bottom was 18–20 m. With the island area of 5.1 km^2, 189 million m^3 of earthen material must have been dumped, followed by the delivery of another 250 million m^3 to build the second island (5.45 km^2) in 1999–2007 (Douglas and Lawson, 2003; Mesri and Funk, 2015). An unexpected problem encountered during the construction phase and experienced until now is accelerated ground subsidence due to a heavy load imposed on the unstable sea floor. The amount of ground settlement for the first island was more than 12 m, causing geological-engineering problems, delays, and the rise of costs to US$14 million (Bowen and Cidell, 2011). According to Mesri and Funk (2015), until the end of the 21st century, the magnitude of ground subsidence may reach 17.6 m for Part I of the island and as much as 24.4 m for Part II. Despite engineering and environmental problems, airport building on artificial islands in Japan continued and currently there

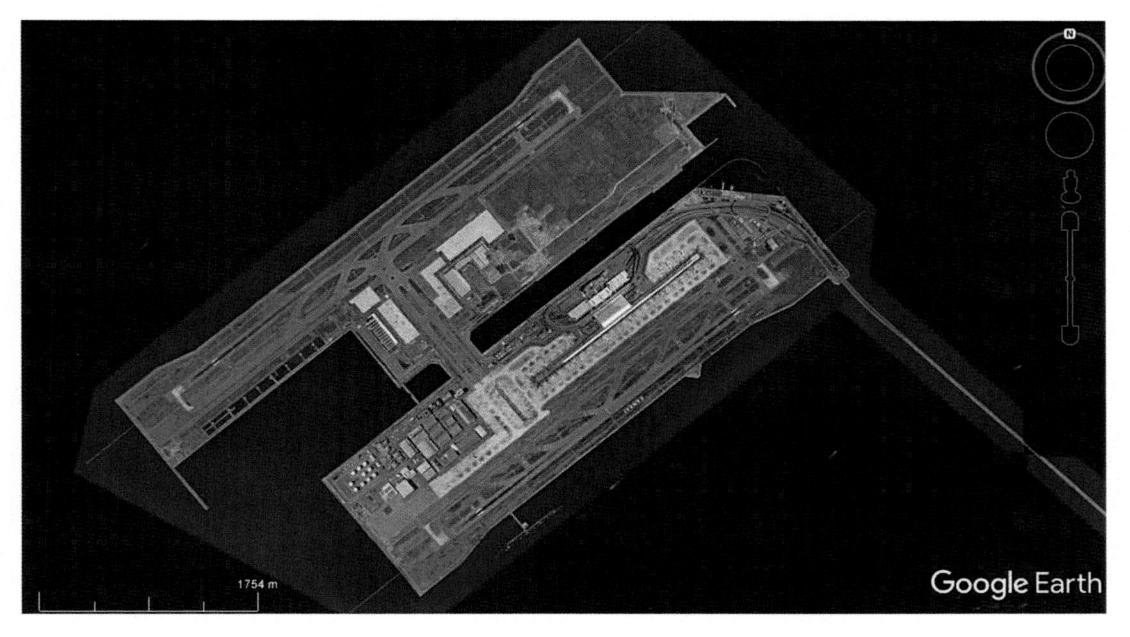

FIGURE 6.6 Artificial islands to host Osaka Kansai International Airport. *Source: © Google.*

are five commercial airports at these locations. Further examples are the nearly 7-km^2 large Chūbu Centrair International Airport in Nagoya (opened 2005), Kobe Airport (opened 2005, circa 2.7 km^2), and the New Kitakyushu Airport (opened in 2006, circa 3.7 km^2). The new Nagasaki Airport opened much earlier (1975) and the smallest among them differs from the others in that an existing island was leveled and significantly reshaped by adding new ground to accommodate the airport.

More complex was the nature of geomorphic change associated with the construction of the new Hong Kong Airport, which replaced the old Kai Tak Airport located right in the city center that was increasingly unsafe. In contrast to Japanese locations, the configuration of the sea bottom, the presence of numerous rocky islets and reefs, and political boundary issues precluded building an entirely artificial island. Instead, the plan involved joining two adjacent islands, Chek Lap Kok and Lam Chan, by a human-made connection that accounts for three-fourths of the airport space, with simultaneous lowering of elevations on these islands by removal of rock, which was then used to gain land. Building of the airport involved one of the largest earthen removals in history; 9.38 km^2 of new land was reclaimed about 69 million m^3 of silt was obtained from the shore seabed, whereas 76 million m^3 of sand and 121 million m^3 of rock was imported for the construction (Dávid et al., 2010).

6.2.5 Small-Scale Alteration of Relief

In the past, especially in the first half of the 20th century, hydroplanes were in use and these required dedicated engineering of the coastline to allow mooring. For example, preparation of the landing station in Szczecin (then Stettin in Germany) in the 1920s, involved

FIGURE 6.7 **Artificial observation hill for visitors at Josef Strauss Munich International Airport.**

dredging of a special basin circa 250 × 250 m and three mooring canals along the shore of Dąbie Lake (Pijet-Migoń and Migoń, 2013). An extension of San Francisco Airport by the end of 1930s included the building of a seaplane harbor, whose dredging provided some material for the parallel extension of the land part of the airfield (https://www.flysfo.com/about-sfo/history-sfo).

At some airports, for example, at Kastrup in Copenhagen, part of the perimeter of the airport area is lined with small earthen embankments, circa 3–5 m high and 20–40 m wide, erected to reduce noise pollution associated with aircraft operations. An interesting example of an artificial hill is present at Munich International Airport, being a part of the Airport Visitor Centre—a group of educational and family entertainment facilities in the area between the two runways. This pyramidal mound 28 m high that is accessible by stairs, hosts a viewing platform on top that allows for observing the apron near one of the terminals and to watch landing and take-off operations (Fig. 6.7).

6.3 CONCLUSIONS

Landform change associated with airport building is rarely visually spectacular, especially if viewed from a ground perspective. This is because construction activities tend to level the terrain rather than increase local relief. Furthermore, activities such as leveling due to filling or emplacement of artificial islands involve significant alterations in the vertical dimension, but these are largely hidden from sight. The spatial scale of change becomes more evident when seen from the air, an opportunity now available to everyone due to easy access to Google Earth. Nevertheless, the magnitude of change is sheer and best expressed in terms of area created or altered. Thus, for example, the artificial twin island that hosts Kansai Airport is nearly 4.8 km long and more than 2.5 km wide, resulting in an area of 12 km^2. The largest inland airports occupy even more

CHAPTER

7

Environmental Contamination by Technogenic Deposits in the Urban Area of Araguaína, Brazil

Carlos A. Machado, Silvio C. Rodrigues***

*Tocantins Federal University (UFT), Araguaína, Brazil; **Uberlândia Federal University (UFU), Uberlândia, Brazil

OUTLINE

7.1 Introduction 115
7.2 Methodology 116
7.3 Technogenic Deposits (TDs) and Soil Contamination 117
7.4 Soil Contamination by TDs in the Urban Area of Araguaína 119
7.5 Conclusion 125
References 125

7.1 INTRODUCTION

Environmental alteration by human activities to produce food and materials has intensified in the last 10 000 years, a period in which human civilization has become sedentary and started developing agriculture and constructing cities in different parts of the world. Among the environmental elements, the pedological layer has been constantly modified by excavations, reallocations, and removal of materials in urban areas and the application of organic (manure) and chemical products (fertilizer) to enhance agriculture. Materials removed from the environment and then transformed into products for human use in the most diverse activities (domestic, commercial, industrial) after being discarded, are generally buried without any treatment and in diverse types of soils, culminating over the years in biological, chemical, and physical activity alterations of these soils by the intense influence of toxic compounds.

Urban Geomorphology. http://dx.doi.org/10.1016/B978-0-12-811951-8.00007-2

In urban areas, mainly construction debris (bricks, tiles, ceramics, paints, metals) and domestic products (packaging, liquid, food waste) are deposited and buried in landfills and erosion spots in poor suburbs of the cities, constituting themselves as technogenic deposits (TDs). The TDs that are created by anthropogenic activity are incorporated into the local environment and covered through sedimentation processes in valleys, so that detection by vegetation growth is difficult and can later result in structural complications for edifications built on these materials. It is notable that public administration does not have any effective control policy or oversight into this matter and the situation has intensified in the last 10 years in the study area according Machado (2014).

The focus of this study is to identify different types of polluting materials and the danger and toxicity risk of the elements that constitute TDs as well as their negative impacts on different types of soils in the urban area of Araguaína, situated in the northern region of Brazil. Studies in technogenic environments in the Amazon region are scarce and poorly detailed and the increasing urbanization of cities has expanded environmental impacts of TDs in the last decade.

7.2 METHODOLOGY

For the identification of soils types, the research of Tocantins (ESTADO) (2004) and Menk et al. (2004) was used, with complementary fieldwork according to procedures outlined by the Brazilian System of Soil Classification of EMBRAPA (1999), the Brazilian Agricultural Research Company (Empresa Brasileira de Pesquisa Agropecuária). The classification of TDs is based on the types of material, as originally proposed by Ter-Stepanian (1988), including terrigenous (soils), chemicals (industrial waste), organics (household waste), and inorganics (construction waste) in addition to the deposition environmental factor (terrestrial, fluvial, lagoon, and marine) according to Machado (2016).

The location and geographical distribution of TDs were obtained using a comparative analysis of SPOT satellite images between 2005 and 2011 available on Google Earth. Satellite-imagery interpretation assisted in locating and identifying TDs and used to perform an environmental impacts evaluation. Due to the large number of small deposits, only areas larger than 500 m^2 were studied by virtue of the difficulty of using satellite images in conjunction to fieldwork. The measurement and mapping of TDs was performed through the use of ArcGIS 9.3 software; and a Garmim Hcx GPS determined the geographical coordinates of deposits in the field.

The collection of technogenic material (TM) followed the standards of Solid Waste Sampling (ABNT/NBR1007) of the Brazilian Association of Technical Standards. For particle-size analysis and the identification of TMs, samples of 1 kg of material were collected. The identification of contaminants in TMs and their hazardousness followed the standards of national resolution of the National Environment Council (Conselho Nacional de Meio Ambiente/CONAMA) 397, 03/04/2008. The classification of solid wastes and their toxicity was based on ABNT/NBR 10004-2004 standards.

With the objective of classifying urban solid waste and considering its potential risks to the environment and public health, we applied Brazilian Association of Technical Standards (Associação Brasileira de Normas Técnicas/ABNT). These regulations (ABNT 2004 NBR

10004-2004) aim to adequately manage waste from its origin to final destination. For the purposes of this standard, solid wastes are classified as:

- Solid Waste Class I: Dangerous
- Solid Waste Class I: Not Dangerous
- Solid Waste Class II A: Not Inert
- Solid Waste Class II B: Inert

The particle-size analysis employed for characterization of sediments in the samples of TDs followed the parameters defined by Suguio (1973), in which the determination of sand fractions (thin, medium, thick), clay, silt, and organic matter are in accordance with the Wentworth scale. Due to the lack of a specific classification for TM size (for fragments and concretions), three classes were defined in consonance with the majority of debris analyzed in the city of Araguaína: small debris (0–5 cm), medium debris (5–10 cm), and large debris (>10 cm). In the last phase of this research, we propose preventive and recovery measures for each specific case, considering social, economic, and environmental aspects.

7.3 TECHNOGENIC DEPOSITS (TDs) AND SOIL CONTAMINATION

The understanding of the current pedological layer behavior, with the insertion of material and the creation of TDs, is vital for comprehension of the dynamics of the physical, chemical, and biological processes in different environments and with anthropogenic materials. According to Ter-Stepanian (1988), TDs are characterized by their large variety, differentiated features, diversified composition, and large variation of thickness determined as an independent genetic class, although analogies with natural deposits can be made. Some deposits accumulate great quantities of organic matter, generally from household waste and others from inorganic matter, derived from construction materials or mining waste.

The USDA/NRCS (2000) has emphasized that the excessive accumulation of heavy metals in soils will, over time, be incorporated in the food chain. The most common problems are cationic metals (metallic elements that in the soil have positively charged cations, for example, the lead ion Pb^{+2}), among which are mentioned mercury (Hg), cadmium (Cd), lead (Pb), nickel (Ni), copper (Cu), zinc (Zn), chromium (Cr), and manganese (Mn). Common anion compounds (elements that form in the soil are combined with oxygen and are negatively charged (e.g. MoO_4^{2-}) are arsenic (As), molybdenum (Mo), selenium (S), and boron (B).

Urban soils contain manipulated, disturbed, and transported materials resulting from human activities. In this context, physical, chemical, and biological properties are generally unfavorable for adequate root growth of exotic or native plants in the urban environment. Various characteristics are common in urban soils, such as spatial and vertical variability, modified soil structure, easy compression, the presence of superficial crusts, high soil reactions, aeration and restricted drainage, interruption of the nutrients cycle, modified biological activity, the presence of contaminating anthropogenic materials, and modification of soil temperature (Craul, 1999).

Anthropogenic interferences on the formation and evolution of the environment strongly impact biological and geological processes, which are dynamic over time. The damage caused to environmental processes and elements may have short-, medium-, or long-term effects, mainly with the incorporation of anthropogenic materials, especially in areas that are more environmentally sensitive. Instability may be reflected in life quality as well as socioeconomic aspects of certain populations (Rogachevskaya, 2006).

Korb (2009), based on collected sediment analysis, verified a great variation of materials and minerals in TDs. Significant differences of organic matter, sand, clay, and ferruginous concretions resulted in oscillations of the water level in dams. Another aspect of the work focused on the concentrations of polluting materials in deposits, mainly Pb and Zn elements, which beside producing physical, biological, and chemical alterations, curtail treatment and damage water quality for public use. According to Korb (2009), the removal of polluting materials is difficult and the best procedure to reduce its concentration is through sewer treatment at these sites.

Vodyanitskii (2014) pointed out that elements like heavy metals present in TDs are considered to be atypical in soils. The best known heavy metals must be considered for studies, such as Mn, Zn, Cu, Pb, Cr, vanadium (V), cobalt (Co), Cd, As, and radioactive elements, such as uranium (U). These elements were present in several industrialized area components and are commonly discarded in the environment. The quantity and time of retention in the soil is variable for each heavy-metal type, as in the case of Zn, which increases significantly in clay textures beyond the unknown behavior of several heavy metals found in organic matter.

The presence of contaminant materials in TDs, as well as their toxicity and permanence in the environment, has been highlighted in many academic works, as for example Bentlin et al. (2009); Domingues (2009); Vodyanitskii (2014); and Ashraf et al. (2015). As an example of heavy metals that are present directly (e.g. paints) or indirectly (e.g. packaging) in composition, Bentlin et al. (2009) explained that

> The paint matrix is complex and difficult to decompose, in view of the presence of various organic and inorganic compounds. Besides that, the remaining salts in the solution sample that was decomposed can interfere in the determination of the analyte. The methods described in technical standards are restricted to some elements (Cd, Co, Cr, Pb and Ti). (***Bentlin et al., 2009, p. 884***)

Meuser (2010), in extensive research about technogenic soils, discussed the problems resulting from each type of material deposited in urban and rural areas; in particular, detailing through chemical analysis in different artificial horizons the types and effects of toxic elements. In some cities, anthropogenic soils presented a diversity of materials (garbage, industrial remains, mining remains, building waste, among others) that form an intricate mosaic of environmental problems with difficult solutions.

According to Ashraf et al. (2015), the destiny of heavy metals will be controlled by means of physical and biological processes that work inside the soil. Metallic ions will enter the soil in solution from various forms and combinations at different rates and may remain in solution, pass in drainage, be absorbed by growing plants, be retained by slightly soluble soils, or remain in insoluble forms. The organic matter of some soils has great affinity with heavy metals that form stable complex cations with a reduced nutrient content. The regeneration process in environments with TDs depends on the type of material present. Generally, deposits with construction waste will need more time as a result of the cement and iron components used in buildings.

7.4 SOIL CONTAMINATION BY TDs IN THE URBAN AREA OF ARAGUAÍNA

Solid waste resulting from the implantation, transformation, and demolition of civic buildings are constantly deposited in ravines, gullies, vacant lands, valley bottoms, or unpaved streets in the city of Araguaína. These areas do not have any necessary structure for waste disposal. This factor will contribute and probably occur with the mobilization of solids and liquids generated from the decomposition of organic and inorganic materials. See the geographical distribution of TDs in Fig. 7.1.

Beyond this factor, Machado (2014) explained that strongly concentrated runoff produced during periods of intense rainfall (January–May), which carries significant amounts of TMs from the top of slopes to river-valley bottoms, increases the accumulation of materials in small sediment plains as well as watercourses. Baird (2002) stated that in fluvial sediments there is a high concentration of heavy metals due to the discharge of sewage.

As Baird (2002) suggested, heavy metals present in various manufactured products frequently gather in the topsoil and their concentrations may increase progressively once they enter the food chain. Combination effects may be important in two aspects. Primarily, the coexistence of contaminants in soils can affect the biological availability of others. Secondly, exposure to a combination of pollutants may be associated with the antagonistic, synergistic, and additive interactions of these pollutants, impacting organisms. A few risks of pollutant mixtures can be predicted based on existing knowledge (Ashraf et al., 2015).

According to the age of the deposit, the types of materials may differ significantly due to either the substitution for new materials or the fact that some materials have become economically viable for recycling or reuse. It is for this reason that the existence of aluminum (Al), Fe, Zn, and Cu is reduced in recent deposits (Machado, 2016). The use of urban soils, as with the expansion of Araguaína, occurred without regulation or inspection by public administration and, for this reason, small and medium industries in peripheral residential areas have become common and led to problems of waste disposal by these companies.

The accelerated urban expansion of Araguaína in the last two decades has favored the advance of occupation into new areas, mainly near the river valleys of several watercourses that cross the city. The current municipal administration does not perform an inspection or regulatory action for environmental planning and urban land use. There is a strong political influence by economic agents on the municipal government. This has facilitated the opening of new allotments and damaged areas of permanent preservation. Only the action of the state environmental agency, the Instituto de Natureza do Tocantins (NATURATINS), is able to constrain environmental damages.

The Araguaína urban area has six types of soils according to the World Reference Base for Soil Resources (2014): ferralic arenosol, ferralsol, nitisol, lithic leptosol, mollic, and gleysol. In the first two types, there have been concentrations of TMs disposal. Considering TMs with a higher concentration of waste originated from the demolition of civic buildings in the city of Araguaína, they are classified as Class II A and may have properties such as biodegradability, combustibility, or water solubility (Machado, 2016).

As an example of the composition of materials in an inorganic TD located in the Cimba sector (see Fig. 7.2) near the city's inner area, with an area of 5.212 m^2 and 2.7 m deep, that has existed for 18 years (Machado, 2011). In this area, there are TMs, including the presence of toxic elements

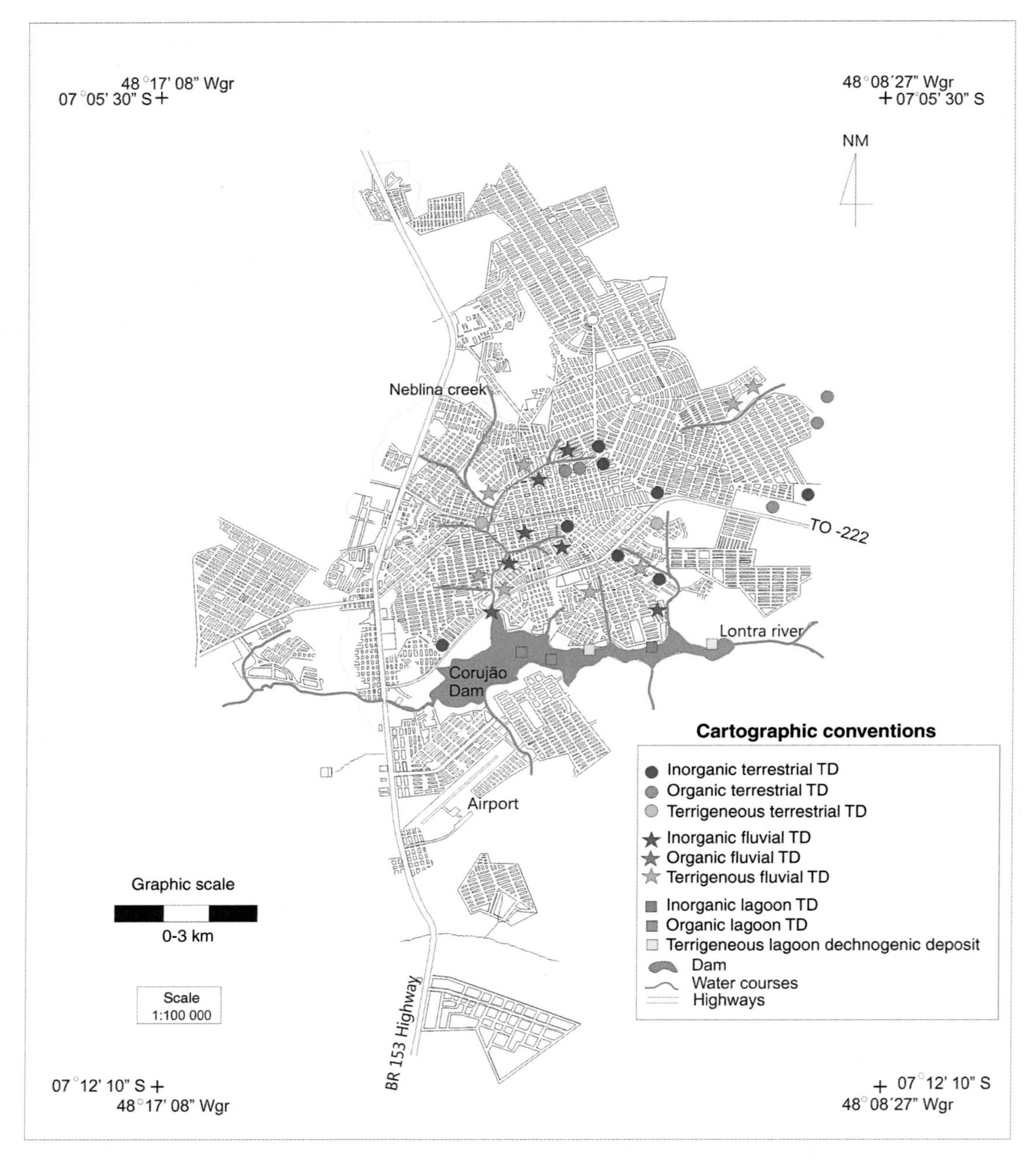

FIGURE 7.1 Geographical distribution and TD types. The highest concentration of TDs in downtown Araguaína city was due to urban land valorization and the floodplain of Neblina Creek being grounded with construction waste and later with the construction of houses and buildings.
Source: Modified from Prefeitura Municipal de Araguaína (PMA), 2005. Plano diretor municipal de Araguaína; Machado, C.A., 2011. Genesis and dynamics of technogenic deposits in the urban area of Araguaína (Brazil). Analls of International Geographic Union (IGU), 2011, Santiago (Chile).

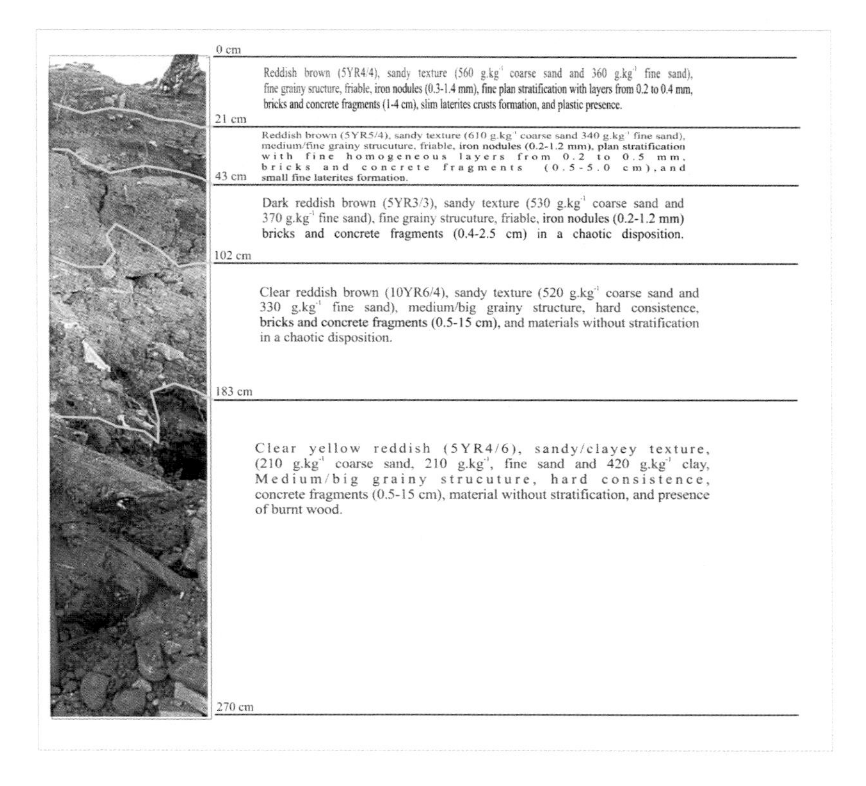

FIGURE 7.2 **Geographical distribution and TD types.** The diversity of materials in this inorganic TD formed in a gully demonstrates the phases of deposition during the last 20 years. The older and deeper layers (183–270 cm depth) exhibited wood and rocks when the area was sparsely inhabited. With the increase of urbanization in the last 10 years, the intermediate layers present a diversity of materials, with residues of construction, plastic, and paper (21–102 cm). The most recent and superficial layers have a large amount of TM inserted in public park construction. *Source: Modified from: Machado, C.A., 2011. Genesis and dynamics of technogenic deposits in the urban area of Araguaína (Brazil). Analls of International Geographic Union (IGU), 2011, Santiago (Chile).*

as fragments of metal and painted wood and fragments of decomposed asbestos tile present in the layer between 102–108 cm in depth. In the other layers, we can only find civic-building waste (bricks, tiles, and small concrete fragments). Due to the presence of contaminating materials, this deposit has been classified as Class I (Dangerous), with high toxicity mainly due to the toxic elements of paint.

According to the topographic profile (Fig. 7.2), the largest verified depths are between 1.9 and 2.7 m in the original deposition area, with approximately 110 m of extension. Because of runoff, most of the fine materials and small debris were carried from the top of the 9% slope into the river valley, with an extension of 40 m and an average depth of 50 cm. This deposit is one of the three largest found in Araguaína and has a large amount of inorganic material, representing a high cost for removal and deposition in a suitable area.

The TD was formed in ferralic arenosol and it is located in the half slope that has a slope of 5%. The superficial layers of sediment between 0 and 102 cm in depth presented greater

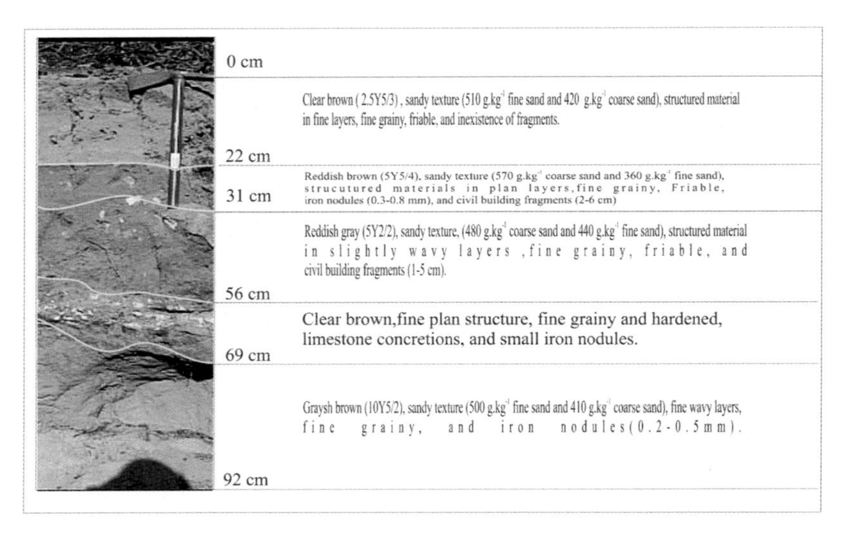

FIGURE 7.3 **Transversal profile of TD in the Cimba neighborhood.** The TDs composed by TMs present the lowest risks to the environment. The superficial layers are composed of sands and small rock fragments stabilized by grass vegetation. In this profile, only the layers between 56 and 96 cm deep are hardened by concrete residue that makes it difficult for water infiltration and the growth of plants roots.
Source: Modified from Machado, C.A., 2011. Genesis and dynamics of technogenic deposits in the urban area of Araguaína (Brazil). Analls of International Geographic Union (IGU), 2011, Santiago (Chile).

fragility due to their friable sandy texture. The material is structured in flat layers, with the presence of iron nodules. Construction and demolition residues of small dimensions between 0.3 and 5 cm have formed small crusts of 0.4 to 0.7 mm in the layers between 0 and 43 cm. Further layers of this deposit, between 102 and 270 cm in depth, present greater consistency and resistance due to the quantity of clay-like material. Construction and demolition wastes are dispersed in a chaotic shape and with large dimensions (>15 cm). Some of these are easily disaggregated by the state of advanced decomposition. A fact that increases instability in the deposit is that at the base there are burned tree trunks that can move under pressure.

The rigid structure of some TMs, such as concrete and metallic structures, increases decomposition time and can possibly cause greater instability in landfills due to the large space among debris. As an example of a TD: it was selected in an area of 4.420 m^2, with an average thickness of 70 cm that has existed for some 12 years and is located in the middle course of Cimba Creek. The material was deposited on ferralic arenosol, containing a medium-to-fine texture in a relief of 5–10% slope covered by thin grass vegetation (see Fig. 7.3).

The TMs present in this deposit consist of small concrete fragments, iron nodules originating from laterites, decomposed bricks, and pieces of rubber in the layer between 22 and 31 cm in depth. No materials containing toxic elements have been identified and this area has been classified as Class II (Nonhazardous), with low toxicity to the environment. Structural and textural

TABLE 7.1 Chemical Composition of TMs

Material type	Composition
Cement[a]	Limestone, sand, calcined clay, and plaster
Concrete[b]	Cement, sand, and rocks
Bricks and tiles[c]	Clay (aluminum-silica), sand, and paint
Iron[d]	Iron, carbon, nickel, and molybdenium
Aluminum[e]	Aluminum
Zinc[f]	Zinc
Asbestos tiles[g]	Chrysotile
Fiber cement tiles[h]	Cement, calcium carbonate, vegetable fibers, and polypropylene thread
Latex paints[i]	Polyvinyl acetate (PCV), zinc oxide, titanium oxide, aluminum powder, mica, barita chrome, and zinc
Oil paints[j]	Alkyd resins, zinc oxide, titanium oxide, aluminum powder, mica, chrome, barite, and zinc
Ceramics[k]	Clay, andahisite, bauxite, calcite, chromite, dolomite, and others
Styrofoam[l]	Ethylene monomer (petrochemical derivative)
Plastics (PCV, PET, PP)[m]	Petrochemical polymers, carbon, and silica
Batteries[m]	Cadmium, lithium, zinc, iodine, and copper

[a] *www.ecivilnet.com/artigos/cimento_portland_composicao.htm*
[b] *www.portaldoconcreto.com.br/cimento/concreto/concretos.html*
[c] *http://www.abceram.org.br/site/index.php?area=4*
[d] *www.acobrasil.org.br*
[e] *www.abal.org.br*
[f] *www.abmbrasil.gov.br*
[g] *www.abrea.com.br*
[h] *www.brasilit.com.br*
[i] *http://187.17.2.135/orse/esp/ES00144.pdf*
[j] *http://187.17.2.135/orse/esp/ES00144.pdf*
[k] *www.abceram.org.br*
[l] *www.quimica.seed.pr*
[m] *www.usp.br/fau/deptecnologia/docs/bancovidros/compplast.htm*
Source: From Machado, C.A., 2016. Depósitos tecnogênicos: gênese, morfologias e dinâmica. Edição do autor. Available from: https://www.researchgate.net/publication/309758050_Depositos_Tecnogenicos_Genese_Morfologias_e_Dinamica?ev(prf_high.

analysis has revealed the existence of five layers. In the first layer, between 0 and 22 cm, the materials present a sandy texture of fine and friable granulation and are free of TMs. The layers between 22 and 56 cm in depth have a sandy, clay-like texture and are slightly plastic, with the presence of iron nodules (0.3–08 cm) and small quantities of building debris dispersed (2–6 cm).

Specifically, in the city of Araguaína, most TDs are formed in river valleys on the embankment of floodplains or in old ravines that were covered by the municipal administration and have since been reopened by heavy rains. In the topographic profile, the greatest depths of verified TD area were between 50 and 92 cm in the original deposition area, with a length of 100 m and width of 68 m. Surface runoff has carried away most of the fine material

TABLE 7.2 Main Materials in Araguaína TDs

Material type	Decomposition time (years)	Toxicity
Concrete	100	Inert
Bricks and tiles	10	Inert
Ceramics	Undetermined	Inert
Iron	150	Inert
Painted wood	13	Toxic
Plastics	100–500	Toxic
Asbestos tiles	30–100	Toxic
Aluminum	100–500	Toxic
Zinc	100–500	Toxic
Paints	50–100	Highly toxic

Source: http://www.deltasaneamento.com.br/, http://www.fec.unicamp.br/~crsfec/tempo_degrada.html, and http://www.set.eesc.usp.br/1enpppcpm/cd/conteudo/trab_pdf/125.pdf, Machado, C.A., 2016. Depósitos tecnogênicos: gênese, morfologias e dinâmica. Edição do autor. Available from: https://www.researchgate.net/publication/309758050_Depositos_Tecnogenicos_Genese_Morfologias_e_Dinamica?ev(prf_high.

and small debris, forming a layer with an average depth of 25 cm that extends to riverbed deposits.

The main TMs found in the soils of the study area include concrete, bricks, tiles, wood, Fe, Al, Zn, plastics, and paint cans, among others. Each material has a specific chemical composition and is associated with other elements, such as cans in the case of paint. Except for concrete, bricks, and tiles (unpainted), the other elements are highlighted in Table 7.1 according to ABNT/NBR 1004 standards and are classified as Toxic and Dangerous when in the environment.

In this study, TMs including metals, plastics, asbestos tiles, and light bulbs, among others, were classified according to the standards of ABNT/NBR 1004 as Class I (Hazardous). Concrete, bricks, and tiles (unpainted) are Class II (Nonhazardous). The definition of TD classification in hazard classes according to their composition is found in Table 7.2. The distribution and geographical concentration in certain areas of Araguaína may increase the contamination of pedological layers, creating problems for the residential population, as in the case of Neblina Creek that crosses the urban area. The decomposition rate of each type of TMs will determine how long the effects of toxic elements will remain active in different soil types. These elements will also influence dispersion and retention. For example, soils with a higher clay content can retain elements like heavy metals that are active for longer in toxic form due to the capacity of their chemical bonds, according to Brady and Weil (2013). Also concerning the activity of toxic elements, Baird (2002) emphasized that humic substances have a great affinity with heavy metals cations.

The Fe element is nontoxic, but in soil Fe and Al oxides strongly absorb Cu^{+2} and Pb^{+2}; also, Mn oxides have high selectivity for Cu^{+2}, Ni^{+2}, Co^{+2}, and Pb^{+2} according to Domingues (2009). The decomposition time of each element may be analyzed in Table 7.2. According to Brady and Weil (2013), Cd and As are extremely toxic; Hg, Ni, Pb, and fluorine (F) are moderately toxic; and B, Cu, Mn, and Zn are slightly toxic.

The solution for the TD area requires the use of public funds to solve the problem. The reuse of solid waste is a reality in some Brazilian cities, with the collaboration of companies that collect and grind solid waste, producing aggregates for various purposes. The removal of TMs for disposal in suitable areas, including solid-waste landfills, or for reuse in civil construction would solve part of the problem of illegal disposal. This measure would be appropriate for 12 deposits in this city, because of their small areas and volume, without incurring high costs for the operation.

7.5 CONCLUSION

Technogenic deposits have increased significantly in area and quantity in the last two decades in the city of Araguaína. The resulting problems have increased and have not been resolved by municipal administrations, so that urgent measures are required to reduce or terminate the environmental impacts. Among possible measures that could be taken to solve the contamination of the pedological layer, we mention the withdrawal of TMs and contaminated soil around them to be discharged in the city's solid-waste landfill once the majority of these deposits have areas smaller than 500 m^2.

The large TDs represent the least amount of these areas, the first measure being the isolation of the area to avoid new waste discards, so that a later study is necessary to define the main types of contaminants and specific measures for decontamination of the area. Deposits containing only construction waste composed of bricks, concrete, and wood will be recycled and used in the manufacture of new bricks and concrete structures like blocks.

There are several methods of soil removal and decontamination with differing complexities and costs. Effective decisions depend on the types of material to be removed and the time that contaminants have acted on each specific soil type. The environmental legislation of Araguaína does not specifically address this issue. It will be necessary to include items that involve the correct disposal, surveillance, encouragement for recycling materials, and penalties for lawbreakers.

Technogenic deposits are largely formed by the unregulated disposal of companies that collect the materials in private works and generally deposit them in peripheral areas of the city. It would, therefore, be necessary for municipal administration to seek an understanding with private companies for disposal in an adequate area.

For the benefit of populations that reside near these contaminated areas, it would be necessary to develop an environmental education program for the entire population that is delivered in nearby schools in order to avoid the accumulation of garbage and burning. Additionally, the population ought to act as inspectors for their respective communities, aiding in the effort to reduce unregulated TDs.

References

Ashraf, M.A., Mohd, J.M., Yusoff, I., 2015. Soil contamination, risk assessment and remediation. Available from: http://cdn.intechopen.com/pdfs-wm/46032.pdf/.

Baird, C., 2002. Química Ambiental. Bookman, Porto Alegre, Rio Grande do Sul, Brazil.

Bentlin, F.R.S., Pozebon, D., Depoi, F., dos, S., 2009. Estudo comparativo de métodos de preparo de amostras de tinta para a determinação de metais e metalóides por técnicas de espectrometria atômica. Química Nova 32 (4), 884–890.

Brady, N.C., Weil, R.R., 2013. Elementos da Natureza e Propriedades dos Solos. Bookman, Porto Alegre, Rio Grande do Sul, Brazil.

Craul, P.J., 1999. Urban Soils: Applications and Principles. John Wiley & Sons, New York, NY, United States.

Domingues, T.C. de G., 2009. Teor de Metais Pesados em Solo Contaminado com Resíduo de Sucata Metálica, em Função de Sua Acidificação. Dissertação de Mestrado Campinas.

Korb, C.C., 2009. A Identificação de depósitos tecnogênicos na barragem Santa Bárbara, Pelotas (RS). Dissertação de Mestrado, Porto Alegre (RS), 2006. Available from: http://hdl.handle.net/10183/1780.

Machado, C.A., 2011. Genesis and Dynamics of Technogenic Deposits in the Urban Area of Araguaína (Brazil). Analls of International Geographic Union (IGU), Santiago (Chile).

Machado, C.A., 2014. Urban expansion and the formation of technogenic deposits in tropical areas: The case of Araguaína city. Available from: http://www.investigacionesgeograficas.uchile.cl/index.php/IG/article/viewFile/32991/34745.

Machado, C.A., 2016. Depósitos tecnogênicos: gênese, morfologias e dinâmica. Edição do autor. Available from: https://www.researchgate.net/publication/309758050_Depositos_Tecnogenicos_Genese_Morfologias_e_Dinamica?ev(prf_high.

Menk, J.R.F., Rossi, M., Bertolani, F.C., Coelho, M.R., Fernández, G.Á.V., 2004. Projeto de Gestão Ambiental Integrada da Região do Bico do Papagaio. Zoneamento Ecológico-Econômico, seconded. Seplan/DZE, Palmas, Tocantins, Brazil, Secretaria do Planejamento e Meio Ambiente (SEPLAN). Diretoria de Zoneamento Ecológico-Econômico (DZE).

Meuser, H., 2010. Anthropogenic Soils: Contaminated Urban Soils. Springer, London, United Kingdom, 121–193.

Prefeitura Municipal de Araguaína (PMA), 2005. Plano diretor municipal de Araguaína.

Rogachevskaya, L.M., 2006. Impacts of technogenic disasters on ecogeological processes. In: Zektser, I.S. (Ed.), Geology and Ecosystems. Springer, New York, NY, United States, 161–169.

Suguio, K., 1973. Introdução a Sedimentologia. Blucher, São Paulo, Brazil.

Ter-Stepanian, G., 1988. The beginning of technogene. Bull. Int. Assoc. Eng. Geol. 38, 133–142.

Tocantins (ESTADO), 2004. Secretaria do Planejamento e Meio Ambiente. Projeto de Gestão Ambiental Integrada da Região do Bico do Papagaio. Zoneamento Ecológico-Econômico. Análise ambiental e socioeconômica: norte do Estado do Tocantins. Palmas, Tocantins, Brazil.

Vodyanitskii, Y.N., 2014. Natural and technogenic compounds of heavy metals in soils. Eurasian J. Soil Sci. 47 (4), 255–265.

Further Readings

Associação Brasileira de Normas Técnicas (ABNT), 2004. Amostragem de Resíduos Sólidos ABNT NBR 10007/2004. Rio de Janeiro, Brazil.

Associação Brasileira de Normas Técnicas (ABNT), 2004. Resíduos Sólidos: Classificação ABNT NBR 10004/2004. Rio de Janeiro, Brazil.

Berger, A.R., 1997. Assessing rapid environmental change using geoindicators. Int. J. Geosci. Environ. Geol. 32 (1), 36–44, July. New York, NY, United States.

Boscov, M.E.G., 2008. Geotecnia Ambiental. Oficina de Textos, São Paulo, Brazil.

Conselho Nacional do Meio Ambiente (CONAMA), 2008. CONAMA Resolução 397, de abril de 2008. Available from: www.mma.gov.br/port/conama/legislacao/CONAMA_RES_CONS_2008_397.pdf.

Empresa Brasileira de Pesquisa Agropecuária (EMBRAPA), 2006. Centro Nacional de Pesquisa de Solo. Sistema Brasileiro de Classificação de Solos, Rio de Janeiro, Brazil.

Goudie, A., 2006. Human Agency in Geomorphology: The Human Impact on Natural Environment: Past, Present and Future. Blackwell, Oxford, United Kingdom, pp. 159–195.

International Union of Soil Science (IUSS), 2014. World Reference Base for Soil Resources (WRB). Available from: http://www.fao.org/nr/land/soils/soil/en.

United States Department of Agriculture/Natural Resources Conservation Service (USDA/NRCS), 2015. Heavy metal soil contamination. Soil Quality–Urban Technical Note no3. Available from: http://www.nrcs.usda.gov/Internet/FSE_DOCUMENTS/nrcs142p2_053279.pdf/.

CHAPTER

8

Transforming the Physical Geography of a City: An Example of Johannesburg, South Africa

Jasper Knight

University of the Witwatersrand, Johannesburg, South Africa

OUTLINE

8.1 Introduction 129

8.2 The Physical Environment of the City 130

8.2.1 Topography, Ecosystems, and Climate 130

8.2.2 Geology and Mineral Resources 134

8.3 Development of the City 135

8.3.1 Precolonial Development 136

8.3.2 The Gold Rush of the 1880s 136

8.3.3 The Apartheid Era (Circa 1948–1994) 138

8.3.4 Post-1994 Development 139

8.4 Discussion: Challenges of the City Today 139

8.4.1 Urban Water Management 140

8.4.2 Industrial Site Rehabilitation and Mine Pollution Management 140

8.4.3 Food Security 142

8.4.4 Urban Greening 142

8.4.5 Sustainable Development 143

8.5 Conclusions and Future Outlook 143

References 144

8.1 INTRODUCTION

Johannesburg (South Africa), like many developing world cities, is currently undergoing rapid change. The drivers of contemporary change in developing world cities include globalization of people-movement, economies, cultures, and the political context in which cities and countries deal with each other in a 21st-century world. As such, the context for development of such cities has changed markedly from the 19 and 20th

Urban Geomorphology. http://dx.doi.org/10.1016/B978-0-12-811951-8.00008-4

centuries, when many (now developing world) countries were administered by European colonial powers, which disproportionately influenced all areas of urban development, planning and governance, socioeconomic and cultural activities, and resource exploitation and management (Winkler, 2012). A key element driving the postcolonial development of developing world cities, especially in Africa, is the expression of local (national) priorities in housing, transport, sustainability, and food/water security; and in issues of heritage and cultural expression and management (Chirisa et al., 2016; Geyer, 2003; Harrison and Rubin, 2016; Todes, 2012a,b). Contemporary changes in these developing world cities, therefore, reflect a combination of both local (national) and global drivers and contexts (Robinson, 2008; Rogerson and Rogerson, 2015; Sihlongonyane, 2016). As such, many developing world cities today can be considered to have had a fractured recent political, socioeconomic, and cultural history of colonial and postcolonial development (Geyer, 2003). Thus, the physical patterns and urban makeup of these cities today can be considered as palimpsests that reflect partially preserved elements of their evolution. Johannesburg, the largest city in South Africa (4.4 million in 2016; circa 9.8 million in the wider metropolitan area), owes its existence to unique features of the physical landscape that have shaped its patterns of development over the last 150 years. Upon this physical background have been superimposed political, socioeconomic, and cultural expressions of power and identity that reflect its colonial, postcolonial, apartheid, and postapartheid histories. This chapter considers the (1) physical landscape; (2) development history of Johannesburg; and (3) discusses relationships between the physical environment and the contemporary city, including water management, food security, urban greening, and sustainability.

8.2 THE PHYSICAL ENVIRONMENT OF THE CITY

8.2.1 Topography, Ecosystems, and Climate

Johannesburg is located in a relatively flat region of northeast South Africa, termed the Highveld (>1700 m altitude). The city today occupies several east-west aligned hills and valleys that form part of the Witwatersrand sedimentary basin. The name Witwatersrand means "the ridge of white water" in Afrikaans and, thus, broadly refers to the relationship between topography and rivers in the landscape (Figs 8.1–8.2).

The topographic variations seen around Johannesburg result from the differential weathering and erosion of rocks of the Witwatersrand Supergroup over long time periods. Harder and more resistant quartzite rocks tend to form east-west aligned elongate ridges (Figs 8.2–8.3) and softer mudstones underlie intervening valleys. The quartzite ridges weather over time to form a thin and nutrient-poor soil, but with high biodiversity, particularly of plant species, and forms part of the rocky Highveld grassland biome (Mucina and Rutherford, 2006). Within this, the presence of ridges and valleys are related to local microclimatic and environmental conditions that give rise to variations in local soil and ecosystem patterns. The ridges host 71% of the province's endemic plant species, many of which are classified as endangered, and also some endangered mammal and bird species (Pfab, 2002). The locations of many of the ridges within the city act as "islands" of high biodiversity and their

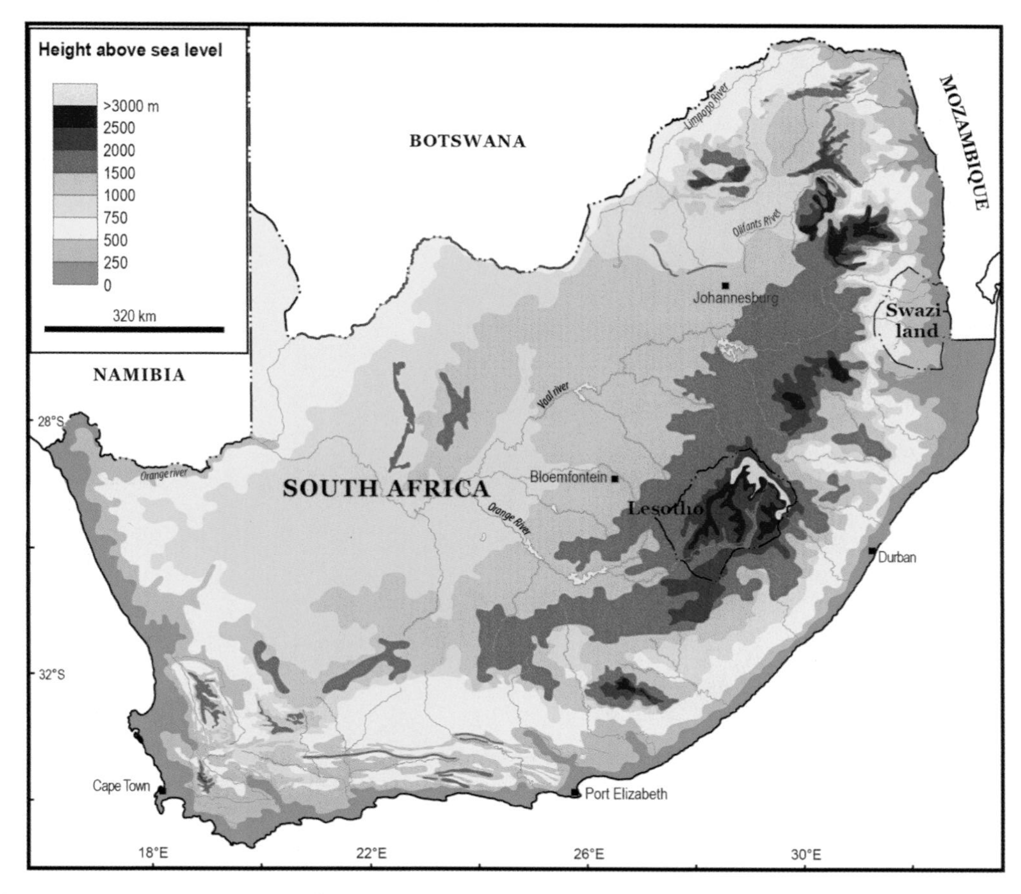

FIGURE 8.1 **Map of South Africa, showing the location of Johannesburg and regional topography.**

linear structure offers migration corridors for these species (Ellery et al., 2001) (Figs 8.2–8.3). By contrast to the built-up areas, on either side the ridges are also climatically cooler and affected by diurnal variations in wind patterns (Goldreich and Surridge, 1988). As the city has developed, these topographic-driven microclimate effects have become more apparent as the urban heat island effect has developed (Goldreich, 1992). In turn, diurnal variations in heating and cooling, amplified by the urban heat island effect, contribute to the generation of katabatic winds as a result of changing the vertical temperature distribution particularly in winter (Tyson et al., 1972).

The regional topography is significant because the continental drainage divide (between the Indian Ocean-flowing Limpopo River system and Atlantic Ocean-flowing Vaal/Orange River system) is located in the middle of the city. Feeding into these major river systems are various smaller rivers and streams, including the Klip and Jukskei Rivers, which drain to the south and north, respectively (Fig. 8.2). Johannesburg has a strongly seasonal climate, with

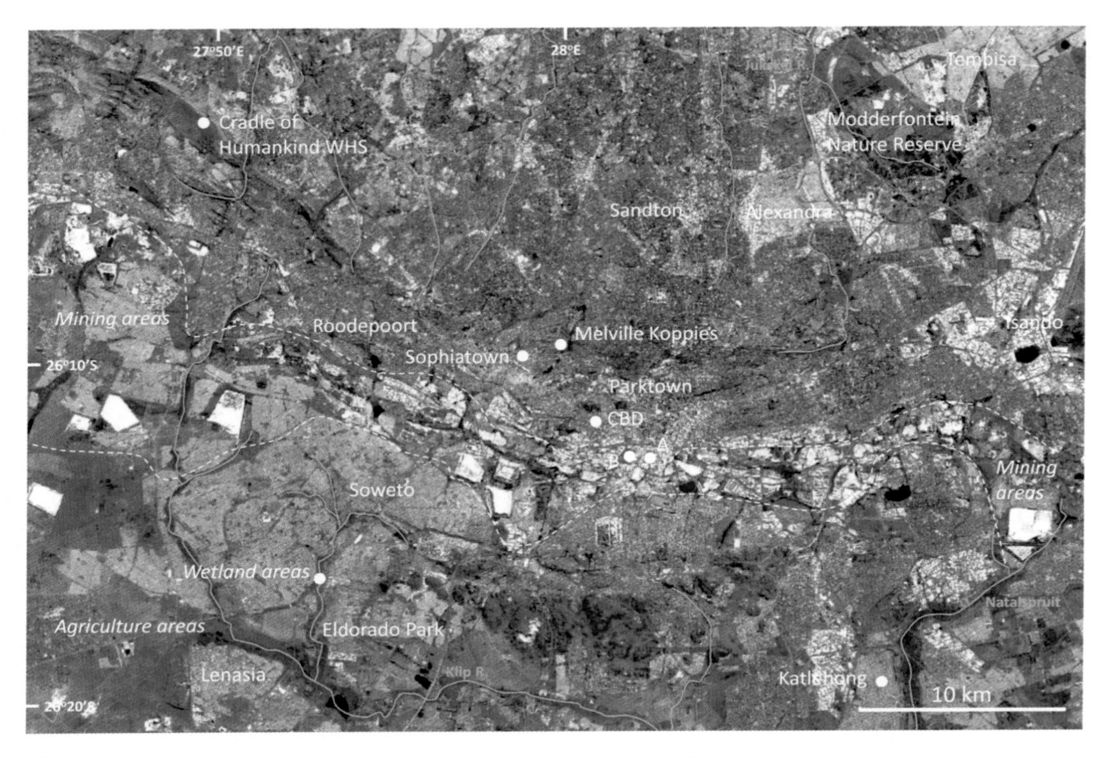

FIGURE 8.2 Regional scale Google Earth image of Johannesburg (image date December 31, 2016), showing the major locations mentioned in the text and the main mining area (defined within the *white dotted lines*) south of Johannesburg central business district or CBD. The locations of Fig. 8.7A and B are marked (south of the CBD). Note the light tones of former apartheid-era nonwhite suburbs of Soweto, Alexandra, and Katlehong, indicative of high-density housing. Contrast this with today's wealthy (and mainly white) suburbs north of the CBD, which show a higher proportion of trees in the urban landscape. Principal rivers are marked with *blue lines*; those north of the mining areas flow north and those south of these areas flow south.

wet austral summers and dry winters. Mean monthly temperature varies from 25°C (January) to 17°C (July) and monthly precipitation from 107 mm (January) to 6 mm (July) (City of Johannesburg, 2009). Annual precipitation is mainly in the range 700–720 mm, which means this environment is marginally semiarid. The influence of topography on rainfall patterns in Johannesburg was examined by Tyson and Wilcocks (1971). Periods of rainfall are associated with moist air masses coming from the northeast to southeast sectors, originating from anticyclonic conditions over the adjacent Indian Ocean. These air masses are forced to rise along their westward track, as they pass over the Great Escarpment of southern Africa toward the Highveld, forcing condensation and thundercloud formation. This is also amplified by urban heat island effects during summer afternoons. The heat island and reduced surface water availability also decrease atmospheric humidity and increase wind gustiness. Natural and urbanized catchments in Johannesburg show differences in runoff ratio (thus, rainfall runoff response) as well as different water uses. Over recent decades, the dramatic

FIGURE 8.3 Photograph of (A) the landscape of Melville Koppies and its view over the city and (B) the quartzite bedrock and grassland of the koppies. *Source: (A) Reproduced with permission from Sian Butcher; (B) Reproduced with permission from Jennifer Fitchett.*

increase in city area by development of satellite suburbs and building of higher structures in the central business district or CBD and in the financial center of Sandton, have contributed to increased urban heat island effects (Goldreich, 1992; Tyson et al., 1972). Goldreich (1992) showed that the city center can be more than 11°C warmer seasonally and has much lower relative humidity (by <43%) when compared to surrounding suburbs. Hardy and Nel (2015), based on remote sensing data, calculated the nighttime average urban heat island effect in the period 2002–2012 to be 2.0–3.5°C, but with significant spatial variations across the city.

8.2.2 Geology and Mineral Resources

The Witwatersrand basin within which Johannesburg is located hosts one of the world's largest gold deposits and more than one-third of all gold that has been extracted globally comes from this region (Tucker et al., 2016). Quartzites, shales, and conglomerates of the Witwatersrand Supergroup were deposited around 2.3–2.9 Ga ago in a shallow marine to braided fluvial plain setting. Total sediment thickness is more than 6000 m and the Witwatersrand basin has dimensions of 350 × 200 km (McCarthy, 2006). Following deposition, these sediments were uplifted, folded, and faulted and mineralization (formation of metal-rich deposits) within the sediments took place in association with several significant high-energy events. These included the Vredefort meteorite impact 2.023 Ga ago, which caused rock fracturing and migration of heated groundwater through the rocks and emplacement of the Bushveld Igneous Complex 2.054–2.059 Ga ago, which also caused groundwater migration, enriching metal concentration within the host conglomerate beds. These host rocks, termed "reefs," outcrop linearly in the landscape for lengths up to 45 km and mines were, therefore, located preferentially in these areas. Reefs of the East Rand and Central Rand Goldfields have been largely mined out (including the 1886 discovery site on Langlaagte Farm, part of the Central Rand Goldfield); fields of the West Rand, Carltonville, and smaller goldfields are still active (McCarthy, 2006) (Fig. 8.2).

The long time period of extraction activities in the Witwatersrand basin today (over 100 years) has meant that most of the remaining resources are at a considerable depth (<4 km), the most accessible and higher grade mineralized layers having already been worked out. Peak gold production from the Witwatersrand basin occurred in the 1970s, with subsequent decline of the industry in the region, with associated environmental and socioeconomic impacts. The surface expression of mining today is more subdued than in the past due to more modern extraction processes and environmental legislation. Activity today has also focused on reworking existing mine waste dumps using more modern processing techniques. Direct negative impacts of mining include groundwater extraction, induced seismicity, and mine shaft abandonment. These can lead to ground-surface subsidence and collapse (Bell et al., 2000; de Bruyn and Bell, 2001). Mine waste dumps cover large areas of the Central Rand Goldfield in the south of Johannesburg and are up to 20 m high with steep (engineered) sides and a flat top (Fig. 8.4). There are significant contemporary environmental issues associated with mine waste dump pollution discussed in the next sections. Many waste dumps today are being redeveloped for industry or are sites of unregulated (and illegal) informal, artisanal mining (Nhlengetwa and Hein, 2015).

FIGURE 8.4 **Photographs of the mine waste landscapes of southern Johannesburg**. (A–B) Abandoned mine waste dump site with secondary scrub encroachment (C–D); waste dump that is now being actively reprocessed (trucks for scale in D); and (E–F) flat top and rectilinear sloping sides of a waste dump. Note the erosional rilling of the dump sides in (F), indicating sediment loss downslope and into water courses. *Source: Photos by Jasper Knight.*

8.3 DEVELOPMENT OF THE CITY

There are many books that have described the colonial history of settlement and the development of the city of Johannesburg since its founding during the gold rush of the late 1880s. These studies have focused mainly on the political, socioeconomic, and racial contexts of city development (Cartwright, 1965; Mandy, 1984; Shorten, 1970). Few studies have considered either the physical landscape generally or the specific relationships between the physical environment and city development. More recently, the geomorphology, ecology, and heritage of postmining landscapes have been discussed (Bobbins, 2013; Toffah, 2013) with respect to rehabilitation and remediation, but these ideas have not yet been fully formulated into management or economic development strategies for the region (GCRO, 2014). Development of Johannesburg city and region since earliest times has, however, been strongly affected by environmental resource availability. This is briefly reviewed.

8.3.1 Precolonial Development

From the 13th century onward, the area later becoming Johannesburg was occupied by San hunter-gatherers, then by Sotho-Tswana farmers who constructed stonewalled enclosures that are present adjacent to hilltops, especially south of Johannesburg (Sadr and Rodier, 2012). During the South African Iron Age (12–18th centuries), metalworking associated with exploitation of Witwatersrand Reefs was evidenced by former mine workings and smelting furnaces, including on Melville Koppies in the center of Johannesburg (Fig. 8.3) and stonewalled enclosures throughout the region (Sadr, 2017). Internal warfare between tribal groups in the 18 and 19th centuries, termed the "mfecane," resulted in large-scale depopulation prior to significant white-settler movement into the region (Hamilton, 1995). Archaeological sites from these periods are still present in the landscape, but many were destroyed as the city expanded, especially in the early 20th century and with the spread of the contemporary city (Sadr, 2017). In the mid-19th century, this area was settled by white farmers; original farm names, such as Doornfontein, Braamfontein, and Turffontein, are still retained in different Johannesburg suburbs today.

8.3.2 The Gold Rush of the 1880s

In March 1886, after around 18 months of informal surveying and exploration on individual farms by mining entrepreneurs, significant gold reserves were discovered within the Witwatersrand Reefs. Within months, hundreds of prospectors arrived, mainly from the diamond-mining town of Kimberley (500 km west) and an official surveying team from Paul Kruger's government in Cape Colony (Cape Town). The resulting surveyor report (on August 12, 1886) recommended that mining leases be issued, but noted the lack of water in the region as a possible limitation on mining development. An announcement in the *Government Gazette* (August 18, 1886) recommended that owners and existing land leaseholders make their claims known. The establishment of mine leases on the adjacent farms of Driefontein and Elandsfontein on September 20, 1886 is commonly taken as the date of establishment of the city of Johannesburg. The city itself was named after one or several prominent mining officials and surveyors whose first name was Johannes (it is not clear precisely who, as there are several possible candidates). The original location of the new settlement to service these mine leases was on a triangule-shaped area of disused land (2 × 3 × 2.5 km dimensions) located between adjacent farms. The original southern edge of the triangle was named Commissioner Street, which is located in the city center today (Fig. 8.5). The northern point of this triangle terminated at the foot of the hill behind (now known as Hillbrow). This area was surveyed between October 19 and November 9, 1886 based on a grid pattern of blocks 50 feet (15 m) that is still

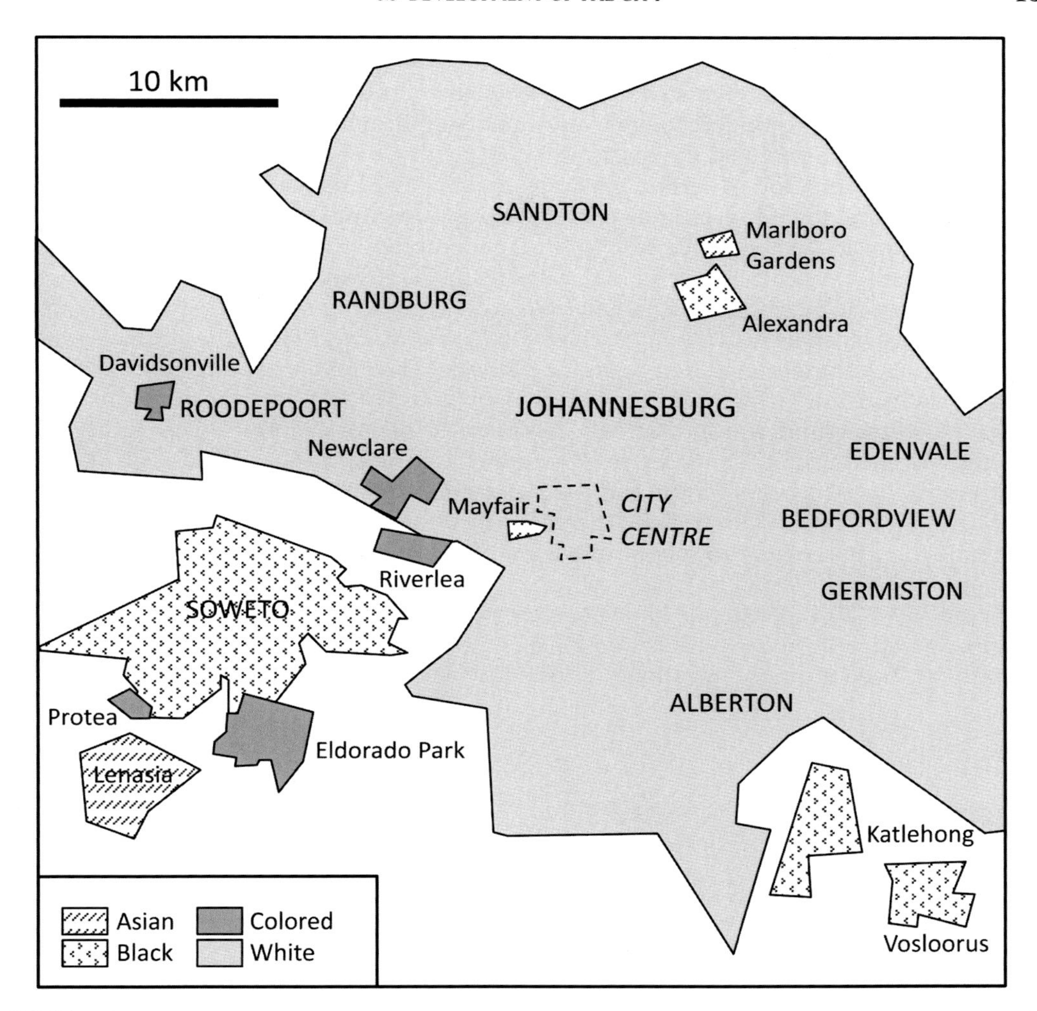

FIGURE 8.5 **Map of central Johannesburg showing the most significant spatial patterns during the first half of the 20th century and the major demographic patterns of the city**. "Open" areas are those with no distinct demographic profile. The *shaded triangle* is the area of original city settlement in 1886. *Source: Modified from http://archive.unu.edu/unupress/unupbooks/uu26ue/uu26ue0g.htm.*

largely in existence today. Prospecting leases and leases for services within the new settlement (bars, shops, banks) were sold on December 8, 1886. Within one year, the population of Johannesburg was 8000 and the Rand Club (social club) and stock exchange were established, followed by a theater (1889), public library (1890), and many others. By 1896, the population had reached 100 thousand (Mandy, 1984). This brief summary shows the speed with which the area was colonized following the discovery of gold and the very rapid formalization of space, social structures, economic activities, and hierarchies of power and privilege (mine owners versus workers) within the new city. Exploitation of and changes to the physical environment was coeval with those of city development.

8.3.3 The Apartheid Era (Circa 1948–1994)

Spatial and socioeconomic structures within the city, starting as a function of colonialism, but continuing to be felt today, were most significantly changed by the Group Areas Act (1950), which formalized the discriminatory principle of apartheid across space. This act involved the designation of certain urban residential and business areas for different ethnic groups and the enforced resettlement of certain groups from one area to another. As such, the socioeconomic, demographic and political geographies of Johannesburg were significantly altered, with implications for spatial patterns in the urban landscape (Christopher, 1997; Pirie and Hart, 1985). These designated areas were commonly separated according to physical boundaries, such as railway lines, rivers, hills, and valleys. Thus, the physical landscape in part influenced these new demographic patterns. The creation and recreation of suburbs along racialized lines was also accompanied by coeval ideas on transport, housing, and urban structures within these suburbs (Chapman, 2015). For example, black residents of the Johannesburg suburb of Sophiatown were forcibly removed to Soweto in 1955, colored and Indian residents to other suburbs, and the area partly demolished and rebuilt for white residents (Fig. 8.6). Soweto (abbreviated from South Western Townships) was adopted in 1963 as a name for different government-built, low-cost housing communities located south of the main mining strip in southern Johannesburg, constrained by wetlands of the Klip River in the south (Phillips, 2014). Wetlands have been areas, historically, of settlement for economically marginalized people and have been areas of higher flood and disease risk. The Klip wetland has also has been strongly negatively affected by acid water drainage from mines, peat ex-

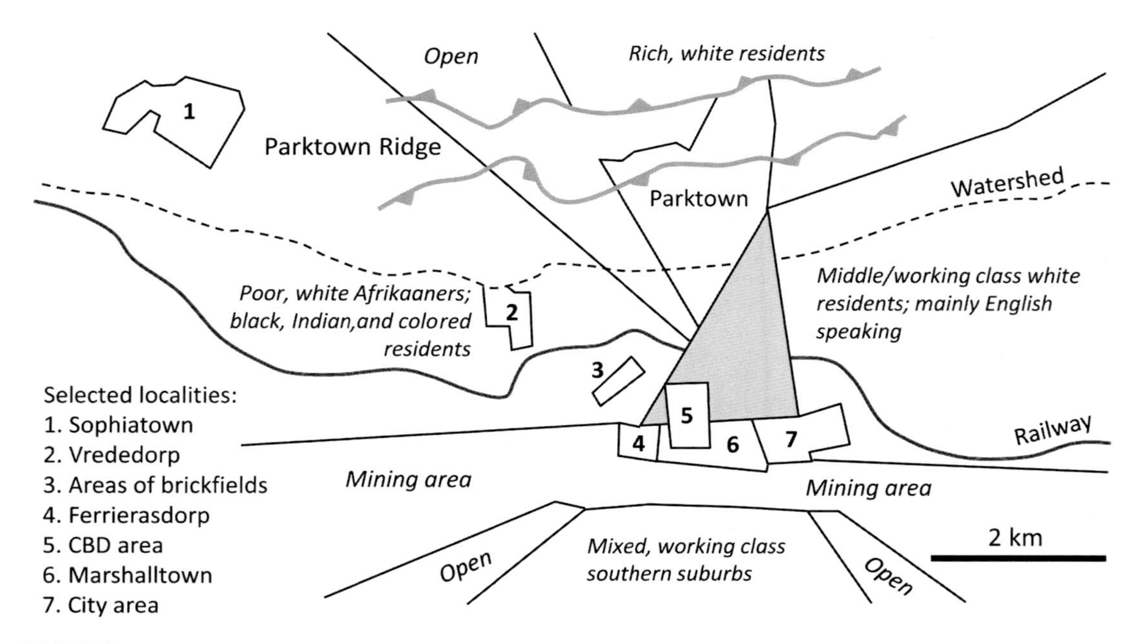

FIGURE 8.6 Map of some of the spatial patterns of Group Areas in the Johannesburg region. *Source: Modified from Parnell, S.M., Pirie, G.H., 1991. Johannesburg. In: Lemon, A. (Ed), Homes Apart. Chapman, London, pp. 129–145.*

traction, pollution from sewage works leading to eutrophication, and a dropping watertable because of groundwater extraction (McCarthy et al., 2007). It could be argued that apartheid decisions in siting these townships were partly based on the disadvantageous physical landscape in these areas (Chapman, 2015; Christopher, 1997).

Demographic patterns toward the end of the apartheid era, and in part continuing today, see a concentration of rich and mainly white residents in northern suburbs of the city. These areas are protected from prevailing winds by Parktown Ridge and other topographic highs, with poorer residents located in lowland areas that are historically more prone to flooding and associated with mining (Fig. 8.5).

8.3.4 Post-1994 Development

Since the abolition of the Group Areas Act in 1991, and the establishment of a democratic government after the fall of the apartheid regime in 1994, spatial constraints on urban development have been loosened and new policies formulated. These have included developing subsidiary suburban centers in a decentralized urban model (Herbert and Murray, 2015; Mabin et al., 2013) and reinventing the purpose of the inner city (Bremner, 2000). The former has included ongoing changes in the demographic makeup of existing suburbs, becoming more mixed in some areas (Rule, 1989), more gated, and securitized in some affluent white areas (Clarno, 2013), with greater socioeconomic inequality especially in historically black areas (Beall et al., 2000; Mathee et al., 2009; Murray, 2009). These changes have also been amplified by the opening of South Africa to international markets and investors and the development of high-end suburban estates, shopping malls, and financial centers. Redevelopment of the inner city has not been a simple process of gentrification, but also by demographic changes (including immigrant communities from other sub Saharan countries), and an emphasis on art, culture, and heritage as economic tools (King and Flynn, 2012; Nevin, 2014). The result of these developments has been a reconsideration of the meaning of space, place, and identity and the emergence of tensions in these meanings. Postapartheid urban planning has, therefore, a strategic and "ethnopolitical" dimension (Winkler, 2012); and the multiple and conflicting needs of the diverse population has meant that no single development strategy has yet been successful (Parnell and Robinson, 2006; Todes, 2012a). This means that, spatially, there are uneven patterns of development in the built environment and relationships to landscape resources, including open spaces and urban greening initiatives.

8.4 DISCUSSION: CHALLENGES OF THE CITY TODAY

Like other climatically stressed regions of the developing world, Johannesburg is presently experiencing constraints on the trajectory and priorities of urban growth that are related in part to constraints exerted by the physical environment, in particular related to the provision of water resources (for domestic, industrial, agricultural, sanitation use) (Dippenaar, 2015). Water availability and provision, especially in poorer areas of the city with deprived populations, is an important contemporary political issue (Nastar, 2014; Turton et al., 2006; Varis, 2006), but are also linked explicitly to flood-risk management (Viljoen and Booysen, 2006), urban food security (Malan, 2015), and sustainable development (Schäffler and Swilling, 2013). These have been highlighted by the City of Johannesburg (2009) as important outcomes of future climate change, and provide the context for the discussion on challenges facing the city today, and the role of the physical environment in mediating these challenges.

8.4.1 Urban Water Management

Urbanization has transformed the hydrological cycle of cities as a result of changed land cover, presence of impermeable surfaces, and modified microclimates (Ferguson, 2016). In South African cities, this tends to result in more intense and frequent rainstorm events and rapid stormwater runoff, with associated flood hazard (Walsh et al., 2012). As such, urban water management is a critical and ongoing issue in South African cities (Adegun, 2014; Knight, 2017; Turton et al., 2006; Varis, 2006). In Johannesburg, flooding in particular affects informal settlements that have a lack of adequate water system infrastructure and with high community vulnerability (Adegun, 2014, 2015; Dippenaar, 2015; Fatti and Vogel, 2011). Rapid overland flow results in stormwater contamination and pollutant transport into and through sewer systems and rivers (Armitage, 2007; Wimberley and Coleman, 1993). Lack of effective water management, resulting in flash floods amid periods of significant water scarcity due to climate, is the most significant management issue (Chirisa et al., 2016; Knight, 2017; Todes, 2012b; Viljoen and Booysen, 2006). Soakways, stormwater harvesting, and biofiltration are means by which water can be used and recycled more efficiently (Knight, 2017).

Topography exerts an important control on overland flow speed and direction. Floodwater routing, soil thickness and vegetation influence infiltration capacity and, in turn, reduce the amount of surface water and resulting flood hazard (Murray, 2009).

8.4.2 Industrial Site Rehabilitation and Mine Pollution Management

Tailings dumps from former mines are common in the Johannesburg landscape and represent the physical expression of the region's industrial heritage (Bobbins, 2013). Some of these sites have been rehabilitated into mixed industrial use, whereas housing has encroached on other sites (Kneen et al., 2015) (Fig. 8.7). Waste contamination from industrial processing means that sites have to be remediated prior to redevelopment, which is commonly not done effectively. The informal, artisanal mining undertaken on many sites today (Nhlengetwa and Hein, 2015) can lead to environmental hazards, such as particulate dust pollution, respiratory diseases, such as silicosis, mine collapse, carbon monoxide and methane emissions, uncontrolled fire, sinkhole formation, and acid mine drainage (Bell et al., 2000; de Bruyn and Bell, 2001; Kneen et al., 2015; McCulloch, 2009; Naicker et al., 2003; Nengovhela et al., 2006). The latter is a significant environmental problem because acidic water can both contaminate groundwater aquifers and spread long distances down-gradient and emerge into rivers, changing water chemistry and water quality. Processing of mineral ores from the host rocks was undertaken in the late 19 and early to middle 20th centuries using mercury and cyanide chemical techniques (Naicker et al., 2003). Rainwater infiltration into mine waste causes oxidation of chemical residues into pyrite (FeS_2) and leaching of metal sulfides, arsenic, and other contaminants through the sediment pile, resulting in acidic conditions at the top and alkaline conditions below (Naicker et al., 2003). These toxic, low-nutrient environments inhibit plant colonization and biodiversity and result in significant water pollution (Bakatula et al., 2012; Lusilao-Makiese et al., 2013). Surface and groundwater samples from

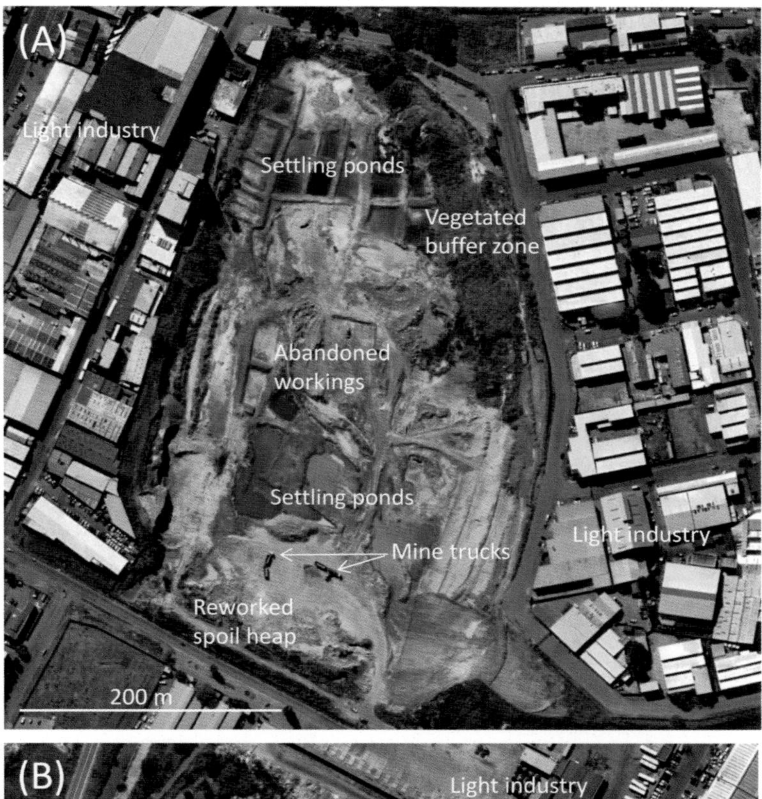

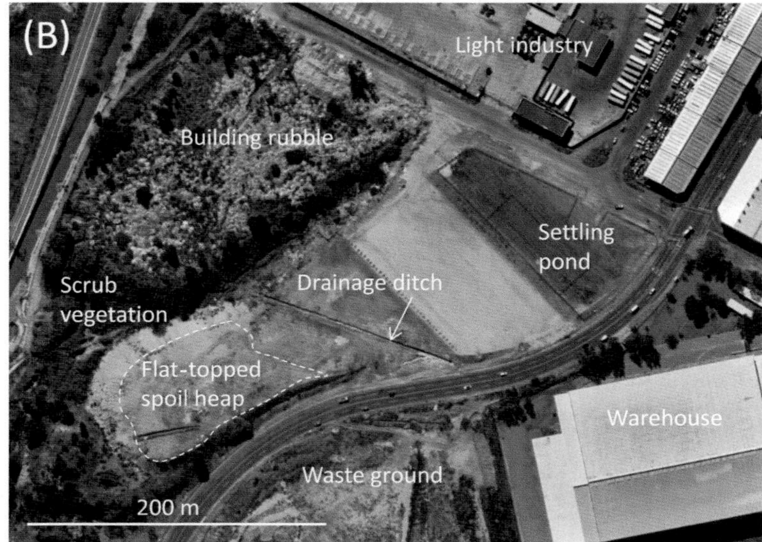

FIGURE 8.7 **Annotated examples of mine site rehabilitation located south of Johannesburg CBD (locations shown on Fig. 8.1).** (A) Enclosed site with steep sides to the former pit, partially covered with scrub vegetation, which acts as a buffer. Light industry surrounds the sites. Note the different colors of the mine water/sediment settling ponds, which reflects variations in water chemistry. A spoil heap south of the site is being actively reworked (e.g., Fig. 8.4D). (B) A more open mining site with a flat-topped spoil heap still intact. Building rubble has been dumped on the site and partly encroached by scrub vegetation. The settling pond and drainage ditches are areas of potential water contamination. *Source: Base images from Google Earth; image date December 31, 2016.*

the Natalspruit River in southern Johannesburg yield low, but variable pH values (3.08–7.90) and very variable conductivity, dissolved solutes, and redox values (Naicker et al., 2003). These values strongly depend on water source (surface versus subsurface) and season (mine water diluted by rainwater during the summer wet season) (Bakatula et al., 2012). Acid water can experience bioremediation when it is filtered through wetland sediments and riparian ecosystems (Abiye, 2015; Ambani and Annegarn, 2015; McCarthy et al., 2007): macrophyte species, such as *Juncus effusus* and *Carex riparia*, can remove dissolved heavy metals from the water as they grow (Ladislas et al., 2013) and, thus, constructed wetlands can be used as natural water filters (Rousseau et al., 2008). Such methods have been used in some places in the Witwatersrand mining belt, but not extensively (Dippenaar, 2015; Toffah, 2013).

Although these site rehabilitation and pollution management methods focus on sediment and water systems, mine waste landscapes are also significant because they were commonly burial locations of mine workers in the late 19 and early 20th centuries; and many of these graves are undocumented. The ownership, occupation, reoccupation, and development of mining landscapes also involve issues of power and socioeconomic exclusion and control of space, but these issues have not been well articulated in discussions of mine site rehabilitation (e.g., Crush, 1994).

8.4.3 Food Security

Formal and informal urban food gardens, collectives, and cooperatives are common in African cities and are extremely important for assuring food security for poor communities (Malan, 2015) and in community cohesion, skills development, and informal trading (Warshawsky, 2014). Making poor communities more resilient to climate change, water scarcity and increased food prices are seen as critical to economic growth and health/well-being (Schäffler and Swilling, 2013). However, the maintenance of food gardens is not well embedded in urban planning strategies in Johannesburg (Hetz, 2016; le Roux and Augustijn, 2017) and areas of disused or underused ground tend to be repurposed for housing or industry, with development of green infrastructure and community food support viewed as low priority (e.g., Schäffler and Swilling, 2013). A key environmental limitation is soil quality, especially in areas of southern Johannesburg, where food gardens are most common but where the soil is nutrient poor through overuse or erosion on made ground and, in particular, on industrially contaminated ground or on mine waste. In addition, a lack of technical expertise, support, and training (in horticulture, agronomy, permaculture, etc.) has limited the potential of urban food gardens (Malan, 2015; Warshawsky, 2014).

8.4.4 Urban Greening

The city of Johannesburg is commonly reported to be "one of the most street tree-lined cities in the world, with the appearance, in satellite images, of a tropical (human)-made forest" (Johannesburg City Parks and Zoo, 2017). This includes 10 million trees planted within the city boundaries, in parks, open spaces, and 1.3 million trees along roads alone. Green, open areas within cities are known to have positive impacts on air and noise pollution, human health and well-being, urban microclimates, and humidity and can buffer flood response and encourage groundwater recharge along soakways (Viljoen and Booysen, 2006). Development of green, open areas can also be seen as a political statement about environmental justice and values (Simon, 2012; Wolch et al., 2014); and it is notable that a strategy of the city of Johannesburg is to develop more of these areas in the historically disadvantaged south of the city as part of a wider environmental remediation and community development program (Johannesburg City Parks and Zoo, 2017). Native trees are most commonly planted, which are better able to survive low nutrient soils and water scarcity, but their wider environmental advantages and ecosystem services have not been well-studied, including their potential for enhancing biodiversity or acting as species migration corridors (Alvey, 2006). There are no specific environmental limitations on tree planting as part of an urban greening policy.

8.4.5 Sustainable Development

Sustainability is a central tenet of future urban development plans in Johannesburg (City of Johannesburg, 2009), similar to many large cities globally (Varis, 2006; Wolch et al., 2014). The context for sustainability here includes adaptation to future climate change, with its associated implications for food and water security, while ensuring equitable provision of services and sustainable economic development (Dippenaar, 2015; Robinson, 2008; Sihlongonyane, 2016; Winkler, 2012). These broad goals are not necessarily compatible (Götz and Schäffler, 2015) and environmental issues, such as water availability, have already been identified as a limiter of future growth in Johannesburg (Hetz, 2016; Nastar, 2014; Varis, 2006; Wolch et al., 2014). Spatial urban planning strategies can also be directly linked to aspects of sustainable development (Le Roux and Augustijn, 2017; Todes, 2012a,b) because they can be designed to increase community resilience to climate change impacts. Strategies that have already been identified include urban greening and reduced carbon emissions through public transport (Bubeck et al., 2014; Götz and Schäffler, 2015). However, as yet, there is no full discussion of the potential role of geoengineering of water, green spaces, and microclimate modification in Johannesburg and, thus far, urban greening strategies have not been applied consistently across the city. Sustainable economic development in Johannesburg is set against a background of globalization, rebalancing the legacy of apartheid socioeconomic policies and inequalities in education, health care, and housing delivery (Harrison and Rubin, 2016; Parnell and Robinson, 2006; Rogerson and Rogerson, 2015). Reconciling the tensions between sustainability of environmental and economic factors is an important future priority (Harrison and Zack, 2014).

8.5 CONCLUSIONS AND FUTURE OUTLOOK

Climate and environmental change is the most pressing 21st-century issue facing developing world cities. This context of climate change frames future development plans for the city (City of Johannesburg, 2009) and are linked to food and water security and sustainable development. Throughout the life history of Johannesburg as a city, there have been complex interplays between the physical landscape and its resources and city development (Shorten, 1970). Today, the city is poised upon transformative changes that are in common with many other cities worldwide, focusing on sustainable and integrated transport systems, transition to a low-carbon and knowledge economy, urban greening, and enhanced biodiversity (Chirisa et al., 2016; Götz and Schäffler, 2015). However, there are significant challenges to these goals, including increased urban population growth; food and water security; immigration/refugees and xenophobia; equity of service provision in housing, education, and health care; and pollution. It is still unclear how these different elements may be reconciled, but there are already examples of heritage conservation and tourism that reflect on the linked physical and human development of the city. Examples include sites of apartheid-era resistance in Soweto and other suburbs and paleontological sites with evidence for human evolution at Swartkrans and Sterkfontein (part of the Cradle of Humankind World Heritage Site situated north of Johannesburg). Mining heritage and tourism can easily fit with this narrative and illustrate the complex and nuanced relationships between the physical and human environments of a rapidly changing developing world city.

References

Abiye, T., 2015. The role of wetlands associated to urban micro-dams in pollution attenuation, Johannesburg, South Africa. Wetlands 35 (6), 1127–1136.

Adegun, O.B., 2014. Coping with stormwater in a Johannesburg, South Africa informal settlement. Proc. Inst. Civil Eng. 167 (2), 89–98.

Adegun, O.B., 2015. State-led versus community-initiated: stormwater drainage and informal settlement intervention in Johannesburg, South Africa. Environ. Urban. 27 (2), 407–420.

Alvey, A.A., 2006. Promoting and preserving biodiversity in the urban forest. Urban Forest. Urban Green. 5 (4), 195–201.

Ambani, A.-E., Annegarn, H., 2015. A reduction in mining and industrial effluents in the Blesbokspruit Ramsar wetland South Africa: has the quality of the surface water in the wetland improved? Water SA 41 (5), 648–659.

Armitage, N., 2007. The reduction of urban litter in the stormwater drains of South Africa. Urban Water J. 4 (3), 151–172.

Bakatula, E.N., Cukrowska, E.M., Chimuka, L., Tutu, H., 2012. Characterization of cyanide in a natural stream impacted by gold mining activities in the Witwatersrand Basin, South Africa. Toxicol. Environ. Chem. 94 (1), 7–19.

Beall, J., Crankshaw, O., Parnell, S., 2000. Local government, poverty reduction and inequality in Johannesburg. Environ. Urban. 12 (1), 107–122.

Bell, F.G., Stacey, T.R., Genske, D.D., 2000. Mining subsidence and its effect on the environment: some differing examples. Environ. Geol. 40 (1–2), 135–152.

Bobbins, K.L., 2013. The legacy and prospects of the Gauteng City-Region's mining landscapes. WIT Trans. Ecol. Environ. 179, 1363–1374.

Bremner, L., 2000. Reinventing the Johannesburg inner city. Cities 17 (3), 185–193.

Bubeck, S., Tomaschek, J., Fahl, U., 2014. Potential for mitigating greenhouse gases through expanding public transport services: a case study for Gauteng Province, South Africa. Transp. Res. D 32, 57–69.

Cartwright, A.P., 1965. The Corner House. The Early History of Johannesburg. Purnell and Sons, Cape Town, South Africa, pp. 293.

Chapman, T.P., 2015. Spatial justice and the western areas of Johannesburg. Afr. Stud. 74 (1), 76–97.

Chirisa, I., Bandauko, E., Mazhindu, E., Kwangwama, N.A., Chikowore, G., 2016. Building resilient infrastructure in the face of climate change in African cities: scope, potentiality and challenges. Dev. S. Afr. 33 (1), 113–127.

Christopher, A.J., 1997. Racial land zoning in urban South Africa. Land Use Policy 14 (4), 311–323.

City of Johannesburg, 2009. Climate Change Adaptation Plan. Johannesburg, South Africa, pp. 96.

Clarno, A., 2013. Rescaling white space in post-apartheid Johannesburg. Antipode 45 (5), 1190–1212.

Crush, J., 1994. Scripting the compound: power and space in the South African mining industry. Environ. Plan. D 12 (3), 301–324.

De Bruyn, I.A., Bell, F.G., 2001. The occurrence of sinkholes and subsidence depressions in the far West Rand and Gauteng Province, South Africa, and their engineering implications. Environ. Eng. Geosci. 7 (3), 281–295.

Dippenaar, M.A., 2015. Hydrological Heritage Overview: Johannesburg. Water Research Commission, Pretoria, South Africa, pp. 61.

Ellery, W.N., Balkwill, K., Ellery, K., Reddy, R.A., 2001. Conservation of the vegetation on the Melville Ridge, Johannesburg. S. Afr. J. Bot 67 (2), 261–273.

Fatti, C.E., Vogel, C., 2011. Is science enough? Examining ways of understanding, coping with and adapting to storm risks in Johannesburg. Water SA 37 (1), 57–65.

Ferguson, B.K., 2016. Toward an alignment of stormwater flow and urban space. J. Am. Water Res. Assoc. 52 (5), 1238–1250.

GCRO (Gauteng City-Region Observatory), 2014. A Framework for a Green Infrastructure Planning Approach in the Gauteng City-Region. GCRO Research Report 04, Johannesburg, South Africa, pp. 132.

Geyer, H.S., 2003. South Africa in the global context: the view from above and below. Ann. Reg. Sci. 37 (3), 407–420.

Goldreich, Y., 1992. Urban climate studies in Johannesburg, a sub-tropical city located on a ridge: a review. Atmos. Environ. B 26 (3), 407–420.

Goldreich, Y., Surridge, A.D., 1988. A case study of low level country breeze and inversion heights in the Johannesburg area. J. Climatol. 8 (1), 55–66.

Götz, G., Schäffler, A., 2015. Conundrums in implementing a green economy in the Gauteng City-Region. Curr. Opin. Environ. Sustain. 13, 79–87.

Hamilton, C. (Ed.), 1995. The Mfecane Aftermath: Reconstructed Debates in Southern African History. Wits University Press, Johannesburg, South Africa, p. 496.

Hardy, C.H., Nel, A.L., 2015. Data and Techniques for Studying the Urban Heat Island Effect in Johannesburg. International Archives of the Photogrammetry, Remote Sensing and Spatial Information Sciences – ISPRS Archives 40 (7W3), pp. 203–206.

Harrison, P., Rubin, M., 2016. South African cities in an urban world. S. Afr. Geograph. J. 98 (3), 483–494.

Harrison, P., Zack, T., 2014. Between the ordinary and the extraordinary: socio-spatial transformations in the "Old South" of Johannesburg. S. Afr. Geograph. J. 96 (2), 180–197.

Herbert, C.W., Murray, M.J., 2015. Building from scratch: new cities, privatized urbanism and the spatial restructuring of Johannesburg after apartheid. Int. J. Urban Reg. Res. 39 (3), 471–494.

Hetz, K., 2016. Contesting adaptation synergies: political realities in reconciling climate change adaptation with urban development in Johannesburg, South Africa. Reg. Environ. Change 16 (4), 1171–1182.

Johannesburg City Parks and Zoo, 2017. A Quick History of Joburg's Trees. Available from: http://www.jhbcityparks.com/index.php/tree-planting/tree-planting-updates/1288-a-quick-history-of-joburgs-trees.

King, T., Flynn, M.K., 2012. Heritage and the post-apartheid city: Constitution Hill, Johannesburg. Int. J. Heritage Stud. 18 (1), 65–82.

Kneen, M.A., Ojelede, M.E., Annegarn, H.J., 2015. Housing and population sprawl near tailings storage facilities in the Witwatersrand: 1952 to current. S. Afr. J. Sci. 111 (11–12), 142–150. doi: 10.17159/sajs.2015/20140186, Art. #2014-0186.

Knight, J., 2017. Issues of water quality in stormwater harvesting: comments on Fisher-Jeffes et al. (2016). S. Afr. J. Sci. 113 (5-6)doi: 10.17159/sajs.2017/a0207, Art. #a0207.

Ladislas, S., Gérente, C., Chazarenc, F., Brisson, J., Andrès, Y., 2013. Performances of two macrophytes species in floating treatment wetlands for cadmium, nickel, and zinc removal from urban stormwater runoff. Water Air Soil Pollut. 224 (1408). doi: 10.1007/s11270-012-1408-x.

Le Roux, A., Augustijn, P.W.M., 2017. Quantifying the spatial implications of future land use policies in South Africa. S. Afr. Geograph. J. 99 (1), 29–51.

Lusilao-Makiese, J.G., Cukrowska, E.M., Tessier, E., Amouroux, D., Weiersbye, I., 2013. The impact of post gold mining on mercury pollution in the West Rand region, Gauteng, South Africa. J. Geochem. Explor. 134, 111–119.

Mabin, A., Butcher, S., Bloch, R., 2013. Peripheries, suburbanisms and change in sub-Saharan African cities. Social Dyn. 39 (2), 167–190.

Malan, N., 2015. Urban farmers and urban agriculture in Johannesburg: responding to the food resilience strategy. Agrekon 54 (2), 51–75.

Mandy, N., 1984. A City Divided. Johannesburg and Soweta. Macmillan, Johannesburg, South Africa, pp. 447.

Mathee, A., Harpham, T., Barnes, B., Swart, A., Naidoo, S., de Wet, T., Becker, P., 2009. Inequity in poverty: the emerging public health challenge in Johannesburg. Dev. S. Afr. 26 (5), 721–732.

McCarthy, T.S., 2006. The Witwatersrand supergroup. In: Johnson, M.R., Anhaeusser, C.R., Thomas, R.J. (Eds.), The Geology of South Africa. Geological Society of South Africa, Johannesburg/Council for Geoscience, Pretoria, South Africa, pp. 155–186.

McCarthy, T.S., Arnold, V., Venter, J., Ellery, W.N., 2007. The collapse of Johannesburg's Klip River wetland. S. Afr. J. Sci. 103 (9–10), 391–397.

McCulloch, J., 2009. Counting the cost: gold mining and occupational disease in contemporary South Africa. Afr. Affairs 108 (431), 221–240.
Mucina, L., Rutherford, M.C. (Eds.), 2006. The Vegetation of South Africa, Lesotho and Swaziland. Strelitzia 19. South African National Biodiversity Institute, Pretoria, South Africa, p. 816.
Murray, M.J., 2009. Fire and ice: unnatural disasters and the disposable urban poor in post-apartheid Johannesburg. Int. J. Urban Reg. Res. 33 (1), 165–192.
Naicker, K., Cukrowska, E., McCarthy, T.S., 2003. Acid mine drainage arising from gold mining activity in Johannesburg, South Africa and environs. Environ. Pollut. 122 (1), 29–40.
Nastar, M., 2014. The quest to become a world city: implications for access to water. Cities 41, 1–9.
Nengovhela, A.C., Yibas, B., Ogola, J.S., 2006. Characterisation of gold tailings dams of the Witwatersrand Basin with reference to their acid mine drainage potential, Johannesburg, South Africa. Water SA 32 (4), 499–506.
Nevin, A., 2014. Instant mutuality: the development of Maboneng in inner-city Johannesburg. Anthropol. S. Afr. 37 (3–4), 187–201.
Nhlengetwa, K., Hein, K.A.A., 2015. Zama-Zama mining in the Durban Deep/Roodepoort area of Johannesburg, South Africa: an invasive or alternative livelihood? Extr. Ind. Soc. 2 (1), 1–3.
Parnell, S.M., Pirie, G.H., 1991. Johannesburg. In: Lemon, A. (Ed.), Homes Apart. Chapman, London, England, pp. 129–145.
Parnell, S., Robinson, J., 2006. Development and urban policy: Johannesburg's city development strategy. Urban Studies 43 (2), 337–355.
Pfab, M., 2002. The quartzite ridges of Gauteng. Veld Flora, 56–59.
Phillips, H., 2014. Locating the location of a South African location: the paradoxical pre-history of Soweto. Urban Hist. 41 (2), 311–332.
Pirie, G.H., Hart, D.M., 1985. The transformation of Johannesburg's black western areas. J. Urban Hist. 11 (4), 387–410.
Robinson, J., 2008. Developing ordinary cities: city visioning processes in Durban and Johannesburg. Environ. Plan. A 40 (1), 74–87.
Rogerson, C.M., Rogerson, J.M., 2015. Johannesburg 2030: the economic contours of a "Linking Global City". Am. Behav. Sci. 59 (3), 347–368.
Rousseau, D.P.L., Lesage, E., Story, A., Vanrolleghem, P.A., De Pauw, N., 2008. Constructed wetlands for water reclamation. Desalination 218 (1–3), 181–189.
Rule, S.P., 1989. The emergence of a racially mixed residential suburb in Johannesburg: demise of the apartheid city? Geograph. J. 155 (2), 196–203.
Sadr, K., 2017. The effect of urban sprawl on archaeological sites between Johannesburg and the River Vaal: a GIS study. S. Afr. Arch. Bull. 72, (in press).
Sadr, K., Rodier, X., 2012. Google Earth, GIS and stone-walled structures in southern Gauteng, South Africa. J. Arch. Sci. 39 (4), 1034–1042.
Schäffler, A., Swilling, M., 2013. Valuing green infrastructure in an urban environment under pressure: the Johannesburg case. Ecol. Econ. 86, 246–257.
Shorten, J.R., 1970. The Johannesburg Saga. Cape & Transvaal Printers, Cape Town, pp. 1159.
Sihlongonyane, M.F., 2016. The global, the local and the hybrid in the making of Johannesburg as a world class African city. Third World Q. 37 (9), 1607–1627.
Simon, D., 2012. Climate and environmental change and the potential for greening African cities. Local Econ. 28 (2), 203–217.
Todes, A., 2012a. Urban growth and strategic spatial planning in Johannesburg, South Africa. Cities 29 (3), 158–165.
Todes, A., 2012b. New directions in spatial planning? Linking strategic spatial planning and infrastructure development. J. Plan. Educ. Res. 32 (4), 400–414.
Toffah, T.N., 2013. Reinstating water, resurrecting the Witwatersrand. J. Land. Arch. 8 (2), 24–31.
Tucker, R.F., Viljoen, R.P., Viljoen, M.J., 2016. A review of the Witwatersrand Basin: the world's greatest goldfield. Episodes 39 (2), 105–133.
Turton, A., Schultz, C., Buckle, H., Kgomongoe, M., Malungani, T., Drackner, M., 2006. Gold, scorched earth and water: the hydropolitics of Johannesburg. Water Res. Dev. 22 (2), 313–335.
Tyson, P.D., du Toit, W.J.F., Fuggle, R.F., 1972. Temperature structure above cities: review and preliminary findings from the Johannesburg Urban Heat Island Project. Atmos. Environ. 6 (8), 533–542.
Tyson, P.D., Wilcocks, J.R.N., 1971. Rainfall Variations Over Johannesburg, Occasional Paper No. 4. Department of Geography and Environmental Studies, University of the Witwatersrand, Johannesburg, pp. 1–14.

Viljoen, M.F., Booysen, H.J., 2006. Planning and management of flood damage control: the South African experience. Irrig. Drain. 55 (Suppl. 1), S83–S91.

Varis, O., 2006. Megacities, development and water. Int. J. Water Res. Dev. 22 (2), 199–225.

Walsh, C.J., Fletcher, T.D., Burns, M.J., 2012. Urban stormwater runoff: a new class of environmental flow problem. PLoS One 7 (9), e45814. doi: 10.1371/journal.pone.0045814.

Warshawsky, D.N., 2014. Civil society and urban food insecurity: analyzing the roles of local food organizations in Johannesburg. Urban Geogr. 35 (1), 109–132.

Wimberley, F.R., Coleman, T.J., 1993. The effect of different urban development types on storm-water runoff quality: a comparison between two Johannesburg catchments. Water SA 19 (4), 325–330.

Winkler, T., 2012. Between economic efficacy and social justice: exposing the ethico-politics of planning. Cities 29 (3), 166–173.

Wolch, J.R., Byrne, J., Newell, J., 2014. Urban green space, public health, and environmental justice: the challenge of making cities "just green enough". Landscape Urban Plan. 125, 234–244.

CHAPTER

9

When Urban Design Meets Fluvial Geomorphology: A Case Study in Chile

Paulina Espinosa, Jesús Horacio**,†, Alfredo Ollero‡, Bruno De Meulder*, Edilia Jaque**, María Dolores Muñoz***

***University of Leuven, Heverlee (Leuven), Belgium; **University of Concepcion, Concepción, Chile; †University of Santiago de Compostela, Galicia, Spain; ‡University of Zaragoza, Zaragoza, Spain**

OUTLINE

9.1 Introduction 150

9.2 Objective and Methods 151

9.3 An Interdisciplinary Dialogue 151

9.4 Study Area Characterization 154
9.4.1 Disaster as an Urban Developer 154
9.4.2 Site-specific Context 156

9.5 Design Exercise: Creating Scenarios 161
9.5.1 Current State of the Fluvial Geomorphology within the Study Area 162
9.5.2 The IHG Index 164
9.5.3 Fluvial Territory: Definition 165
9.5.4 Design Actions 166

9.6 Discussion around Feasibility Issues 169

9.7 Conclusions 169

References 171

Urban Geomorphology. http://dx.doi.org/10.1016/B978-0-12-811951-8.00009-6

9.1 INTRODUCTION

Before the Spanish colonization of Chile in 1520, the country's geomorphology and its human development were tightly interwoven. Unable to domesticate the country's wild and inhospitable environment, indigenous communities adapted themselves to the landscape. As a result, the different macrogeomorphological units that define the landscape of Chile's central area almost exactly overlap with the settlements of its indigenous populations (Fig. 9.1).

This historical relationship between landscape and development invites one to take a closer look at the current relationship between Chile's geomorphology and its urbanization. This relationship proves to be deeply fraught with misunderstanding and misuse. This rift is to a large extent responsible for the natural disasters that punctuate Chile's urban history, ranging from floods to landslides and earthquake liquefaction. The catastrophic urban history of Chile is related to this broken relationship. Between 1960 and 1991, 16 catastrophic floods were registered, 63% of which were situated in Chile's central area, where 73% of the population lives (Rojas, 2012). The majority of these events were related to the combination of intense rainfall and anthropogenic intervention (Rojas, 2012).

In order to mend the historical rift between environment and human occupation and use, this study focuses on the key role of rivers in both Chile's natural landscape and cities. It is proposed to keep urban rivers "alive." River restoration–rehabilitation (R–R) is tested to reinvigorate life in rivers, allowing one to learn about the natural behavior of the environmental system. This action would deliver the knowledge necessary to envision new forms of development that are based on adaptation to nature instead of drastic alteration of the same.

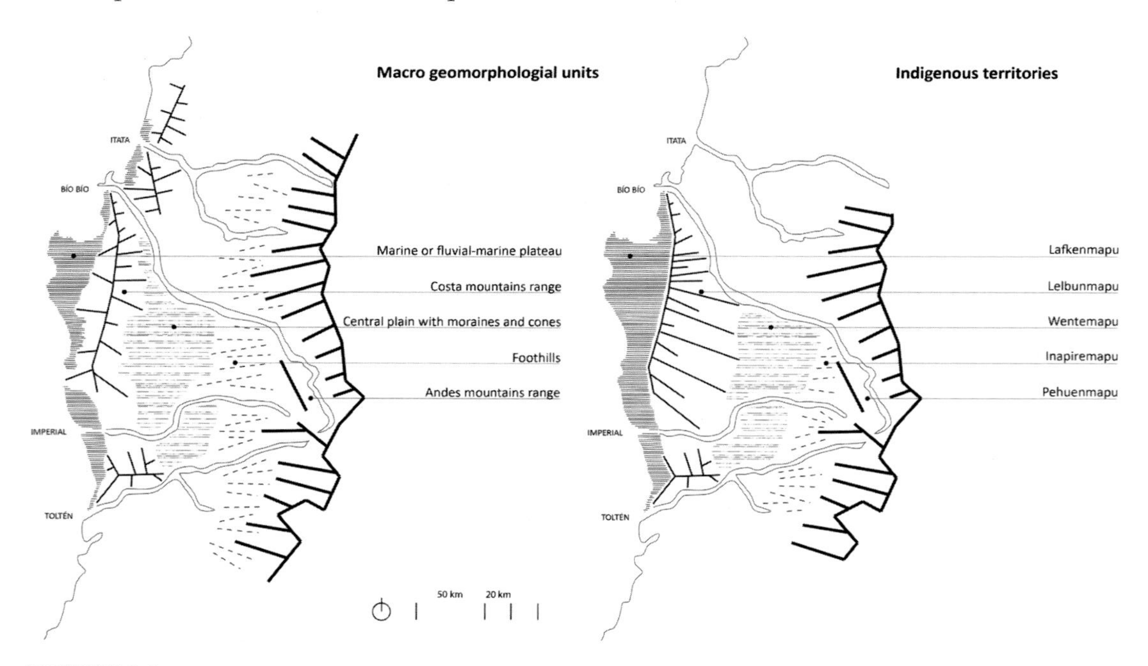

FIGURE 9.1 Comparative map between the national macrogeomorphological units and the indigenous Mapuche territory in Chile called "Wallmapu". *By the author Espinosa modified from: (*Left*)* Mapa de Chile de macro-unidades geomorfológicas in Red Cedeus (2015)*; (*Right*)* "Territorio Mapuche hacia 1540" *in Duquesnoy (2012).*

Geomorphology is one of the key dimensions of both the natural system and production model and, more specifically, features as its base and substrate. By developing a practical design exercise, this study rejects the traditional dichotomy between landscape and urbanization in favor of an alternative and more integrative scenario. On the one hand, this chapter deals with fluvial geomorphology; the interactions between river channel forms; processes on a space and timescale—in other words, the study of erosion, transportation, and sedimentation processes (Charlton, 2008), which are essential for the proper ecological functioning of rivers. On the other hand, the chapter deals with the increasing lack of building space; an extensive growth pattern; and the major forces driving the processes of urbanization.

9.2 OBJECTIVE AND METHODS

A review of fluvial geomorphology and landscape-oriented urban design allows for establishing an interdisciplinary dialogue between the two disciplines, with the aim of using design as a tool to guide a balanced and recalibrated urban development. The urban design exercise is carefully assessed with regard to its compatibility with the goals of R–R, testing how far is it possible to go in this particular urban context. The idea is to move beyond the merely theoretical and to suggest a concrete model of urban growth that factors in the complex and multifaceted reality of urban development.

The starting point of the design is the definition of a fluvial territory. This is defined by using fieldwork and desk-based analysis, including cartography, historic aerial photographs, technical reports, flood management works, and environmental impact assessments literature from the Ministry of Public Works of Government of Chile. Photo-interpretation analysis allows one to structure the geomorphological dynamics of the river in three different periods (1955, 1992, 2016). However, the area of the calculated fluvial territory is compared to the flood hazard maps elaborated by Rojas (2015) for different periods of return in the lower section of the Nonguén Stream, as this area is bigger than the flood area of a return period of 10 years.

After that, the design proposes landscape-based urban solutions for redefining the border area and recovering rugosity, permeability, and natural functions of the fluvial system. This newly created scenario also functions as a communication tool, allowing various stakeholders—academic, community, corporate, governmental, and otherwise—to, in the future, articulate and interrogate their respective interests and concerns and to open a dialogue between the same.

9.3 AN INTERDISCIPLINARY DIALOGUE

Rivers are extremely dynamic and complex systems, with permanent adjustments in space and time of their liquid and solid flow fluctuations (Werritty, 1997). Rivers are also self-constructed, that is their processes and forms are developed to efficiently transport liquid, solid, and biological flow along their drainage network (Horacio, 2015; Ollero, 2010). The lateral and vertical mobility of the river is, therefore, a mechanism of self-regulation and reflection of intense ecological dynamics and guarantor of bio- and geodiversity

(Elosegi et al., 2010, 2017; Gray et al., 2013; Malavoi et al., 1998). In this way, geomorphological processes are crucial to the correct functioning of fluvial courses as ecosystems (Ollero, 2007).

Despite their pivotal role as a natural system, rivers have been increasingly exploited for human use. In the last decades, this use has only intensified. This exploitation by occupying and polluting river territory, modifying land uses in the basin, changing river courses, and consuming river resources not only drastically interferes with a river's natural processes and forms; ultimately, it also risks depriving the river of its wider social value (Honey-Rosés et al., 2013; Palmer and Richardson, 2009). Demographic projections for the coming decades, moreover, predict a rapid and substantial growth of the urban population (Angel et al., 2011; Seto et al., 2011). This will only increase the strain on riparian ecology. Add to this the potentially devastating impact of global warming (Davies, 2010) and it is easy to see why rivers in Chile, and elsewhere, face a bleak future. In the light of this bleak prognosis, urban river management should: (1) develop a model far from current river-based approaches as a channel (Vietz et al., 2016); (2) emphasize the value and key functions of river systems in territorial planning (Brierley and Fryirs, 2005) and urban design; (3) address the causes of river degradation at all scales starting from the scale of the river basin, instead of merely patching up and working on the symptoms; and (4) focus more efforts on hydrogeomorphological processes (Fryirs and Brierley, 2012; Newson, 2002), given the key functions of river systems in territorial planning.

It is by now common knowledge that the increasing occupation of river banks and floodplains has led to a rise in floods. Important legal texts, such as *Floods Directive 2007/60/ EC* (for Europe), recognize the fundamental role of floodplains as necessary spaces for rivers to discharge water, store it, and, subsequently, return it to the bed at a slower rate, thereby moderating downstream flooding. Traditionally, flood-risk management has relied heavily on large-scale structural measures, such as levees, dikes, dams, deviations, and channelizations. While such interventions may prove effective occasionally, they almost always come at great environmental cost because they distort the river's biodiversity, disturb its ecological function, and most often turn the stream into a mere channel. Addressing the causes and not simply the symptoms of flooding requires the recovery of floodplains and working together with the river's flooding dynamics (Ollero, 2014).

In recent years, many urban design researchers have attempted to understand and repair the broken relationship between nature and city. Increasingly, their attention has turned toward the concept of landscape urbanism, which was introduced at the landscape urbanism symposium and exhibition in Chicago in 1997. Yet, to be fully defined and commonly understood as an approach that uses landscape structure as a substrate of urban design, landscape urbanism is rooted in the work of Scottish landscape architect Ian L. McHarg, whose seminal work *Design with Nature* (1969) rejected the orthodox dichotomy between urban design and natural environment and, instead, promoted a type of architecture that sought to incorporate the natural landscape as an integral component of the design process. In the 1990s, McHarg's ecological ideas found resonance in the work of Waldheim (2010), who even more firmly positioned human activities in the natural landscape, envisioning the voids as commons (Shane, 2003). Once this vision became a systemic approach, larger scales appeared on the scene, so that infrastructure, water systems, and natural systems started to be revealed in urban design practice. Corner (1999) deepened the concept of landscape urbanism by reintroducing mapping as a fundamental tool. In this way, the landscape scale, with its own rules and tools (such as interpretative mapping), became part and parcel of urban design theory and practice. In view of this and consequently nowadays landscape urbanism incorporates a long-term perspective and, therefore, uncertainty and complexity of the natural systems as a necessary transdisciplinary layer in the design process (Mostafavi and Doherty, 2010; Redeker, 2013; Waldheim, 2010).

A closely related concept in current urban design research and practice is water urbanism (Shannon et al., 2008), which ties urban design to hydrology and a wide range of water issues going far beyond the simply aesthetic or hygienic. Under this construction, Shannon (2013) emphasized water as a key element playing a main role in forging a harmonious yet dynamic balance between city and nature. Claiming water as a common and/or a public domain and concern (de Meulder and Shannon, 2013a,b), this study, with a focus on rivers, uses water urbanism to provide a program framework necessary to reclaim and restore what is simultaneously one of the most destructive and most vitalizing forces in the city. Consequently, this study is in-line with Mathur and da Cunha's ideas expressed in "Design in the Terrain of Water," about understanding water as a subject matter that challenges assumptions, reminds us of our fallibility, accommodates complexity, and sets our horizon (Mathur et al., 2014).

As there is no broad or coherent consensus about how to relate new methods of flood defense and water management to urbanization and economic development (Meyer et al., 2010), what is especially relevant for this study is water urbanism's treatment of natural water spaces (river spaces, in particular) as dynamic processes rather than static entities. Therefore, if urban design aims to include river space, it needs to be process-oriented. This new way of planning is performed in terms of options, follow-up measures, and responses that not always allow designers to foresee a fixed result for a project (Prominski et al., 2012).

Currently, in practice, there have been several urban design cases that include the concept of restoration in their formulation, but only a few cases are in-line with this study, considering a processes-oriented design that is focused on geomorphology restoration. In the US, interesting initiatives are being undertaken to return freedom to the river. Projects, such as "daylighting," express the need for rivers to discover and recover their old course. In the area of Berkeley (California), there are a large number of creeks on which these types of restoration measures have been applied, such as Strawberry Creek, Baxter, and Cerrito Creeks, Quail Creek, or Blackberry Creek. Subsequently, these projects have been followed and studied by authors, such as Asher and Atapattu (2005); Pinkham (2000); and Riley (2016). Although they established the clear intention to give more space to the river, the interventionist measures that are applied at riversides, after eliminating the main impacts, conceptually conflict with the approach presented in this chapter.

Three other projects are particularly relevant for our case study because of their comparable conditions, the morphodynamic activation techniques they each employed, and their respective outcomes. The first case is the Isar River in Munich, with the Isar-Plan, which started in 2000. The second is the Emsher River in Dortmund, with a 30-year plan started in 1990. The third is the Shunter River in Braunschweig, with a general plan started in the mid-1990s and a more structural plan developed between 2009 and 2011 (Prominski et al., 2012). These three cases were similar in several ways: they all embraced a long-term approach focused on self-recovery; they allowed channel migration; and they encouraged channel dynamics so as to improve the structure of the watercourse. More specifically, in all three cases, riverbed and riverbank reinforcements were removed, the channel-cross section was reprofiled, and seminatural riparian management was performed. Both in the Isar and Shunter Rivers, disruptive elements were introduced as part of seminatural channel management (Prominski et al., 2012).

9.4 STUDY AREA CHARACTERIZATION

Chile has a context of sociospatial capital adjustments. Chilean rivers, both their sources and their management, have advanced in their "commodification." This privatization, based on evicting peasant and indigenous communities, has been the key to the expansion of mining, energy, agro-export activities, and the inclusion of transnational capital in drinking water after neoliberal transformations.

This study focuses on the confluence of the lower Andalién River and its subbasin, the Nonguén Stream, in the urban area of Concepción. With approximately 1 million inhabitants (Pérez and Hidalgo, 2010), Concepción is the second most populated city in the country, after the capital of Santiago (Fig. 9.2). Its main economic activities are manufacturing, forestry, and fishing. It is also a service and transportation hub (Pérez and Hidalgo, 2010).

9.4.1 Disaster as an Urban Developer

When we look at the historical trajectory of urban planning in Concepción in the 20th century, we can identify four different stages, which reflect the city's different economic and sociopolitical developments. The first stage is linked to the reconstruction of the city after the earthquake of 1939 and the subsequent formulation of the Regulatory Plan of Concepción (PRC-40) of 1940. The PRC-40 included proposals to (1) rebuild damaged areas; (2) organize the urban expansion resulting from state fostered industrialization; and (3) solve housing scarcity facing the existence of large peripheral areas of precarious and spontaneous development.

Covering the 1960s and 1970s, the second stage is marked by the 1960 earthquake and the ensuing need to regulate the development of sprawling housing districts for workers in emerging industries. In 1960, a new Regulatory Plan (PRC-60) was introduced that aimed to (1) meet the need for postearthquake reconstruction; (2) consolidate the city as a service center for the metropolitan area; and (3) solve the housing problem through the densification of the central area. In the PRC-60 plan, the Andalién River and the Nonguén Estuary were designated as natural areas adjacent to the urban area (Muñoz, 1995). This plan was succeeded by the Intercommunal Regulatory Plan (IRP) in 1963. This plan aimed to solve housing needs and boost the economy. These objectives, however, were not met and, in the early 1970s, the original plan underwent significant modifications. In the midst of these comings and goings, in 1965, the Ministry of Housing and Urbanism was founded.

The third stage was initiated by the 1973 military coup, which overthrew Allende's socialist government and aligned with neoliberal economic systems. In practice, this meant moving from one model based on national industrialization to another focused on giving the private sector freedom to act. In this neoliberal system, state-owned enterprises were privatized and forced to compete on an international market, instituting a radical change that seriously damaged the domestic industry, causing large segments of the manufacturing industry to even disappear altogether (Agosin, 2001).

In 1979, the government introduced the National Policy for Urban Development, which—wholly in-line with the logic of the free-market—reduced land to its mere economic exchange value and largely disregarded its wider social and ecological importance. Needless to say, this neoliberal policy had direct consequences for Chile's urban landscape and communities,

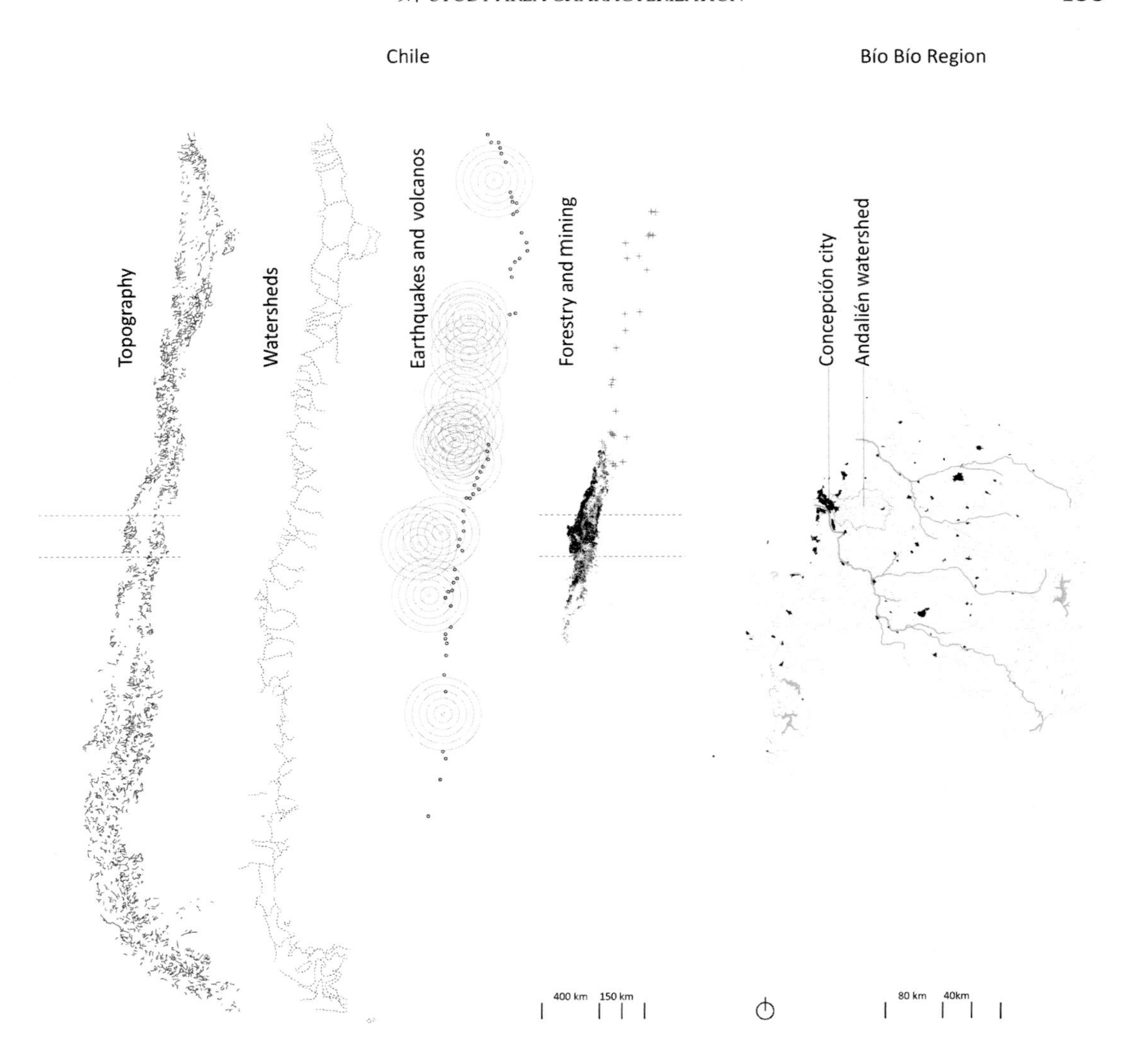

FIGURE 9.2 Mapping explorations of Chile characterization and Andalién watershed location. *By the author Espinosa based on: Chile Map, Cereceda et al. (2011); cartographic material,* Intendencia Regional de Concepción *(2013); and* Carta Base Andalién, *Jaque (1996).*

encouraging as it did private investment and urban growth without attempting to understand the social, cultural, or natural value of cities and landscapes. The Metropolitan Regulatory Plan of Concepción (PRMC) of 1980, framed in the policy described above, included expansion areas that coincided with those of the IRP of 1963, but with fewer regulations. It is also noteworthy that IRP afforestation zones were classified in PRMC as areas of natural value, although this designation did not result in an assessment of landscape or its ecological characteristics.

After returning to a democratic system, a fourth stage began in 1996, with the updating of the PRMC of 1980 (PRMC-96). Three urban issues guided the update: (1) improving intercommunal connectivity; (2) encouraging and facilitating infrastructure and productive activities; and (3) balancing natural and built environment. The update of the PRMC redesignated 26% of all land as residential area, 13% as area for urban expansion, and 46% as ecological protection zones, including the Bio-Bio and Andalién Rivers along with their floodplains, lagoons, wetlands, and dunes. The PRMC-96 also designated development-restricted zones, which were at risk of flooding or landslides. The Natural and Cultural Heritage protection areas were identified in accordance with the legal framework (Articles 2.1.7 and 2.1.8 of the General Ordinance of Urbanism and Constructions), integrating them into intercommunal urban development with the objective of "achieving an adequate safeguard of resources of natural and scenic value that characterize the metropolitan area, (p. 4)" set out in the explanatory section presentation, within the framework of current legality and the competencies of the PRMC-96. However, Article 5.1.7 of the PRMC-96 Ordinance states that areas of natural value may be modified by management plans that provide a sound background to redefine the land use of these areas. As a result of this loophole, areas of natural value remain open to interpretation and vulnerable to development due to the elasticity of planning instruments (Arenas et al., 2010).

The different regulatory plans led to a tentacular expansion of Concepción's urban center, moving along the city's main roads and waterways and occupying the rivers' floodplains. This growth pattern can be explained by the fact that regulatory plans are deeply affected by a neoliberal agenda, which has allowed corporate rather than public interest to shape the city's expansion. Indeed, since the 1980s, territorial planning instruments have been dictated by private investors and the fluctuations of the land market (Fig. 9.3). As a result, regulatory plans have been all too often tailored to incorporate new zones, thus, increasing real estate speculation and changing land value. This subservience to private and corporate interests and the general failure to safeguard ecological protection and risk areas from development has significantly undermined the validity of planning instruments.

9.4.2 Site-Specific Context

The Andalién River system is an area that is predominantly granite and mountainous; and occupies more than 60% of its watershed, with slopes of more than 20° in the surrounding urban areas (Mardones and Vidal, 2001). It is an exoreic system of rainfall regime. Its average annual flow is 14.3 $m^3\ s^{-1}$, reaching 565 $m^3\ s^{-1}$ in a return period of 50 years (Rojas, 2015). Its main tributary is the Nonguén Stream, which contributes 13% of the flow (Rojas, 2015). This tributary is relatively short in length. Its steep slopes and brief but intense bursts of rainfall (1200–1600 mm concentrated in just 4 months) make it, and specially its confluence with the Andalién River, highly susceptible to flooding (Jaque, 1996; Rojas, 2015).

Due to urban development and agricultural production, the fluvial system has suffered the destruction of its floodplain and changes and channeling of its course. In addition to this, the management of massive monocultures of eucalyptus and pine, which occupy a main part of the watershed, entail clear-cutting practices, which leave behind bare soil during an entire season (1 year), causing pluvial erosive processes in steep slopes and consequently unbalanced processes of sedimentation in the watershed (Jaque, 2010). Other impacts related to this monoculture production are

The area of the Nonguén Stream floodplain was mainly a wetland and, for this reason, the land was qualified as poor quality and cheap. The global rise and success of grassroots movements at that time inspired the local population to set up cooperatives in order to improve access to affordable housing. These cooperatives relied on a system that would later be formalized into a governmental development policy called "Operación Sitio," or "site operation" (Hidalgo, 1999; Quintana, 2014). It enabled the local population to buy land with a basic urban layout, which they should develop by themselves (without any government involvement). An entire urban infrastructure including private dwellings, public buildings, streets, bridges were built. They established a drinking water network in 1968, an electrical grid in 1970, and a sewage system in 1990 (Urrutia et al., 2012). These grassroots associations continued to operate until being converted into social partnerships to address common concerns, such as flood damage or the corporate destruction of natural resources and territory by private initiatives.

Around the 1980s, neoliberalism had firmly established itself as the dominant ideology in politics and urban policymaking. The landscape was increasingly reduced to a mere commodity that could be exploited for private gain without concern for public safety or environmental protection. The neoliberal logic led to unregulated urban development and expansion. The image of 1978, for instance, clearly demonstrates the alteration of floodplains and wetlands through artificial fillings (Fig. 9.6C). Between 1955 and 2007, urban land added to the Andalién River system increased seven-fold, which ultimately dangerously reduced its rainwater storage potential (Romero and Vidal, 2010).

Finally, from the second half of the 1990s until now, the landscape scale started to play an important economic role at a transnational level, as large storage facilities began to appear along the highway. At the same time, real estate developers began to build two-story houses on the Andalién River floodplain at the confluence with the Nonguén Stream, creating even more danger for the population, as confluences are known risk zones for flooding, and this area is now categorized as the main flood-risk zone (Rojas, 2015) based on the 2006 flooding (Fig. 9.6D). After the 2006 flood, the government aimed to improve the safety of the local population of this area by channeling 5.6 km of the lower reaches of the Nonguén Stream—35% of its total length—an intervention that was completed in 2012 (SEA, 2016).

9.5 DESIGN EXERCISE: CREATING SCENARIOS

Drawing on both urban design and RR, the next exercise will explore alternative ways of dealing with flooding, which is conceptually disregarded as a natural and essential component of the river's ecological dynamics, mainly because it is a potentially damaging force. Untangling this conceptual difference, it is the first step to be addressed as a way of positioning the design focus of this exercise (Fig. 9.7).

The main geomorphological elements of the fluvial system are analyzed through cartography, photo-interpretation, and the application of the hydrogeomorphological index (IHG) to lay out a starting point for our alternative design proposal. After gathering the data, the minimum area needed to develop an R–R process is demarcated as the "fluvial territory." By retracing and contrasting urban voids as part of the plan, the generation of a new negotiated area is enabled in order to implement this restoration and recalibrate urban development.

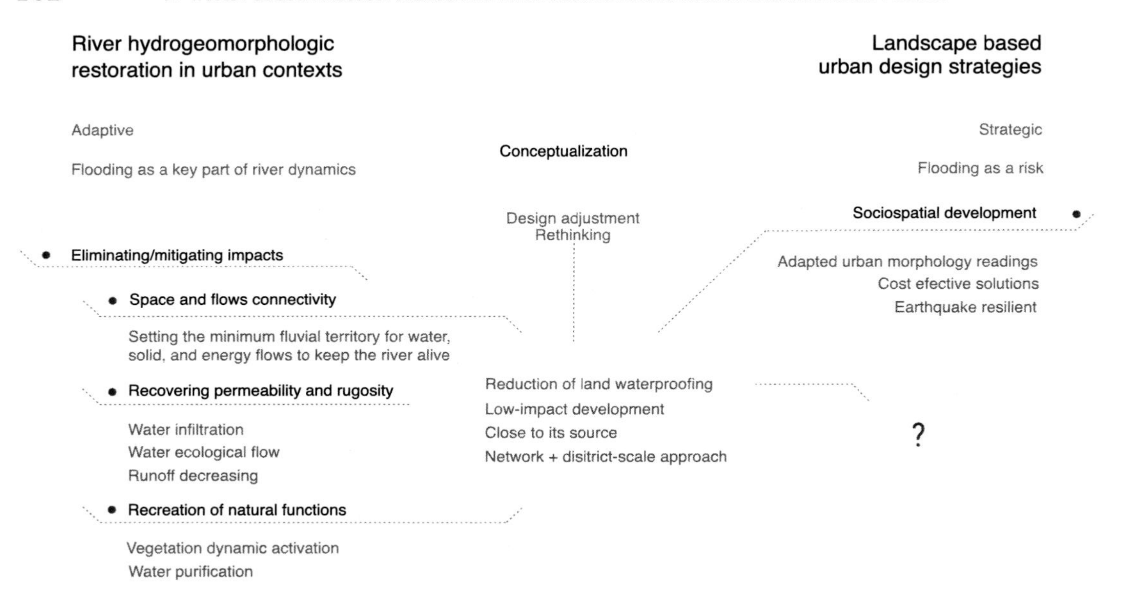

FIGURE 9.7 **Conceptualization problem by Espinosa.**

By performing RR, the aim is to restore the functions of the river system as well as the quality of the riverbed and riversides, activating lateral and vertical processes, giving continuity to longitudinal processes, restoring some naturalness to the layout and morphology and, consequently, improving the fluvial corridor. The social and urban context of this study imposes the question of feasibility. In this particular case, two main facts suggest that this approach is possible. One is that the full occupation of the area that Rojas (2015) designated as the most hazardous started with its urbanization in the 1980s and constitutes an ongoing process that can or should be reoriented. The other is that the designated budget to build the channelization project in the lower course of the Nonguén Stream was around €30.5 million (SEA, 2016). Besides this, last year a project worth €385 000 was launched to reinforce the channel revetment (MOP, 2016). This means that important budgets continue to be designated to solve these problems. An alternative approach is consequently needed to propose solutions for other urban rivers and streams that are likely to be channelized as if we were back in the most 19th-century hygienist programs.

9.5.1 Current State of the Fluvial Geomorphology within the Study Area

The diachronic analysis carried out from 1955 to the present-day has revealed that the geomorphological structure of the lower stretch of the river has been increasingly modified and that the floodplain, which in 1955 still occupied an area of 321 ha, was reduced by 72% in 2017 (Fig. 9.8).

Three main forces forged the development of this area: the cooperative organizations, a powerful private sector, and a high-risk and difficult landscape. After studying the urban

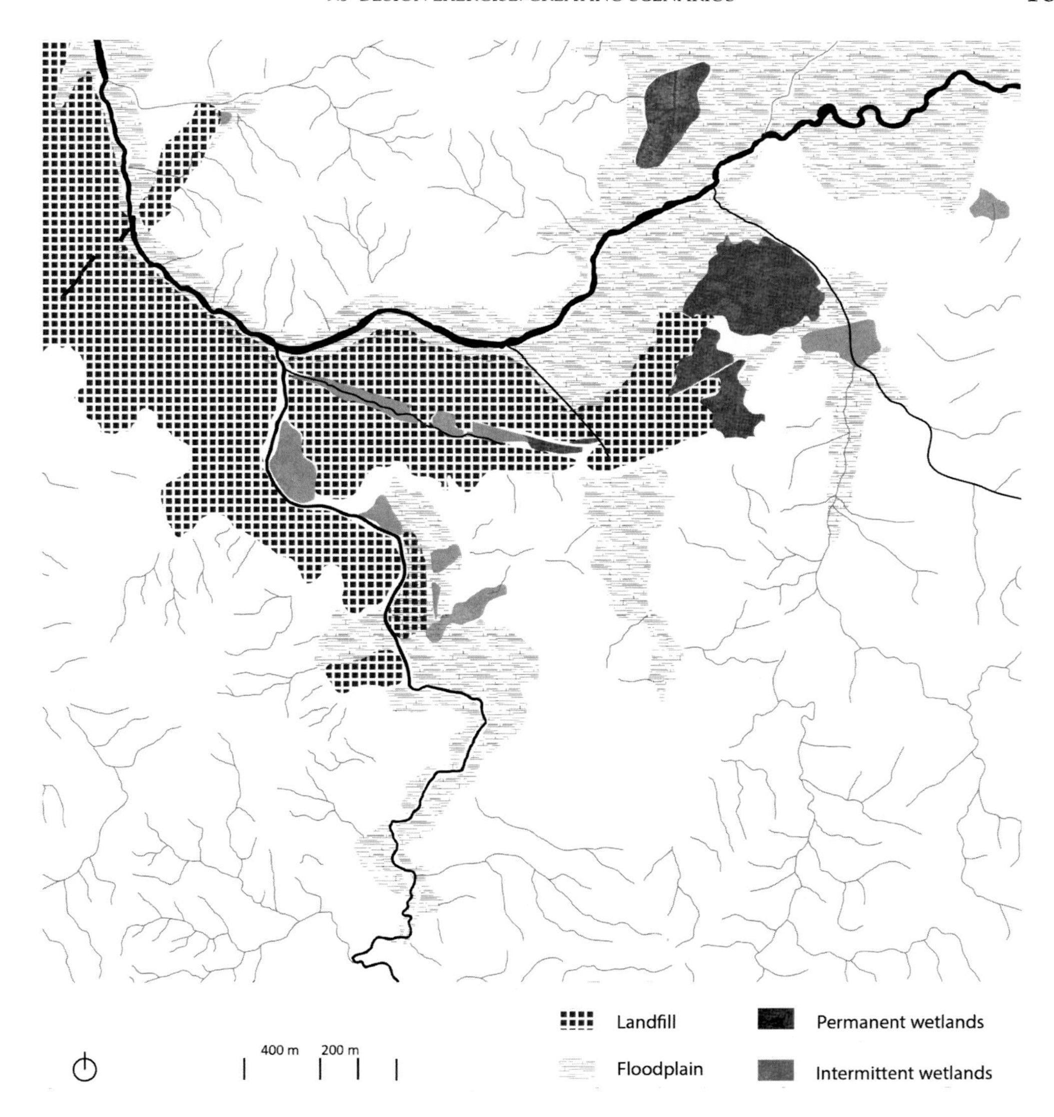

FIGURE 9.8 Geomorphological scheme. *By the author modified from Jaque (1996) and PRMC (2013).*

formation of the area, one of the most relevant findings is that the right bank of the Nonguén Stream still contains some remnants of former wetlands (Fig. 9.9). It is no coincidence that this area was mainly built by the cooperative organizations' limited means. The resulting fabric of the study area meant that the plan would start out by reinforcing a hybrid urban design between landscape and urbanization, based on restoring the wetlands in order to reincorporate them into the river system.

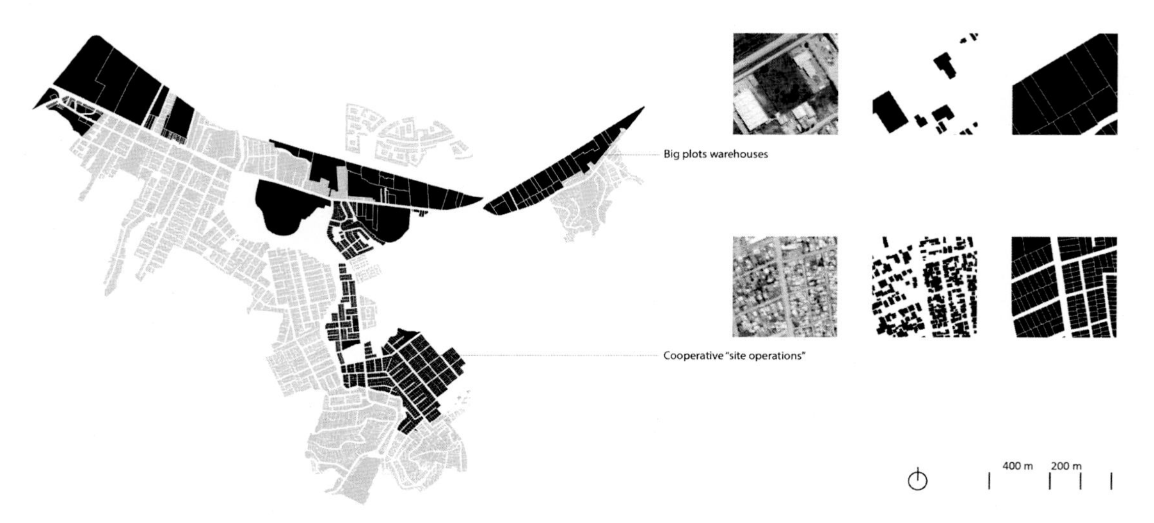

FIGURE 9.9 Interpretative map of the urban tissue. *By the author Espinosa modified from PRMC (2013) and photo-interpretation in 2013.*

9.5.2 The IHG Index

The hydrogeomorphological quality of the Nonguén Stream is diagnosed by applying the IHG index (Ollero et al., 2011). The index rates nine parameters: (1) functional quality, including (a) flow regime naturalness, (b) sediment supply and mobility, and (c) floodplain functionality; (2) channel quality, including (a) channel morphology and planform naturalness, (b) riverbed continuity and naturalness of longitudinal and vertical processes, and (c) riverbank naturalness and lateral mobility; and (3) riparian quality, including (a) longitudinal continuity, (b) riparian corridor width, and (c) structure, naturalness, and cross-sectional connectivity of the riparian corridor. Each parameter has an initial score of 10, corresponding to the natural state and functionality of the system. The pressures on the system are analyzed as well as their consequent impacts, so points are subtracted from that initial value. As pressures and impacts have different origins and consequences, their evaluation is therefore variable.

The Nonguén Stream was divided into four reaches from the source (upper reach) to the river mouth (lower reach): Nonguén Nature Reserve reach (R1), the ritron reach (R2), the transition reach (R3), and the potamon reach (R4). This division is based on the geomorphological characteristics of each reach as well as on the biological characteristics of associated fauna (Correa-Araneda and Salazar, 2014; Illies and Botasaneanu, 1963). Fig. 9.10 shows the hydrogeomorphological state of the reaches after applying the IHG index (we applied the IHG index at 11 sampling points distributed in the four reaches). The results indicate a marked demographic density gradient, causing an impoverishment of the ecological quality of the river, moving from quasipristine conditions (R1: very good quality) to completely urbanized sector and heavily altered at the river mouth (R4: very bad quality).

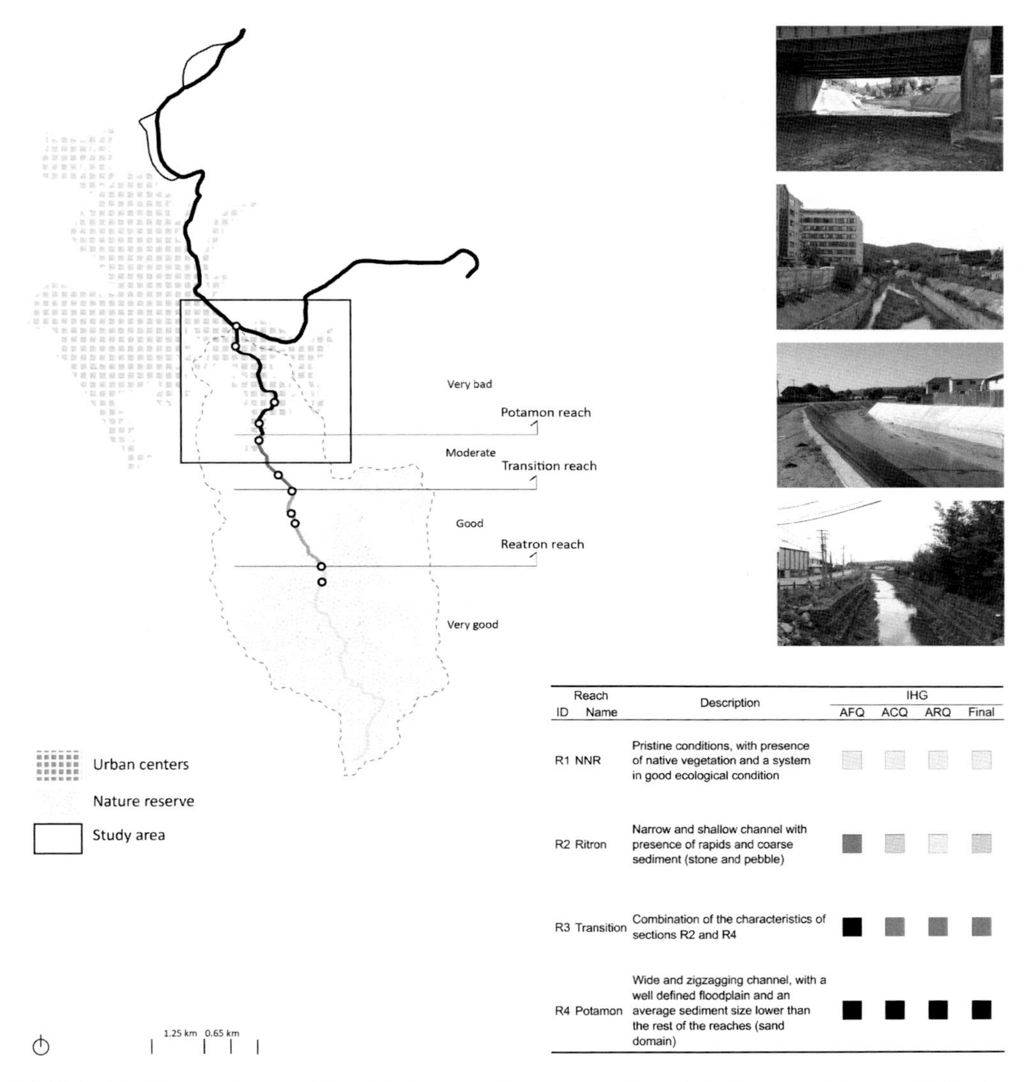

Reach ID	Reach Name	Description	IHG AFQ	IHG ACQ	IHG ARQ	IHG Final
R1	NNR	Pristine conditions, with presence of native vegetation and a system in good ecological condition				
R2	Ritron	Narrow and shallow channel with presence of rapids and coarse sediment (stone and pebble)				
R3	Transition	Combination of the characteristics of sections R2 and R4				
R4	Potamon	Wide and zigzagging channel, with a well defined floodplain and an average sediment size lower than the rest of the reaches (sand domain)				

FIGURE 9.10 Characteristics of the IHG index in the study sections of the Nonguén Stream. *NNR*, Nonguén Nature Reserve; *AFQ*, assessment of the functional quality; *ACQ*, assessment of the channel quality; *ARQ*, assessment of the riparian quality. The colors indicate the quality: very good (black) > good > moderate > poor > very bad (light gray). Photograph series along the channelized river. *By the authors Horacio and Espinosa.*

9.5.3 Fluvial Territory: Definition

A proposal of a so-called fluvial territory has been designated in the urban river area (low reach: R4); (see also "free space for rivers," "room for rivers," "space to move," or "river widening"). Fluvial territory refers to a space required to naturalize the running waters of the river and restore its basic hydrogeomorphological and ecological functions of erosion and sedimentation (Dister et al., 1990; Ollero et al., 2015; Piégay et al., 2005; Rohde et al., 2006). Fluvial territory included the minor riverbed, the riverside corridor, and, either partially or totally, the floodplain, and has been delimited following these criteria (Ollero and Ibisate, 2012): (1) erodible river corridor (Piégay et al., 2005), which includes sectors that are susceptible to lateral erosion in the coming decades and, which, therefore, include the current and past meander-train developed by the river; (2) the whole river corridor, as well as snippets thereof; and (3) disconnected fluvial annexes that are like vestiges of the river's past geomorphological dynamics (e.g., oxbow lakes, abandoned channels, remains of isolated river forests). The flood area with return periods of 5–10 years is a fourth criterion (Malavoi et al., 2002), but this information was not available in the ministry's technical reports.

Three options of fluvial territories are proposed: (1) one meeting the minimum requirements; (2) a compromise option; and (3) an ideal option. While the first option stakes out the minimum space that is required to restore some of the stream's basic ecogeomorphological functions, the third option outlines the ideal space for the stream to recover its natural vitality with a certain guarantee. Option 2 entails a compromise, reached after mediating with a multidisciplinary criterion in order to enhance feasibility. The main goal of this negotiation was to minimize resettlement and, at the same time, maximize the use of available vacant space. This option has a surface of 2000 m (N–S) and 670 m (W–E) and 0.40 km^2 and covers 2400 m of the river (15%). The delimitation of the fluvial territory was not an easy task due to a lack of good quality of historical material and the absence of geomorphological remains in the ground as a result of the highly altered condition of the study area. This supposes that the ideal option of fluvial territory should possibly have an extension superior to the one shown in Fig. 9.11.

9.5.4 Design Actions

9.5.4.1 Redefinition of the Boundaries (the Edge)

Allowing flooding, while simultaneously protecting the population, the "negotiated area" of the fluvial territory proposed is the backbone of the design proposition. The priority is to work within the most flood-prone areas, which are coincidentally also the most recently urbanized areas (Fig. 9.12A).

To restore fluvial geomorphology, it is first suggested that the river's edge be redefined by removing the anthropogenic impact at the riverside and transversely on the riverbed. This redefinition requires three actions: (1) removal of the channel, where possible; (2) rebuilding a hidden dyke that provides the river with more space; and (3) leaving parts of the channel untouched when the first two actions cannot be carried out. The carrying out of cut-and-fill earthwork operations is also suggested to restore the damaged geomorphology of the fluvial system within the "negotiated" area, this along with the design of buildings adapted to flooding. The second part of the redefinition of the river's edge involves a resettlement of the population to collective buildings that conform to the landscape and that are located no further than 200 m from their original dwellings. Public safety is to be guaranteed by the design of flood-resistant buildings.

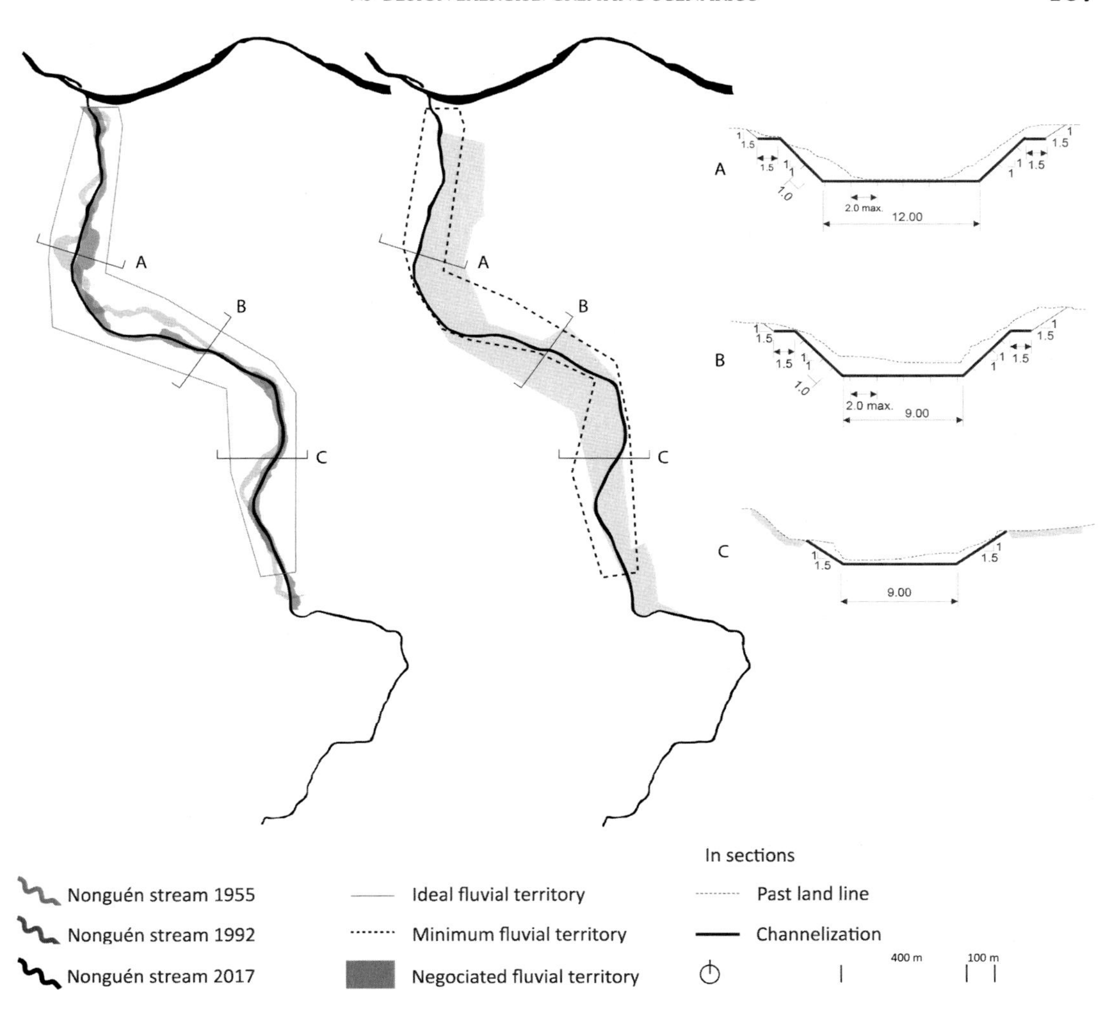

FIGURE 9.11 **Several fluvial territories alternatives designed for the study area.** The "negotiated fluvial territory" includes different kinds of urban voids: former wetlands, sports fields, green areas, vacant spaces, and large private surfaces. *The cross-sections represent the current channelization and the old riverbed (modified images from the studies of the Project "Diseño de obras fluviales río Andalién, esteros Nonguén y Palomares. VIII Región del Bío Bío, 2008" of the Chilean Government). By the authors Horacio and Espinosa based on fieldwork and photo-interpretation in 2017.*

9.5.4.2 Water Management System

This alternative design proposes multifunctional measures, starting with Natural Water Retention Measures (http://nwrm.eu/), with the aim of "protecting and managing water resources and addressing water-related challenges by restoring or maintaining ecosystems as well as natural features and characteristics of water bodies using natural means and processes." This is proposed as a system that aims to control flood damage and at the same time manages to preserve and improve the water-retention capacity of aquifers, soil, and ecosystems in order to restore their status. This multifunctionality is achieved by three key actions: (1)

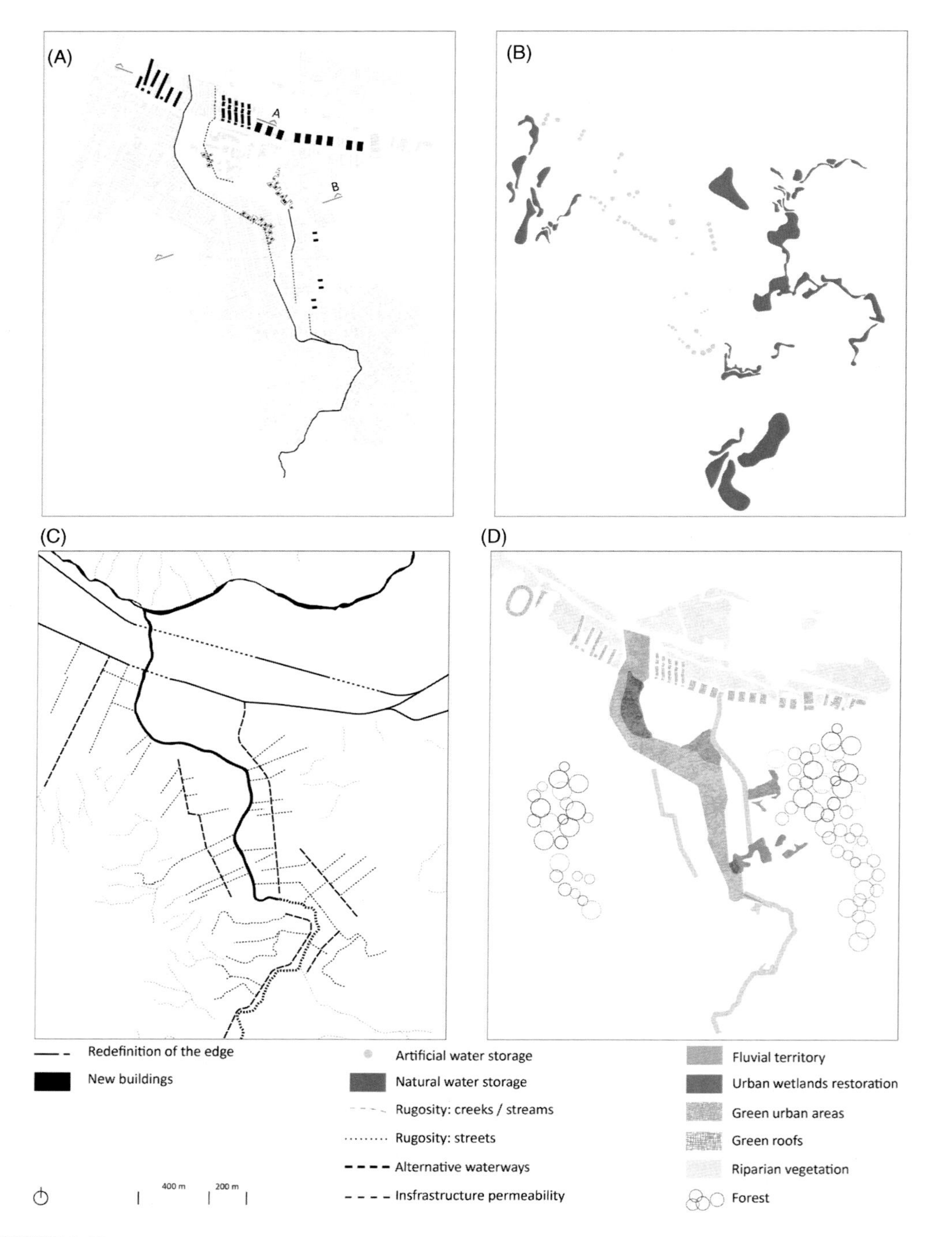

FIGURE 9.12 Systematic mapping of the design exercise. (A) Current urban tissue and redefinition of the edge, typology, and new urbanization proposal; (B) water management/storage; (C) water management rugosity and permeability; and (D) multipurpose green system. *By the author Espinosa.*

water storage via an artificial and natural system created by the restoration of wetlands; (2) run-off control that is generated by adding rugosity to the main flows in the study area, such as key streets in the floodplain, creeks, the upper part of the Nonguén Stream, and large-surface roofs; and (3) the creation of new waterways along some streets in order to improve waterflow-where necessary (Fig. 9.12B, C).

9.5.4.3 Multipurpose Green System

To reestablish the natural functions of the fluvial green system, we propose restoring (1) the river corridor; (2) wetlands; (3) native forest areas and creating public green areas in correspondence with alternative waterways; (4) building green roofs over large surfaces; and (5) planting native plants in nurseries all along the newly created sites to expand green spaces with suitable plants (Fig. 9.12D). Sections A and B, show the development of the system of multipurpose blue and green infrastructure, allowing more space for the river and its overflows within a mixed green area to increase the biodiversity and rugosity of the system (Fig. 9.13).

9.6 DISCUSSION AROUND FEASIBILITY ISSUES

In order to succeed, it is vital to create an adaptive management program. This should accompany the objectives of naturalness, functionality, connectivity, and self-regulation (Ollero, 2015) and should be applied to water and solid flows. Central to the proposal made by this study is the removal of anthropogenic impacts along the riverside and transversally across the riverbed. Future actions should focus on (1) purifying wastewater by filtering discharges and (2) recognizing the crucial role played by deadwood and sediments as part of the restored dynamics. As the headwaters and the main part of the Nonguén watershed is native-forested and has no significant deferrals of flow, it does not require actions to manage hydrological regulation of extreme flows.

A monitoring program that constantly tracks changes in the fluvial system ensures an adequate response to each change and allows for the process of continuous adaptation. For the management plan to be effective, it should be accompanied by environmental legislation that regulates all the actions in the watercourses of the basin and needs to find support among its inhabitants to validate an effective, long-term process. In this context, and within the city production forces, grassroots cooperatives transformed into new organizational units can play an active role in the management program. The designation of the Nonguén as a nature reserve would be a good starting point in this regard too because it can set a regulatory framework supporting the management program.

9.7 CONCLUSIONS

Three fundamental contributions to fluvial geomorphology restoration in urban contexts as a flood-risk management tool and landscape-based urban design strategy were formulated: (1) a new interdisciplinary methodology that forges a dialogue between RR and urban design; (2) a mapping of different scenarios to function as communication tools between

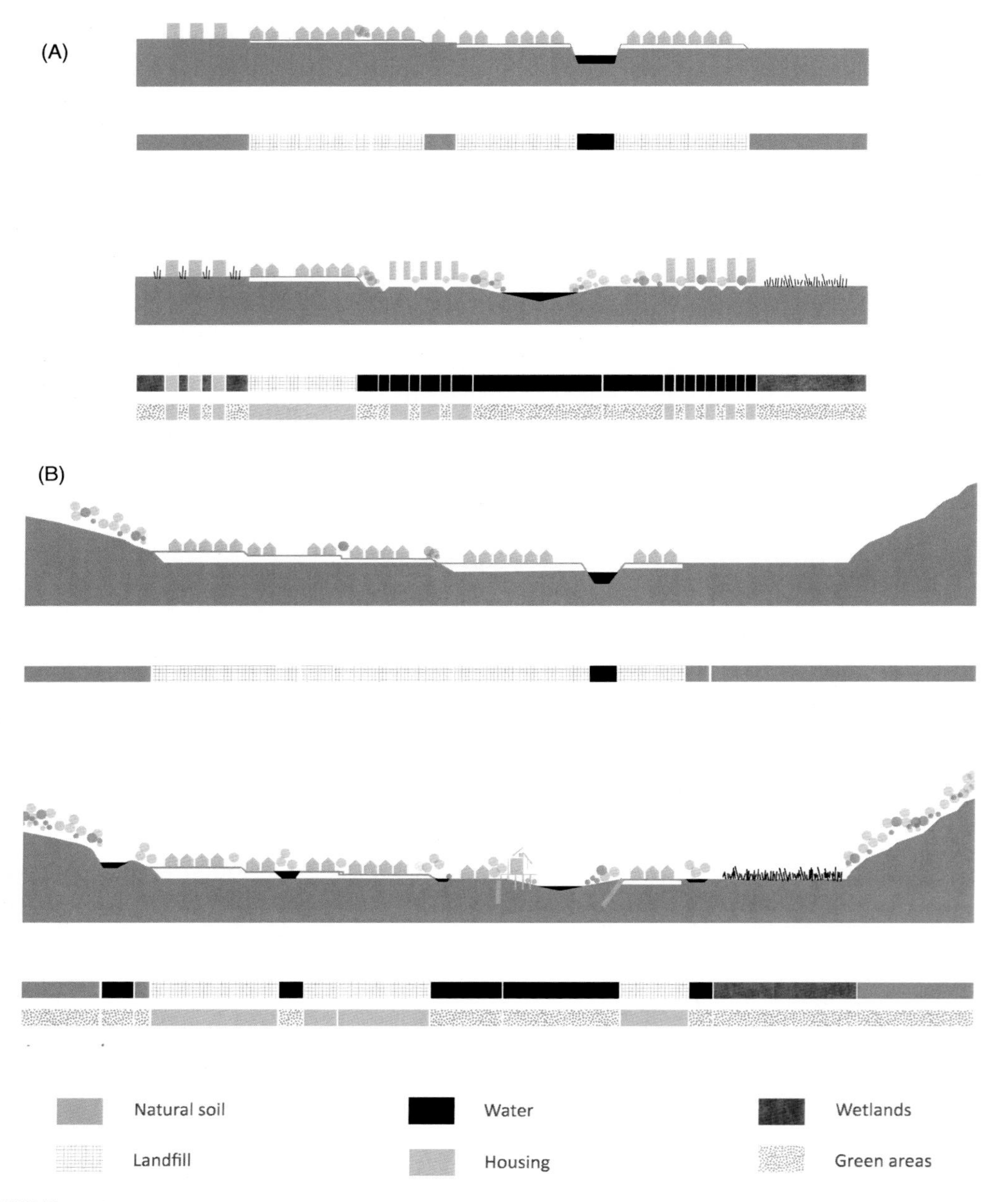

FIGURE 9.13 Schematic cross-section A: the confluence Nonguén Andalién. Schematic cross-section B: the middle section. *By the author Espinosa*

various disciplines and stakeholders; and (3) the study has opened up several promising avenues for future research based on multiple possibilities that appeared in the dialogue between two disciplines.

This study offers a new methodology that integrates insights from both urban-design and RR. We established this interdisciplinary dialogue by drawing up an urban-design exercise that elaborates the conceptual differences between these two fields concerning rivers and their dynamics. So, by using the definitions from RR, we proposed a "negotiated" fluvial territory, where the river can recover some of its vital dynamics, instead of setting any urban-design aim as a priority. In short, the practical design exercise is a key starting point for developing new paradigms in looking for landscape-based answers to urban design questions.

In-line with the methodology of landscape urbanism, we depicted different scenarios to function as a communication tool. The spatial dimension of the fluvial territory allowed us to combine the two disciplines and their respective complexities, thereby creating a valuable communication tool and establishing a fruitful dialogue. The scenarios did not merely represent possible solutions, they also generated these solutions, thus, functioning as a dynamic problem-solving tool, rather than as a static measuring instrument.

In practice, interdisciplinary approaches can encounter many difficulties and it is necessary to point them out in order to improve the final result of these kinds of processes. In this particular case, two main limitations were found. The first one was understanding the conceptual language differences of each discipline and the second one pertained to the need for rebalancing the expectations of both disciplines, considering that one can act as a restriction to the other and vice versa.

This study has opened up several promising avenues for future research. In the coming years, an interesting aim would be to take a closer look at restorative techniques within the context of urban design, exploring the potential of the spatial dimension on a microscale, or examining the benefits that a cross-scalar scenario production might yield on a watershed or city scale. The particular case of the Andalién River and Nonguén Stream confluence is highly representative of the Chilean context, which simultaneously remains one of the main problematic areas with flooding problems in Concepción city. This condition allows the study to project the methodology onto other urban cases in the future.

Acknowledgments

We would like to thank Jorge Félez for his generosity in providing us with various cartographic data; and Octavio Rojas for proving us with historical aerial photography from the project FONDECYT N° 11150424. We are also very grateful to Javiera de la Peña, Darío Almendra, Alberto Reichelt, and Sebastián Delgado for their data and field work.

References

Agosin, M., 2001. Reformas comerciales, exportaciones y crecimiento. In: French-Davis, R., Stallings, B. (Eds.), Reformas, Crecimiento Y Políticas Sociales En Chile Desde 1973. LOM Ediciones, Santiago, pp. 99–132, Chapter 3.

Aliste, E., Almendras, A., 2010. Trayectoria territorial de la conurbación Concepción-Talcahuano: Industria, asentamientos humanos y expresión espacial del desarrollo, 1950–2000. Pérez, L., Hidalgo, R. (Eds.), Concepción Metropolitano, Evolución y Desafíos, vol. 14, Editorial Universidad de Concepción. GEOlibros Series, Concepción, pp. 123–149.

Aliste, E., Almendras, A., Contreras, M., 2012. La dinámica del territorio en la conurbación Concepción-Talcahuano: huellas urbanas para una interpretación de las transformaciones ambientales durante la segunda mitad del siglo XX. Rev. Geog. Norte Grande. 52, 5–18.

Angel, S., Parent, J., Civco, D.L., Blei, A., Potere, D., 2011. The dimensions of global urban expansion: Estimates and projections for all countries, 2000–2050. Prog. Plan. 75 (2), 53–107.

Arenas, F., Lagos, M., Hidalgo, R., 2010. Los Riesgos Naturales En La Planificación Territorial. Centro de Políticas Públicas Pontificia Universidad Católica de Chile.

Asher, M., Atapattu, K, 2005. Post Project Appraisal of Village Creek Restoration, Albany, CA. Restoration of Rivers and Streams (LA 227). eScholarship UC Berkeley, Berkeley, CA.

Brierley, G.J., Fryirs, K.A., 2005. Geomorphology and River Management: Applications of the River Styles Framework. Blackwell Publishing, Oxford.

Cereceda, P., Errázuriz, A.M., Lagos, M., 2011. Terremotos y Tsunamis en Chile. Origo Ediciones, Santiago.

Charlton, R., 2008. Fundamentals of Fluvial Geomorphology. Routledge, London.

Chilean Government, 2008. Project "Diseño de obras fluviales río Andalién, esteros Nonguén y Palomares. VIII Región del Bío Bío. Chile.

Corner, J., 1999. The agency of mapping: speculation, critique and invention. In: Cosgrove, D. (Ed.), Mappings. Reaktion Books, London, pp. 151–213.

Correa-Araneda, F., Salazar, C., 2014. Caracterización fisicoquímica del agua del estero Nonguén y su confluencia con el río Andalién, región del Bío Bío. Variación en relación a los distintos usos de suelo en su cuenca. Sustain. Agri Food Environ. Res. (SAFER) 2 (2), 33–46.

Davies, P.M., 2010. Climate change implications for river restoration in global bio-diversity hotspots. Restor. Ecol. 18, 261–268.

De Meulder, B., Shannon, K., 2013a. Water urbanism east. Emerging practices and age-old traditions. In: De Meulder, B., Shannon, K. (Eds.), Water Urbanism East. Park Books, Zurich, pp. 4–9.

De Meulder, B., Shannon, K. (Eds.), 2013. Water Urbanisms East. Park Books, Zurich.

Dister, E., Gomer, D., Orbdlik, P., Petermann, P., Schneider, E., 1990. Water management and ecological perspectives of the Upper Rhine's floodplains. Regul. River Res. Manag. 5 (1), 1–15.

Duquesnoy, M., 2012. The tragedy of the utopia of the Mapuche of Chile: territorial vindications in the times of applied neoliberalism. Revista Paz y Conflictos 5, issn: 1988-7221.

Elosegi, A., Díez, J., Mutz, M., 2010. Effects of hydromorphological integrity on biodiversity and functioning of river ecosystems. Hydrobiology 657, 199–215.

Elosegi, A., Gessner, M.O., Young, R.G., 2017. River doctors: learning from medicine to improve ecosystem management. Sci. Total Environ. 595, 294–302.

Fryirs, K.A., Brierley, G.J., 2012. Geomorphic Analysis of River Systems: An Approach to Reading the Landscape. Wiley-Blackwell, Hoboken, NJ.

Gray, M., Gordon, J.E., Brown, E.J., 2013. Geodiversity and the ecosystem approach: the contribution of geoscience in delivering integrated environmental management. Proc. Geol. Assoc. 124, 659–673.

Hidalgo, R., 1999. Políticas de vivienda social en Santiago de Chile: la acción del Estado en un siglo de planes y programas. Scripta Nova Geo Crítica 45 (1).

Honey-Rosés, J., Acuña, V., Bardina, M., Brozovic´, N., Marcé, R., Munné, A., Sabater, S., Termes, M., Valero, F., Vega, Á., Schneider, D.W., 2013. Examining the demand for ecosystem services: the value of stream restoration for drinking water treatment managers in the Llobregat River, Spain. Ecol. Econ 90, 196–205.

Horacio, J., 2015. Reflexiones y enfoques en la conservación y restauración de ríos: georrestauración y pensamiento fluvial. Biblio3W 1, 142.

Huber, A., Iroumé, A., Mohr, C., Frêne, C., 2010. Efecto de plantaciones de Pinus Radiata y Eucalyptus Globulus sobre el recurso agua en la Cordillera de la Costa de la región del Bío Bío, Chile. Bosque 31 (3), 219–230.

Illies, J., Botasaneanu, L., 1963. Problemes et methodes de la clasification de la zonation ecologique des eaux courantes, considerees surtout du point de vue faunistique. Mitteilungen Internationale Vereiningung fuer Theoretische und Angewandte Limnologie 12, 1–57.

Instituto nacional de estadísticas (INE), 2002. http://www.ine.cl/.

Intendencia Regional de Concepción, 2013. Official thematic cartography.

Jaque, E., 1996. Análisis integrados de los sistemas naturales de la cuenca hidrográfica del río Andalién: Proposición para el manejo integral de la cuenca (Ph.D. thesis). Universidad de Concepción, Bío Bió Region, Chile.

Jaque, E., 2010. Diagnóstico de los paisajes mediterráneos costeros. Cuenca del río Andalién, Chile. Boletín de la Asociación de Geógrafos Españoles 54, 81–97, ISSN: 0212-9426.

Jaque, E., 2017. Construyendo riesgo de incendios forestales en el Área Metropolitana de Concepción, Chile. XVI EGAL, Encuentro de Geógrafos de América Latina. La Paz, Bolivia, April 25–26, 2017.

Malavoi, J.R., Bravard, J.P., Piégay, H., Héroin, E., Ramez, P., 1998. Determination de l'Espace de Liberté des Cours d'Eau. SDAGE Rhône-Méditerranée-Corse, Lyon.

Malavoi, J.R., Gautier, J.N., Bravard, J.P., 2002. Free space for rivers: a geodynamical concept for a substainable management of the watercourses. In: Bousmar, D., Zech, Y. (Eds.), In: Proceedings of the International Conference on Fluvial Hydraulics River Flow 2002. Swets and Zeitlinger, Lisse.

Mardones, M., Vidal, C., 2001. La zonificación y evaluación de los riesgos naturales de tipo geomorfológico: un instrumento para la planificación urbana en la ciudad de Concepción. EURE 27 (81), 97–122.

Mathur, A., Da Cunha, D., Meeks, R., and Wiener, M., 2014. Design in the Terrain of Water. Applied Research + Design Publishing with the University of Pennsylvania School of Design. Philadelphia.

Mazzei, L., Pacheco, A., 1985. Historia del Traslado de la Ciudad de Concepción. Editorial de la Universidad de Concepción, Concepción.

McHarg, I., 1969. Process as Values, Design with Nature. On Landscape Urbanism. University of Texas, Austin, TX, Reprinted by Austin School of Architecture, 1021.

Meyer, H., Bobbink, I., Nijhuis, S., 2010. Delta Urbanism. The Netherlands. American Planning Association, Chicago, Washington, D.C..

Ministerio de obras públicas (MOP), 2016. Dirección general de Obras públicas.

Mostafavi, M., Doherty, G. (Eds.), 2010. Ecological Urbanism. Harvard University Graduate School of Design Lars Müller Publishers, Baden.

Muñoz, M.D., 1995. El Plan Regulador de Concepción 1960. Arquitecturas del Sur 24, 22–28.

Newson, M., 2002. Geomorphological concepts and tools for sustainable river ecosystem management. Aquat. Conserv. 12, 365–379.

Ollero, A., 2007. Territorio Fluvial. Diagnóstico y Propuesta para la Gestión Ambiental y de Riesgos en el Ebro y los Cursos Bajos de sus Afluentes. Bakeaz y Fundación Nueva Cultura del Agua, Bilbao.

Ollero, a., 2010. Channel changes and floodplain management in the meandering middle Ebro River, Spain. Geomorphology 117, 247–260.

Ollero, a., 2014. Guía Metodológica Sobre Buenas Prácticas en Restauración Fluvial. Manual para Gestores. Contrato de río del Matarraña. ECODES, Zaragoza.

Ollero, a., 2015. Guía Metodológica Sobre Buenas Prácticas en Restauración Fluvial. Manual para Gestores. Contrato de río del Matarraña. ECODES, Zaragoza.

Ollero, A., Ibisate, A., 2012. Space for the river: a flood management tool. Wong, T.S.W. (Ed.), Flood Risk and Flood Management, vol. 9, Nova Science Publishers, New York, NY, pp. 199–217.

Ollero, A., Ibisate, A., Gonzalo, L.E., Acín, V., Ballarín, D., Díaz, E., Domenech, S., Gimeno, M., Granado, D., Horacio, J., Mora, D., Sánchez Fabre, M., 2011. The IHG index for hydromorphological quality assessment of rivers and streams: updated version. Limnetica 30 (2), 255–262.

Ollero, A., Ibisate, A., Granado, D., Real de Asua, R., 2015. Channel responses to global change and local impacts: perspectives and tools for floodplain management (Ebro River and tributaries, NE Spain). In: Hudson, P.F., Middelkoop, H. (Eds.), Geomorphic Approaches to Integrated Floodplain Management of Lowland Fluvial Systems in North America and Europe. Springer, New York, NY, pp. 27–52.

Owens, P.N., 2005. Conceptual models and budgets for sediment management at the river basin scale. J. Soils Sediments 5, 201–212.

Owens, P.N., Slob, A.F.L., Liska, I., Brils, J., 2008. Towards sustainable sediment management at the river basin scale. In: Owens, P.N. (Ed.), Sustainable Management of Sediment Resources: Sediment Management at the River Basin Scale. Elsevier, New York, NY, pp. 217–259.

Palmer, M.A., Richardson, C.D., 2009. Provisioning services: a focus on fresh water. In: Levin, S.a. (Ed.), The Princeton Guide to Ecology. Princeton University Press, Princeton, NJ, pp. 625–633.

Pérez, L., Hidalgo, R., 2010. Concepción metropolitano: realidad preterremoto 27/f 2010 y desafíos de su reconstrucción. Pérez, L., Hidalgo, R. (Eds.), Volución y Desafíos, vol. 14, Editorial Universidad de Concepción. GEOlibros series, Bío Bío Region, pp. 7–22.

Piégay, H., Darby, S.E., Mosselman, E., Surian, N., 2005. a review of techniques available for delimiting the erodible river corridor: a sustainable approach to managing bank erosion. River Res. Appl. 21, 773–789.

Pinkham, R., 2000. Daylighting: New Life for Buried Streams. Rocky Mountain Institute, New York, NY.

PRMC (Peninsula Regional Medical Center), 2013. Thematic Cartography (Preliminary information). Ministerio de Vivienda y Urbanismo, Región Metropolitana.

Prominski, M., Stokman, A., Zeller, S., Stimberg F. D., Voermanek, H., 2012. River. Space. Design. Planning Strategies, Methods and Projects for Urban Rivers. Brikhauser, Basel.

PROT, 2011. Thematic Cartography (Preliminary Information). Regional Government Bío Bío, Bío Bío Region.

Quintana, F., 2014. Urbanizando con tiza. Revista ARQ 86, 30–43.

Red Cedeus, 2015. Unidades geomorfológicas chile geo-node cedeus. Available from: http://datos.cedeus.cl/layers/geonode:cl_unidades_geomorfologicas_geo.

Redeker, C., 2013. Rhine Cities/Urban Flood Integration (UFI) (Ph.D. Research). Delft University of Technology, Delft, The Netherlands.

Riley, A., 2016. Restoring Neighborhood Stream, Planning, Design, and Construction. Island Press, Washington, DC.

Rohde, S., Hostmann, M., Peter, A., Ewald, K.C., 2006. Room for rivers: an integrative search strategy for floodplain restoration. Landsc. Urban Plan. 78 (1–2), 50–70.

Rojas, O., 2015. Cambios Ambientales y Dinámica de Inundaciones Fluviales en una Cuenca Costera del Centro Sur de Chile (Tesis Doctoral). Universidad de Concepción, Concepción, Chile.

Rojas, O., Mardones, M., Arumí, J., Aguayo, M., 2014. Una revisión de inundaciones fluviales en Chile, período 1574-2012: causas, recurrencia y efectos geográficos. Revista de geografía Norte Grande 57, 177–192.

Romero, H, Vidal, C, 2010. Efectos ambientales de la urbanización de las cuencas de los ríos Bío Bío y Andalién sobre los riesgos de inundación y anegamiento de la ciudad de Concepción. Pérez, L., Hidalgo, R. (Eds.), Concepción metropolitano, evolución y desafíos, vol. 14, Editorial Universidad de Concepción. GEOlibros series.

Servicio de evaluación ambiental (SEA) del gobierno de Chile, 2016. Available from: http://www.sea.gob.cl/.

Sernageomin, 2010. Risk cartography. Concepción, Talcahuano, Hualpén y Chiguayante, Chile.

Seto, K.C., Fragkias, M., Guneralp, B., Reilly, M.K., 2011. A meta-analysis of global urban land expansion. PLoS One 6, 9.

Shane, G., 2003. The emergence of "landscape urbanism": reflections on stalking Detroit. Harvard Design Magazine, Fall 2003/Winter 2004, 1–8.

Shannon, K., 2013. Eco-engineering for water: from soft to hard and back. In: Pickett, S.T.A., Cadenasso, M., McGrath, B. (Eds.), Resilience in Ecology and Urban Design: Linking Theory and Practice for Sustainable Cities. Future Cites Series. Springer, New York, NY, pp. 163–182, vol. 3.

Shannon, K., De Meulder, B., D'Auria, V., Gosseye, J. (Eds.), 2008. Water Urbanisms. UFO 1. Sun, Amsterdam.

Urrutia, H., Manchileo, D., Sanhueza, C., Jara, D., Vidal, C., 2012. Evolución Urbana del Sector Nonguén entre los años 1950–2012. Análisis de Problemáticas de Riesgos Naturales. Universidad San Sebastián, Concepción.

Vietz, G.J., Rutherfurd, I.D., Fletcher, T.D., Walsh, C.J., 2016. Thinking outside the channel: challenges and opportunities for protection and restoration of stream morphology in urbanizing catchments. Landsc. Urban Plan. 145, 34–44.

Waldheim, C., 2010. On landscape, ecology and other modifiers to urbanism. Topos 71, 21–24.

Werritty, A., 1997. Short-term changes in channel stability. In: Thorne, C.R., Hey, R.D., Newson, M.D. (Eds.), Applied Fluvial Geomorphology for River Engineering and Management. Wiley, Chichester, pp. 47–65.

SECTION IV

DEVELOPING GEOMORPHOLOGICAL HAZARDS DURING THE ANTHROPOCENE

10 *Urban Geomorphology of an Arid City: Case Study of Phoenix, Arizona* 177

11 *Bivouacs of the Anthropocene: Urbanization, Landforms, and Hazards in Mountainous Regions* 205

12 *Pokhara (Central Nepal): A Dramatic Yet Geomorphologically Active Environment Versus a Dynamic, Rapidly Developing City* 231

CHAPTER

10

Urban Geomorphology of an Arid City: Case Study of Phoenix, Arizona

Ara Jeong, Suet Yi Cheung, Ian J. Walker, Ronald I. Dorn

Arizona State University, Tempe, AZ, United States

OUTLINE

10.1 Sonoran Desert Setting of the Phoenix Metropolitan Area 177

10.2 Common Desert Geomorphic Processes in the Phoenix Metropolitan Area 179
10.2.1 Rock Decay, Rock Coatings, and Soils 180
10.2.2 Interplay of Aeolian, Fluvial, and Anthropogenic Processes 187
10.2.3 Mass Wasting 189
10.2.4 Pedimentation 190

10.3 Desert Geomorphic Hazards 195
10.3.1 Alluvial Fan Flooding 195
10.3.2 Street Flooding in Planned and Unplanned Housing Developments 195
10.3.3 Debris Flows 199
10.3.4 Haboobs and Dust Storms 199

10.4 Summary Perspective on Human Influences on the Arid Geomorphic System in the Urbanizing Sonoran Desert 200

References 201

10.1 SONORAN DESERT SETTING OF THE PHOENIX METROPOLITAN AREA

The present-day climate and vegetation of the Phoenix metropolitan region (Fig. 10.1) resembles much of the rest of the Sonoran Desert in central Arizona, USA. Annual precipitation displays a bimodal distribution with summer and winter maxima. Summer convective thunderstorms occur during the July–September Mexican Monsoon. Winter frontal rainfall derives from Pacific cyclones. Mean annual precipitation tends to be evenly split between winter and summer, averaging about 200 mm (Arizona State Climatologists office, https://azclimate.asu.edu/). The arid climate is typified by a distinct biogeography typical of the

Urban Geomorphology. http://dx.doi.org/10.1016/B978-0-12-811951-8.00010-2

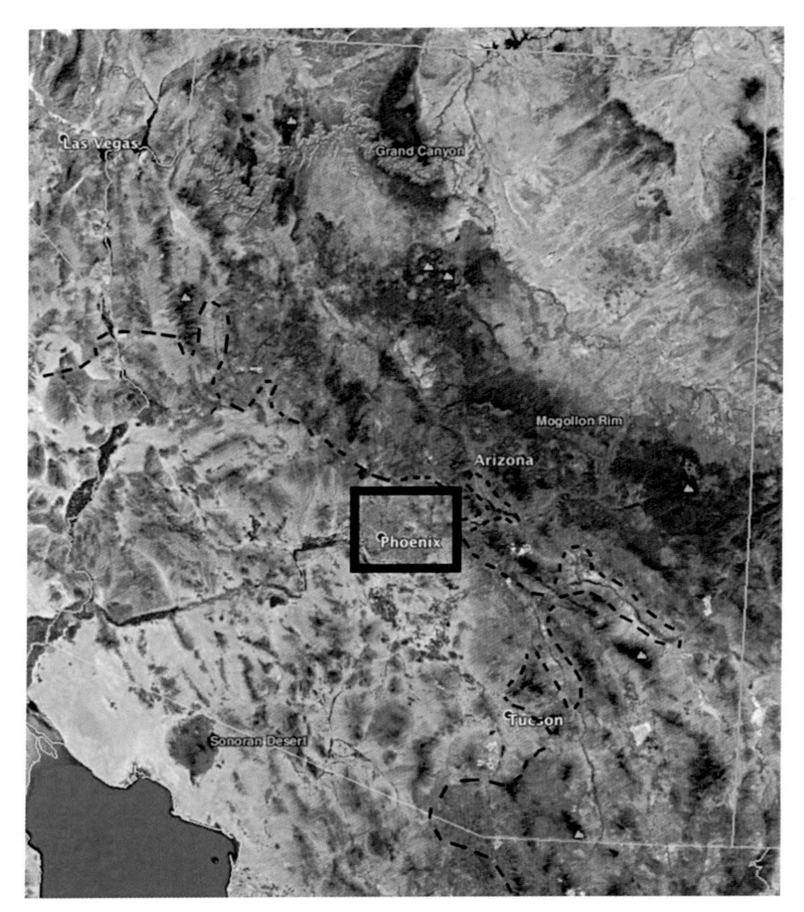

FIGURE 10.1 The state of Arizona as seen in Google Earth. The *black box* identifies the metropolitan Phoenix region as framed in Fig. 10.2. The *dashed line* indicates the boundary of the Sonoran Desert, where Phoenix is situated in the northeastern corner. The forested highlands of the Mogollon Rim to the northeast provide much of the water for metropolitan Phoenix, funneled by the Verde and Salt River drainages. *Source: The image is used following permission guidelines for Google Earth (http://www.google.com/permissions/geoguidelines.html).*

Sonoran Desert. Trees include palo verde (*Parkinsonia microphylla*), ironwood (*Olneya tesota*), and elephant tree (*Bursera microphylla*). Common desert scrub vegetation includes creosote bush (*Larrea tridentata*), brittlebush (*Encelia farinosa*), triangle-leaf bursage (*Ambrosia deltoidea*), catclaw acacia (*Acacia greggii*), desert globe mallow (*Sphaeralcia ambigua*), and ocotillo (*Fouquieria splendens*). Succulents occur in great variety, notably the iconic saguaro (*Carnegiea gigantea*), barrel (*Ferocactus cylindraceus*), and hedgehog (*Echniocereus engelmannii*) cacti that are abundant throughout the natural landscape.

Pollen and packrat midden studies in the Sonoran Desert and adjacent areas suggest the region did not become a desert until the Holocene and it was not a desert during the Pleistocene. Pollen records in northern Baja California from 44 to 13 ka reveal the presence of pines, junipers, and sagebrush in that area indicating more humid and cooler conditions

(Lozano-García et al., 2002). Packrat midden sequences in the Sonoran Desert indicate the presence of the dwarf conifers *Juniper osteosperma* and *Pinus monophylla* in the lower Sonoran Desert in this same late Pleistocene time range (Allen et al., 1998; McAuliffe and van Devender, 1998; van Devender, 1990). Thus, abundant evidence of a wetter and cooler time generated more extensive vegetation in the last glacial period and perhaps previous glacial cycles.

The geology of the Phoenix area underwent major crustal extension during the mid-Tertiary. This crustal extension resulted from the release of compressional stress after the Laramide mountain building period (Coney and Harms, 1984; Holt et al., 1986; Nations and Stump, 1981). This extension generated basin and range topography that developed between about 25 and 8 Ma. As a part of this extension, major rhyolite caldera eruptions started about 20 Ma, and metamorphic core complexes also domed up before most volcanic activity and extension finally ceased around 8 Ma (Reynolds, 1985; Spencer, 1984). The net result is a bedrock geology that mixes intrusive igneous granitic rocks, foliated metamorphic blocks, as well as extensive outcrops of rhyolitic welded tuff from major eruption episodes during extension (Fig. 10.2).

The geomorphic landscapes of the Phoenix area contain a mixture of classic desert landforms. Bedrock landforms depend greatly upon the rock type. Granitic forms include classic domed inselbergs or bornhardts where jointing is far apart, but more complex landscapes where the jointing density increases and influences biotic communities (Seong et al., 2016b). Metamorphic slopes tend to host debris-flow chutes and levees (Dorn, 2012). Rhyolitic welded tuff deposits of the Superstition volcanic field (Stuckless and Sheridan, 1971) develop more massive cliff faces.

The nature of piedmont slopes in front of the ranges depends on drainage area. Larger ranges have sufficient drainages to develop alluvial fans or alluvial slopes (Applegarth, 2004), whereas bedrock pediments form in front of smaller mountain masses (Kesel, 1977). The low-relief areas now occupied by the urbanscape of the Phoenix metropolitan area (Fig. 10.2) consist of the distal ends of pediments, the distal end of alluvial fans, aeolian sand sheets, and alluvial deposits.

An individual walking on these landforms, before massive land-use change associated with cattle grazing and urban expansion, would have experienced very different surface conditions than found by the average hiker today. Extensive areas once hosted desert pavements, biological soil crusts (Nagy et al., 2005), and interlocking colluvium on steeper slopes, providing a net armoring effect (Bowker et al., 2008; Granger et al., 2001; Seong et al., 2016a). Today, only patches of such armored surfaces remain, providing glimpses into the original land surfaces. Thus, this chapter attempts to give the reader a sense of the geomorphology that once was (and still exists in a few places) and a desert geomorphology influenced by the Anthropocene (Waters et al., 2016) and its urban footprint.

10.2 COMMON DESERT GEOMORPHIC PROCESSES IN THE PHOENIX METROPOLITAN AREA

This section presents some of the more important desert geomorphic processes that occur in the Phoenix area. The section starts with processes related to rock decay (weathering) (in general, we prefer the use of the term rock decay instead of weathering for reasons elaborated elsewhere; Hall et al., 2012), rock coatings, and soils. The second section explores how aeolian,

FIGURE 10.2 **Geological map of bedrock ranges in the metropolitan Phoenix overlaid on a Google Earth oblique image.** The mapping units derive from the Arizona Geological Survey (Richards et al., 2000), but are generalized here to help visualize isolated bedrock mountainous areas: intrusive igneous (pink), extrusive igneous (red), metamorphic (green), and mid-Tertiary sedimentary rocks (gray). Most urban space, thus, rests on Quaternary sediment ranging from Pliocene to Holocene in age. *Source: The image is used following permission guidelines for Google Earth to Source: The image has been modified from a Google Earth image (http://www.google.com/permissions/geoguidelines.html) with mapping units derived from Richards, S.M., Reynolds, S.J., Spencer, J.E. Pearthree, P.A., 2000. Geologic Map of Arizona. Arizona Geological Survey Map M-35, 1 sheet, scale 1:1,000,000.*

fluvial, and human activities interact in the fringe of the Phoenix urban area. The third section overviews mass-wasting processes on the steep slopes of desert mountain ranges in the middle of Phoenix. Much of the Phoenix metropolitan region is built on pediments and the fourth section explains that the Phoenix area is truly unique in terms of the occurrence of pediments in several different rock types.

10.2.1 Rock Decay, Rock Coatings, and Soils

10.2.1.1 Dominant Rock Decay Process

Dirt cracking is the dominant process of physical rock decay in the Phoenix area. Any random rock fracture, when pried open, reveals evidence of the dirt-cracking process

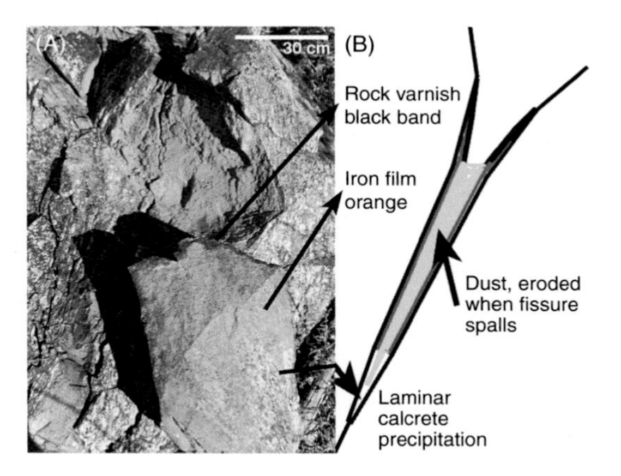

FIGURE 10.3 **Dirt cracking wedges open joint faces in two ways: laminar calcrete precipitation and the wetting and drying of dust in the joints.** (A) A 0.5 mm fracture was manually pried open at South Mountain, Phoenix. Dust filled the fracture. The two sides of the fracture display rock coatings diagnostic of dirt cracking. (B) Idealized diagram of rock coatings associated with fractures opened by dirt cracking.

(Dorn, 2011; Ollier, 1965). A combination of laminar calcrete precipitation and the wetting and drying of dust accumulated in fractures gradually opens fissures to the point where spalling occurs. Unlike other forms of physical rock decay, dirt cracking leaves behind visual evidence of the process (Fig. 10.3). Laminar calcrete coats the walls along the narrowest parts of a fissure, which is a space wide enough for capillary water to penetrate and precipitate calcium carbonate. Eventually, the fracture widens enough to allow dust to infiltrate. Iron (Fe) films typically less than 10 μm-thick coat the walls, where dust remains in contact with rock surfaces. Black rock varnish forms a rim around the margins of the fracture, where rainwater has washed away the dust and the removal of this alkaline dust allows manganese-enhancing bacteria to develop and form a coating of rock varnish over the Fe film.

10.2.1.2 Granitic Landforms Generated by Rock Decay

Granitic rocks underlay extensive areas of metropolitan Phoenix (Fig. 10.2). Thus, classic forms of cores stones, tors, domed inselbergs (bornhardts), and kopje occur throughout the Phoenix area (Fig. 10.3). Jointing is particularly important in the morphogenesis of granitic terrains: "Here we speculate that a corollary to the arguments given above about the role of tectonics as a crusher of rock is that in those places where rock has dodged the rock crusher, it may be stronger and less easily removed by erosive agents" (p. 10) (Molnar et al., 2007). Jointing is particularly important in arid weathering-limited landscapes (Abrahams et al., 1985; Howard and Selby, 2009; Viles, 2013).

Core stones and tors are common in the wealthier areas of metropolitan Phoenix, such as north Scottsdale, Fountain Hills, and east Mesa. Core stones are the spheroidal less-decayed boulders that emerge at the surface as grus erodes (Fig. 10.4B) (Twidale, 1982). Domed inselbergs, also known as bornhardts, are bald, and steep-sided domes with a range of shapes and size (Twidale, 1981) (e.g., Fig. 10.4A). Bornhardts like those seen in metropolitan Phoenix (e.g., Fig. 10.4A) maintain a lower joint density than the surrounding granite. Meanwhile, the

FIGURE 10.4 Landforms resulting from decay of granite in a desert setting in Scottsdale create an aesthetic setting for the wealthy in metropolitan, Phoenix. (A) Troon Mountain, a bornhardt, is surrounded by large home and mansions. (B) Golf courses place greens adjacent to spheroidal core stones and tors. (C) Large homes and mansions surround the collapsed bornhardt known as Pinnacle Peak, a kopje. *Source: (A) The image is used following permission guidelines for Google Earth (http://www.google.com/permissions/geoguidelines.html).*

surrounding granitic rocks with high joint-density experience more active mineral decay to grus. Eventually, the existing joints in bornhardts separate, leading to rock slides and collapse into a landform known as a "kopje" (e.g., Fig. 10.4C), similar to those studied in Africa and Australia (Michael et al., 2008).

10.2.1.3 Desert Pavements

Desert pavements consist of a smooth surface with closely packed, interlocking pebbles, cobbles, and sometimes with scattered boulders (Fig. 10.5A). Pavements provide insight into the antiquity of the underlying landform (Seong et al., 2016a), aid in the preservation of ancient artifacts (Adelsberger et al., 2013), and can provide information about environmental change (Dietze et al., 2016) and desert soils (Peterson et al., 1995). Although introductory textbooks often attribute desert pavements to deflation winnowing of fines, the desert pavements in the Phoenix area are not a result of wind erosion. In the case of the entire Sonoran Desert, very little evidence of aeolian abrasion exists (Seong et al., 2016a). For example, desert pavements at South Mountain Preserve, Phoenix, have: coatings of rock varnish that deflation would have abraded away (Fig. 10.5A); vesicular Av soil horizons from the accumulation of dust underneath surface clasts (Fig. 10.5B and C); closely spaced clasts separated by silt and clay surfaces; and no evidence of ventifacts.

The desert pavement in the Phoenix area initiate when floods or debris flows deposit loose and unconsolidated clasts on the surface (Fig. 10.5D). Aeolian fines slowly move into the matrix between the large clasts deposits (Fig. 10.5C) and the size of clasts decreases mostly from dirt-cracking processes (Fig. 10.3). As dust accumulates and clast size decreases, the relief of the original "bar-and-swale" topography gradually reduces (Fig. 10.5E). The keys to stable pavements (e.g., Fig. 10.5A) in the Phoenix area are a combination of: a relatively flat surface; the accumulation of allochthonous dust; a lack of headward retreating swales or gullies (Seong et al., 2016a); and most critically a minimal amount of human activity, as even just one vehicle driving over a pavement surface can do damage (Fig. 10.5F).

10.2.1.4 Biological Soil Crusts

Biological soil crusts (BSCs) consist of assemblages of living organisms on soil or rock surfaces in arid and semiarid areas. Typically composed of cyanobacteria, fungi, lichens, and algae, they cover a wide variety of undisturbed Sonoran Desert soils (Fig. 10.6) and protect desert surfaces from erosional shear stresses imposed by overland flow and strong winds (Allen, 2005, 2010).

When soil is wet, the mucilage of cyanobacteria swell and filaments of cyanobacteria move up toward the soil surface (Belnap et al., 2001). This repeated swelling and frequent movement leaves copious sheath material in the uppermost soil layers that, in turn, maintains soil structure after BSCs are dehydrated and soil particles become loose (Belnap, 2003). Thus, BSCs, and especially filamentous cyanobacteria, adhere to and aggregate with soil particles and their cohesion increases surface stability and inhibits erosion in arid and semiarid lands (Belnap, 2003; Bowker et al., 2008).

Although BSCs are extremely well adapted to the harsh growing conditions in deserts, they can be significantly altered by disturbances, such as grazing, recreational activities (hiking, biking, and off-road driving), and military activities (Belnap and Gillette, 1998) (Fig. 10.6). Faist et al. (2017) examined BSC hydrologic responses to disturbance at different crustal development stages on sandy soils on the Colorado Plateau through a simulated rainfall experiment. They found that trampling well-developed dark cyano-lichen-dominated crusts increased total sediment loss by nearly four times in comparison to intact controls during a 30 min simulated precipitation event, suggesting that well-developed, intact dark BSCs generally decrease runoff and sediment loss and considerably increase aggregate stability (Faist et al., 2017). While BSCs are extremely vulnerable to disturbance, their recovery time can be

FIGURE 10.5 Desert pavement surfaces at South Mountain, Phoenix. (A) Pleistocene desert pavement showing weathered clasts that are closely packed and interlocked, where the sunhat provides scale. (B) Dust accumulation underneath surface clasts seen after removal of the pavement clasts with rock hammer for scale. (C) Vesicular soil horizon termed the Av horizon and the underlying Bk horizon with carbonate-covered clasts, where the Av/Bk boundary is only about 5 cm beneath the pavement. (D) Flood deposits typically maintain a rough bar and swale form; the sunhat provides scale. (E) Over time, the relief of the bar-and-swale topography decreases as swales fill in and bars erode; desert scrub vegetation with a height of 70 cm provides scale. (F) Desert pavement impacted by just one vehicle (*arrow*).

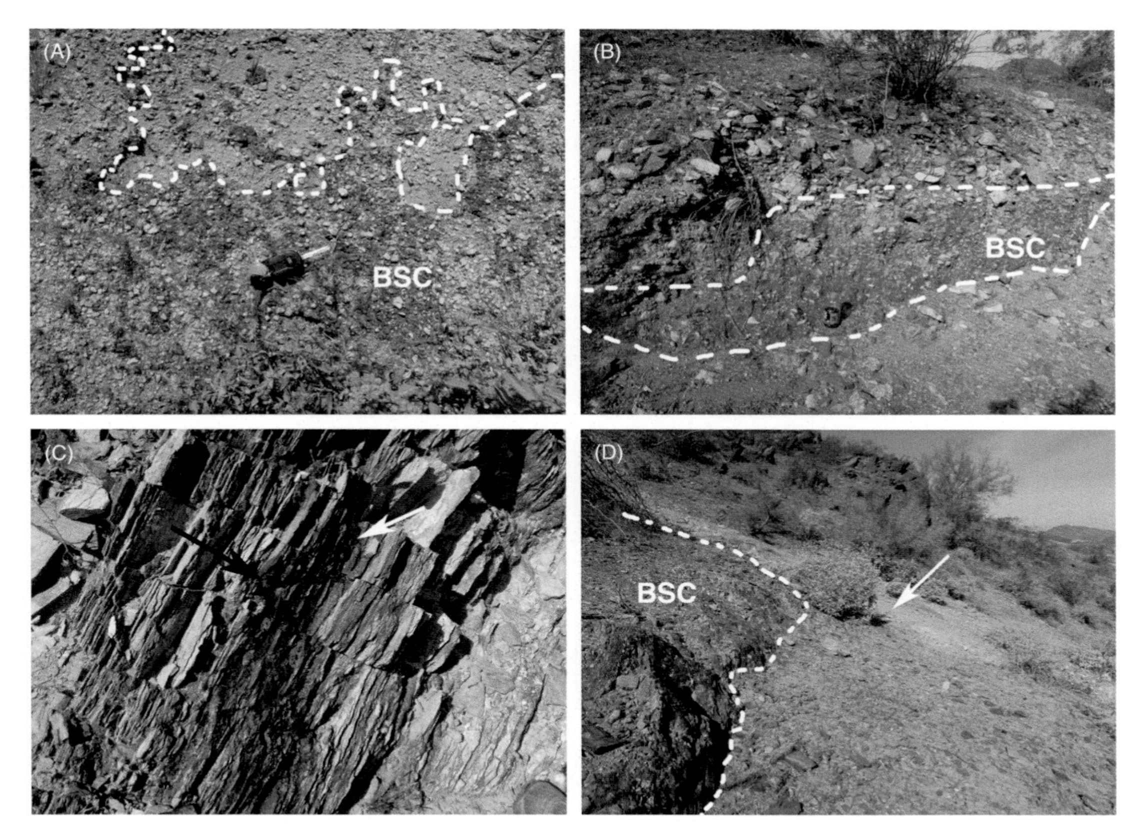

FIGURE 10.6 **Biological crusts on soil and rock surfaces in metropolitan Phoenix.** (A) Biological crusts on a soil surface with car keys for scale. The upper portion of the image was disturbed by the compressive force of cattle grazing, although some recovery has occurred as seen in the lower half of the image (underneath *dashed line*) over the past 12 years since cattle grazing ceased. (B) Biological soil crusts growing on the side of the east-facing side of a desert wash, an aspect that reduces exposure to directly sunlight in the warmest part of the day. (C) Biological soil crusts on a rock surface, where the surface was wetted resulting in a greening up by the algae (*white arrow*); however, BSCs dominated by fungi did not green up (*black arrow*). (D) Biological soil crusts disturbed along a hiking trail, but still evident away from the trail to the left of the *dashed white line*.

relatively slow. No growth of BSCs occurred in the central Namib, for example, over an 8-year period of observation (Viles, 2008).

10.2.1.5 Rock Coatings

A variety of different rock coatings occur throughout the metropolitan Phoenix area, but manganese(Mn)-rich rock varnish darkens the vast majority of exposed rock faces. Fig. 10.7A illustrates the typical appearance of rock varnish. Fig. 10.7C presents an ultrathin section of varnish, revealing the presence of fine micrometer-scale laminations or "microlaminations." These layers form as a result of Holocene and Pleistocene climatic changes (Liu and Broecker, 2007, 2008). Where these layers have been calibrated by independent ages (Liu, 2017), it is possible to assign millennial-scale ages to landforms (Liu and Broecker, 2013).

FIGURE 10.7 Rock varnish as the dominant natural rock coating in metropolitan Phoenix. (A) Colluvial boulder field at Shaw Butte darkened by rock varnish. The occasional orange Fe film indicates rocks spalled by the dirt cracking process (Fig. 10.3). (B) Urbanization tends to create scars across rock faces, but developers in an affluent Phoenix neighborhood applied "artificial varnish" to minimize the aethestic impact of this road cut. (C) Microlaminations form discrete black, orange, and yellow layers in rock varnish thin sections. (D) Back-scattered electron microscope image of artificial varnish from image B that is experiencing ongoing dissolution, generating a granule-like appearance.

In the case of Fig. 10.7C, a varnish started to form about 8.1 ka, indicated by wet holocene (WH) layer WH9 at the base of the varnish.

Human activity has left a distinct chemical imprint on rock varnishes. The Fe and Mn hydroxides that provide the varnish color scavenges Pb, for example Pb additives used in gasoline in the early part of the 20th century. This Pb accumulates in the surface-most micron of the varnish (Fig. 10.7C), as evidenced by electron microprobe measurements (e.g., 0.44% PbO in Fig. 10.7C). Such "spikes" in Pb abundance are far greater than background levels seen in natural varnish of <0.03% PbO. Although this Anthropocene signal may seem negative at first, representing widespread contamination, the Pb actually provides a useful chronometric marker able to identify purely 20th-century flooding surfaces as well as authentication of rock engravings (Dorn et al., 2012).

The dominance of rock varnish makes it easy to spot anthropogenic disturbances associated with urban construction because the underlying rock is always much lighter in color when disturbed. Some developers in affluent communities decided to try an experimental treatment of artificial varnish (Elvidge and Moore, 1980) to reduce the aesthetic impact of road construction, as seen in the road cut shown in Fig. 10.7B. This artificial varnish is slowly dissolving into granules (Fig. 10.7D). In contrast to natural varnish, artificial varnish has not been binded with clay minerals that, in turn, help cement natural varnish to the underlying rock (Dorn and Oberlander, 1982).

10.2.2 Interplay of Aeolian, Fluvial, and Anthropogenic Processes

The interaction between aeolian and fluvial processes can be an important factor in the shaping of dryland environments (Bullard and Livingstone, 2002). Source-bordering dunes represent a common landform in many dryland environments, such as dunes closely bordering a river (Page et al., 2001). When these fluvial sediments are exposed to the air during a prolonged dry period, winds are more likely to affect sediment transport and wind velocity and particle size are critical factors to entrain sediments.

Two large exoreic river systems cross the Phoenix metropolitan area. The Salt River runs through the center, while the Gila River flows along the southern boundary. A small area of source-bordering dunes and a sand sheet occurs north of the Gila River (Fig. 10.8).

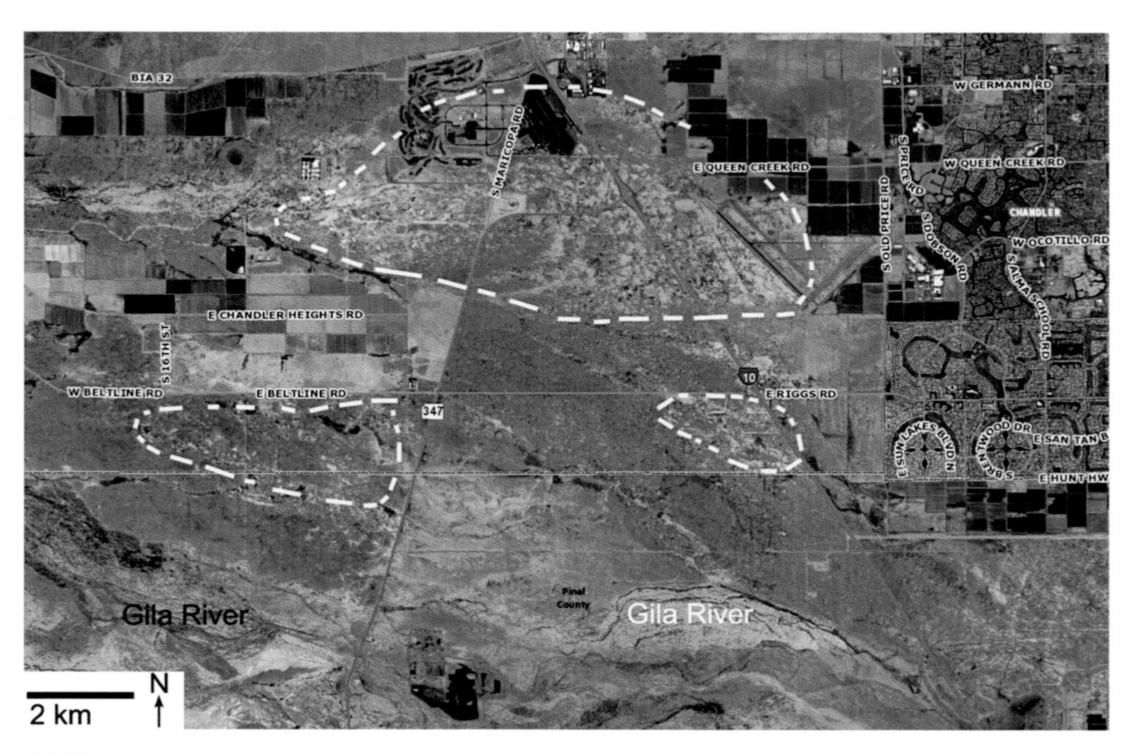

FIGURE 10.8 **Source bordering dunes near the Gila River in the southern part of the Phoenix Metropolitan area.** Areas denoted by the *dashed lines* locate areas with distinct dune forms. However, the land between these areas is covered by a sand sheet.

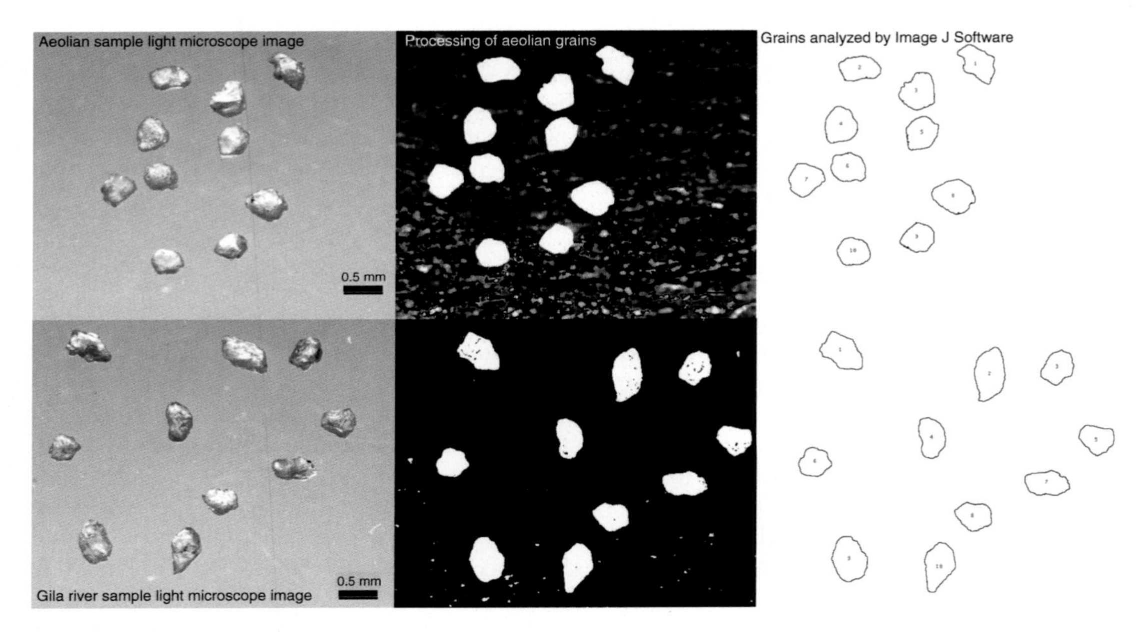

FIGURE 10.9 Grain size analysis of aeolian and fluvial sediments using the ImageJ software. The aeolian sample (*upper row*) from source-bordering dunes display more rounding than the fluvial sediments from the Gila River (*lower row*). The grains are first imaged with light microscopy (*left column*), then subject to digital image processing (*middle column*), and the resultant grain perimeter is used by ImageJ software to generate shape parameters. The result of grain shape analysis shows that aeolian sediments have a higher value in roundness (0.78–0.79) and a lower ratio in aspect ratio (1.29–1.30) than fluvial sediment that has 0.70–0.72 in roundness and 1.45–1.50 in aspect ratio.

Using the method developed by Eamer et al. (2017), sediments collected from the Gila River show considerably less rounding than sediments collected from source-bordering dunes (Fig. 10.9).

This area experienced a variety of land uses, including cattle grazing, irrigated agriculture, road construction, and the building of subdivisions. Concomitantly, human activities along the Gila River altered the natural river system, hydrological processes, and, thus, sediment supply to these source-bordering dunes.

Construction of the Coolidge Dam in the upper course of Gila River in 1928 greatly reduced the flood frequency and flood magnitude, leading to a decline in sediment supply. At the same time, an invasive species, Tamarisk, invaded the riparian zone of the Gila River and spread rapidly in the 1900s (Graf, 1988). The presence of Tamarisk affects both aeolian and fluvial processes in terms of reducing the sediments in transport and shear stress on the soil surface. These anthropogenic effects likely decreased sediment transport to the Gila source-bordering dunes. However, extreme weather events, such as the 1993 flood in Phoenix, triggered by El Niño, reactivated the formation of sandbars and changed the channel form to a braided stream pattern along the Gila River due to increased sediment supply. Fig. 10.10 summarizes some of the major controls influencing the potential supply of sediment along the Gila located next to the dunes.

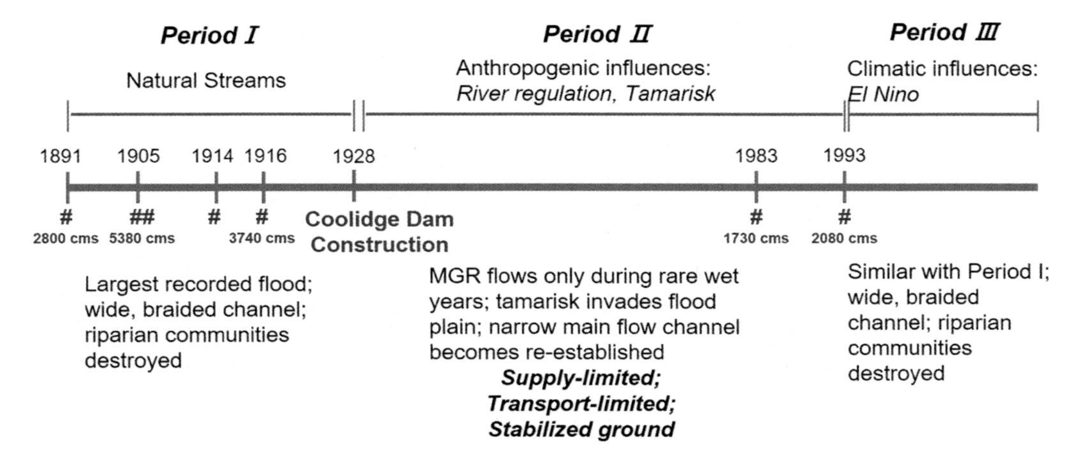

FIGURE 10.10 Timeline of anthropogenic alteration and large flood events along the Gila River. *Period I*: Gila River was a natural channel with frequent seasonal flooding. *Period II*: Anthropogenic activities altered the natural channel system of Gila River, largely reducing sediment supply for both aeolian and fluvial transport as well as flood frequency. *Period III*: Extreme climatic events caused large floods along the Gila River reactivated aeolian and fluvial-transport processes due to the increased sediment availability.

Extensive areas of the cities of Chandler, Gilbert, Mesa, and Tempe in the metropolitan area are covered by several meters of fine sandy material. This material could have been derived in part by aeolian transport from the Gila River. In addition, there are lenses of river-transported gravels and cobbles derived from bordering mountains. Thus, it is likely that aeolian and fluvial processes resulted in a mixture of a sand sheet intercalating with low-energy rivers. Fig. 10.11 illustrates an anthropogenic excavation into this mixed, interdigitated aeolian and fluvial deposit.

10.2.3 Mass Wasting

Talus from rockfalls and rockslides covers steep slopes of desert mountain ranges throughout the southwestern USA (Melton, 1965; Parsons et al., 2009). Urban expansion in arid regions globally (Cooke et al., 1982) continues to thrust infrastructure at the base of steep desert slopes. This is certainly the case in the Phoenix area (Dorn, 2014; Harris and Pearthree, 2002), where the wealthy build homes right on the margins of mountain preserves (Ewan et al., 2004) and often beneath steep bare rockfaces (Fig. 10.12). Chronometric studies of rockfall in the Phoenix area reveal that rockfalls occurred throughout the Holocene (Dorn, 2014) and historically in the Anthropocene.

Debris flows are one of the most hazardous landslide types in any region with steep terrain and precipitation, including the Phoenix area (Dorn, 2012, 2016). Debris flows occur when slopes fail to maintain the equilibrium between gravitational drivings and frictional resisting forces (Iverson, 2005). Thus, they typically occur on steep-slope areas between 20 and 45° after prolonged or particularly intense wetting events (Jakob and Hungr, 2005); in the case of the Phoenix area, from thunderstorms, soaking hurricane moisture, or a series of winter frontal storms (Dorn, 2016).

FIGURE 10.11 **An excavation exposed the mixed fluvial and aeolian deposit located between the Salt and the Gila Rivers in metropolitan Phoenix.** Some sediments are clearly fluvial gravels, while sandy units show evidence of both fluvial and aeolian transport. Carbonate cementation could be related to groundwater processes or pedogenic processes. 4-m tall Paloverde trees provide scale. The uppermost 1.5 m consists of rock and sand from construction activities.

Debris flows have three major zones, including initiation, transportation, and deposition (Fig. 10.13) (Jakob and Hungr, 2005), and geomorphic features can be identified in each zone. Distinct head scarps indicate the initiation zone, where slope failures start. Once initiated, chutes develop along the debris-flow channel, and debris-flow materials are transported down slope. Finally, debris flows produce levees and alluvial fans at the mouths of drainages (Fig. 10.13A and B) (Webb et al., 2008; Youberg et al., 2008). In the case of most debris-flow contexts in Phoenix, the chutes are only a few hundred meters long and the alluvial fans that result exist only at the base of the slopes, as illustrated in Fig. 10.13.

Rockslides are another type of mass wasting that involves the displacement of rock materials along a sliding plane, such as a bedding plane, and the interface between two different rock types. Granitic rocks experience sheeting and produce pressure-release shells once the overlying materials have been removed (Bahat et al., 1999). The Phoenix neighborhood of Awhatukee illustrates three different types of mass-wasting events associated with pressure release shells: debris flows (Fig. 10.14B), rockfalls (Fig. 10.14C), and rockslides (Fig. 10.14D). What may be surprising is that the homeowners at the base of these steeply dipping joint faces have little to no understanding of the potential hazard just meters from their homes.

10.2.4 Pedimentation

The Phoenix metropolitan area hosts iconic pediments with a variety of rock types that makes this area the rival of other well-known pediment sites on Earth (Fig. 10.15). Pediments with the classic sharp piedmont angle exist on Forest Service and city preserve lands, allowing

FIGURE 10.12 Historic rockfalls place wealthy homes in potential danger in such areas as Camelback Mountain (A) and the Phoenix Mountains (B).

for study of the entire inselberg-pediment landscape (Fig. 10.15A, B, D, and H). The largest expanse of pediments, however, rests under urban sprawl (Fig. 10.15C, F, and G).

The pediment literature maintains an extensive bias toward granitic study sites (Dohrenwend and Parsons, 2009), including central Arizona (Kesel, 1977; Pelletier, 2010). However, the Phoenix area contains pediments in four broad rock types (Larson et al., 2016): granitic (Fig. 10.15A, B, and D); foliated metamorphic (Fig. 10.15C, G, and H); sedimentary

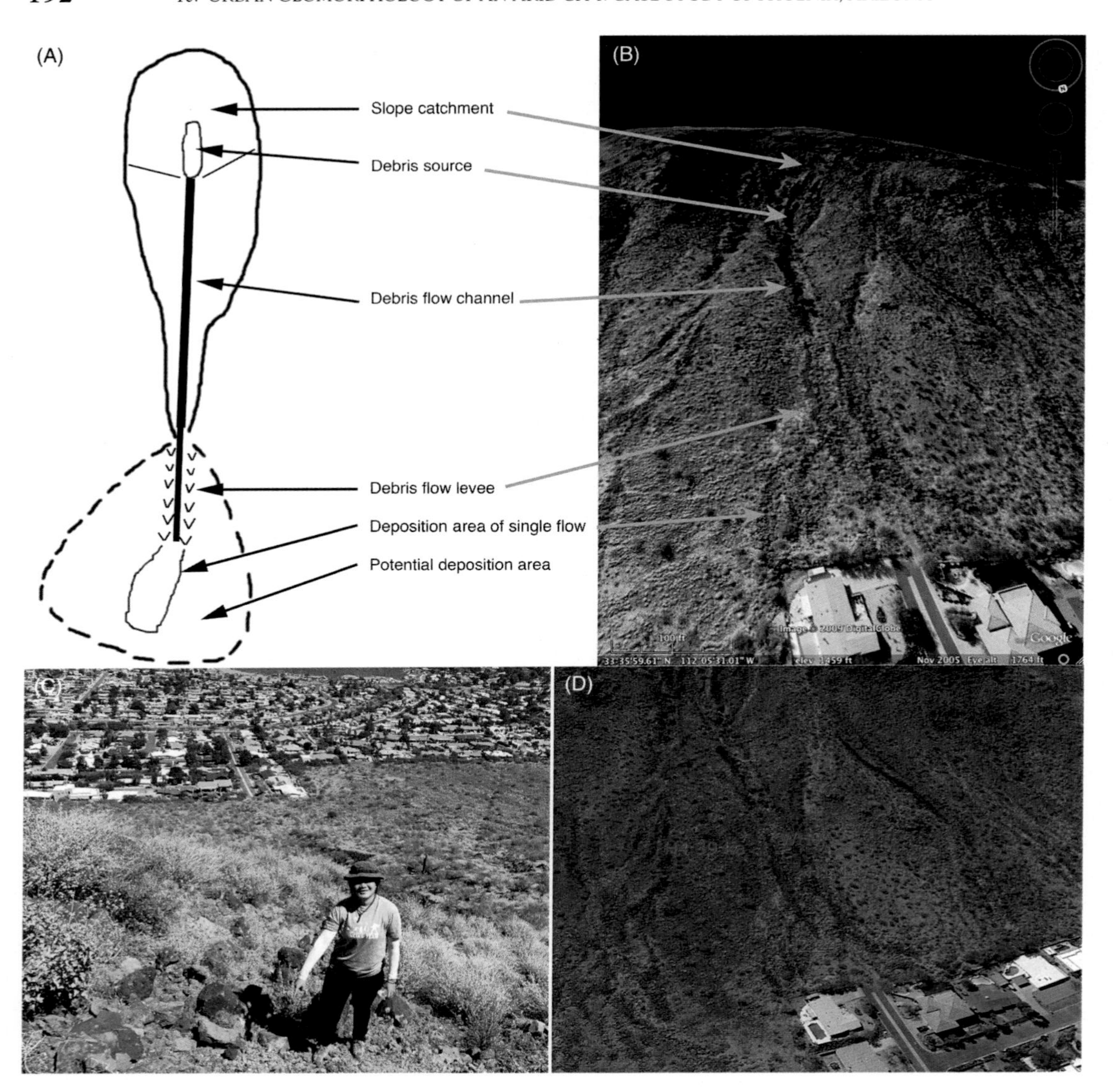

FIGURE 10.13 Shaw Butte in the Phoenix Mountains illustrates how debris flows interface with urbanization. (A) and (B) illustrate the debris-flow system, where small catchments generate debris flows that move down chutes a few hundred meters long, resulting in debris-flow deposition near the mountain front. (C) illustrates the source region of the most recent Little Ice Age flow that occurred about 0.65 ka. (D) identifies debris-flow deposits about 8.1, 16.5, 24, and 39 ka. However, an unknown number of other debris flows occurred, with evidence destroyed by subsequent events. *Source: The images in B and D are used following permission guidelines for Google Earth (http://www.google.com/permissions/geoguidelines.html).*

FIGURE 10.14 (A) is from the perspective of looking at the mountain slope from the resident's back fence, where (B) debris flow levees, (C) talus from rockfalls, and (D) a rockslide all represent hazards above a suburban neighborhood. Note the massive rock in (D) that is situated on the pressure-release joint of granitic surface with little to no support to inhibit the next mass-wasting event.

breccia with extensive sandy facies (Fig. 10.15E and F); and ignimbrite. Given this mixed lithology, explanations of the pediment form requiring differential decay of granitic rocks and fossilized landscapes (Oberlander, 1989) do not work, since similar forms exist side-by-side in rock types other than granite. Furthermore, given the central Arizona evidence that pediment forms are able to adjust to base-level change in the timeframe of the last glacial cycle (Larson et al., 2016), also removes the need for complicated explanations of form requiring two-stage etching (Twidale, 2002).

Our view of pedimentation as a process in the Phoenix area returns to early German geomorphological thinking (Penck, 1924); G.K. Gilbert's classic observations (Gilbert, 1877); and more modern process-geomorphic interpretations (Applegarth, 2004; Larson et al., 2016; Parsons and Abrahams, 1984). Pediments function as transport surfaces (conveyor belts) of materials detached and eroded from small mountain masses. Pediments form where drainage

FIGURE 10.15 Pediment-inselberg landscapes of metropolitan Pheonix, illustrating planar pediments in front of small inselberg ranges. (A) granitic eastern McDowell Mountains; (B) granitic northern Usery Mountains; (C) foliated metamorphic Mummy Mountain; (D) granitic San Tan Mountains; (E) breccia Red Mountain; (F) breccia Camelback Mountains; (G) massive metamorphic North Mountain; and (H) foliated metamorphic Phoenix Mountains.

areas are too small to develop alluvial fans. The classic piedmont angle, which is seen as a fairly dramatic slope break, results from the greater resistance to detachment and transport of larger slope colluvial particles and bedrock that leads to the generation of steeper slopes. The slopes of once-graded pediments led to closed basins throughout the late Miocene and Pliocene, but pediments have been experiencing ongoing adjustment to fluctuating base level throughout the Quaternary (Larson et al., 2014)

10.3 DESERT GEOMORPHIC HAZARDS

10.3.1 Alluvial Fan Flooding

The basic ephemeral channel morphologies of the Sonoran Desert (Sutfin et al., 2014) include bedrock channels in the upper interior of the drainage that transitions to bedrock mixed with alluvium and ultimately to incised alluvium at an embayment that merges into an alluvial-fan piedmont. For much of the Sonoran Desert, including Phoenix, the alluvial-fan piedmont is incised. The risk for flooding only comes where the channel emerges from the incised area and is then able to experience an avulsion (Fuller, 2012). Such alluvial-fan avulsions only occur below the hydrological apex and not on older abandoned alluvial-fan surfaces.

The Federal Emergency Management Agency (FEMA) uses a procedure for delineating flood-hazard zones on alluvial fans, with fiscal implications for those building on surfaces so delineated on flood insurance-rate maps based on the FEMA approach. Put simplistically, FEMA treats as potentially hazardous all surfaces beneath the topographic apex of a fan. However, vast tracks of land in the 100-year flood-hazard zone are not truly flood prone if they exist above the hydrological apex. Such a condition occurs when there is a "fan-head trench" that delivers water and sediment in a naturally incised water conduit toward the toe of the fan form. Prior research indicates that the FEMA approach simply does not work in places like Laughlin, Nevada (House, 2005); Tucson, Arizona (Pearthree et al., 1992); and certainly not in the Phoenix area (Fuller, 1990). Fig. 10.16 illustrates the offset between the FEMA procedure and reality in the community of Scottsdale, Arizona.

10.3.2 Street Flooding in Planned and Unplanned Housing Developments

Street flooding occurs throughout the Phoenix metropolitan area, typically during the summer monsoon season, when short, but intense, downbursts result in localized overland flow. The local Maricopa County Flood Control District receives property tax funding to route water efficiently through the metropolitan area, working with local municipalities (Fig. 10.17). The county and cities can take different cost-benefit strategies to dealing with street flash flooding. Over-engineering has often been done by the county. However, local communities have made other choices at times.

The homes, infrastructure, and retail space of the Phoenix suburb of Fountain Hills rests on an eroding alluvial fan landform known as a "ballena" (Fig. 10.18). However, the roads cross a series of incised ephemeral washes that experience frequent flooding (Rhoads, 1986). The management challenge rests in the cost-benefit trade-off of engineering for decadal events,

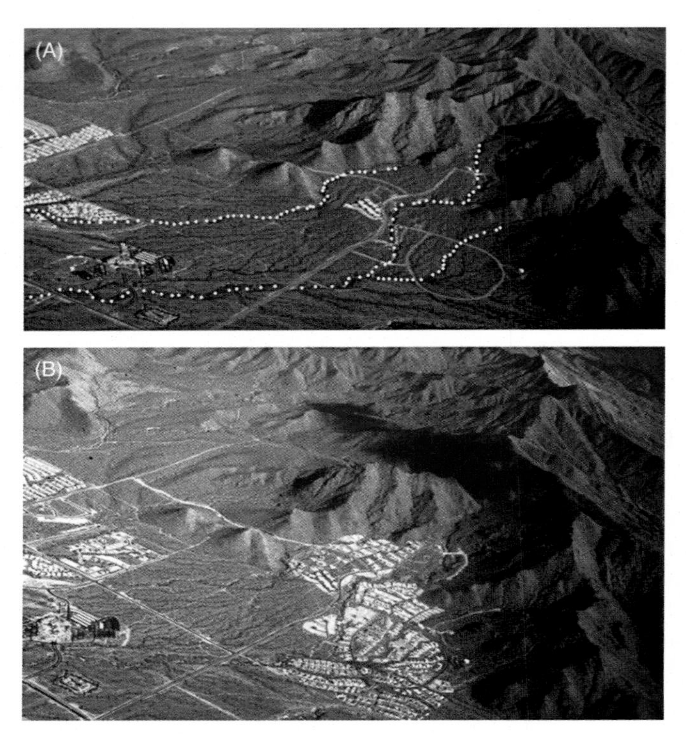

FIGURE 10.16 Development burgeoned on alluvial-fan surfaces during the 1990s, as shown in a comparison of 1991 (A) and 1995 (B) aerial photographs of the southern McDowell Mountains, Scottsdale. In the 1991 image, *dots* delineate the presence of entrenched channels transporting water and sediment almost 8 km downstream from this fan. Thus, with the hydrological apex located at distance from the topographic apex of the alluvial fan, all of the development is safe from fan-related flooding.

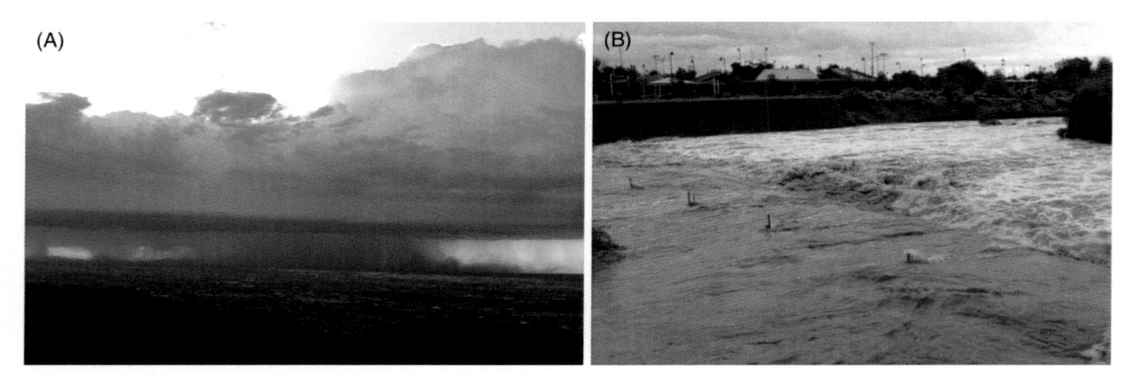

FIGURE 10.17 Monsoon downburst taking place over the western portion of metropolitan Phoenix (A) and corresponding localized flooding being routed through flood-control structures (B).

and then let century or millennial-scale flash flooding require infrastructure replacement. This leads to ongoing construction at problem locations, where initial engineering structures have repeatedly failed.

Much of Phoenix has been built on pediments with low slopes (Fig. 10.19). The engineering associated with this development ranges considerably in terms of the investment to deal

FIGURE 10.18 **Fountain Hills is a community built on a ballena, or eroding alluvial fan. Although structures are safe because homes and businesses are placed on ballena tops and side slopes, roads must be engineered to survive occasional flooding in the washes between the ridges.**

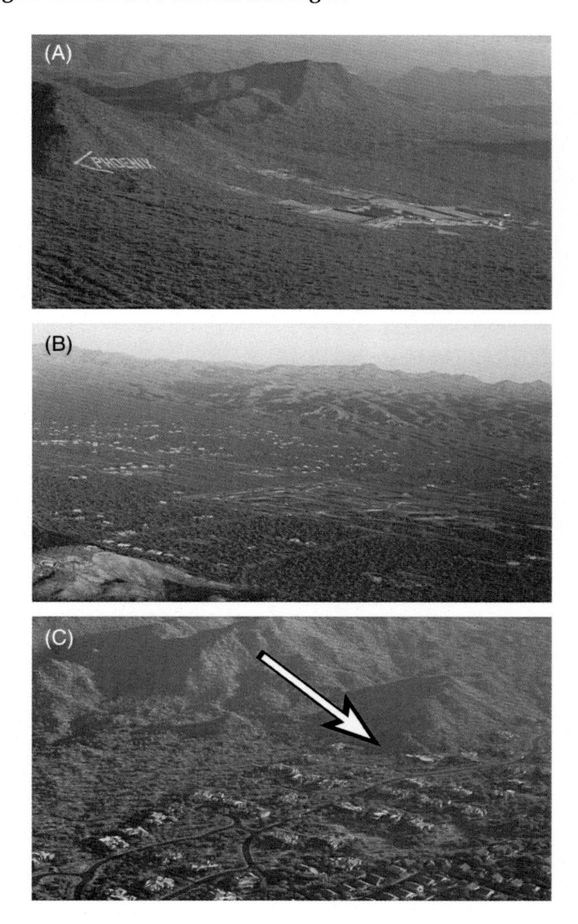

FIGURE 10.19 **Development on pediments takes different strategies in dealing with ephemeral flooding.** (A) A gun club simply built large levees to divert flow into the surrounding desert. (B) A "wildcat" development called Rio Verde continues to experience localized flooding, since little or no effort focuses on water routing during home or road construction. (C) Affluent subdivisions do consider flash flooding issues and route water into natural or human-enhanced washes. The arrow identifies the classic sudden break in slope between inselberg and pediment.

with ephemeral flooding. In Fig. 10.19A, a gun club has built simple berms to deflect flooding around the complex. In contrast, Fig. 10.19B displays a wildcat community known as "Rio Verde." This development is almost entirely unplanned with respect to dealing with runoff. An individual purchasing a plot of land will build, quite often, without any concern for the flooding issues caused by upstream neighbors or that they may cause for downstream property owners. More wealthy communities, such as the exclusive Las Sendas neighborhood of Mesa (Fig. 10.19C), include structures to deal with the routing of water.

The largest river running through metropolitan Phoenix has ceased to pose a flooding hazard. Throughout the 20th century, the Salt River flooded repeatedly, causing considerable losses to property and sometimes lives (Gober, 2005). However, the last time that the Salt River experienced destruction associated with flooding took place in 1993 during a major El Nino Southern Oscillation (ENSO) event (Fig. 10.20C) that corresponded with construction at Roosevelt Dam (Fig. 10.20A), requiring the release of water from the reservoir behind. The combination of a

FIGURE 10.20 During the winter of 1993, a major ENSO event led to (A) the release of water from the reservoir behind the Roosevelt Dam. The released water destroyed the bridge at Mill Avenue in Tempe (B) as well as other infrastructure along the course of the river through metropolitan Phoenix. *Source: (A) and (C) are copyright free and made available courtesy of the Bureau of Reclamation (A) and NASA (B).*

very wet winter and a lowered dam level destroyed bridges and a lot of other infrastructure (Fig. 10.20B). However, since the Roosevelt Dam's 1993 construction resulted in increased reservoir capacity, flooding has not been an issue since along this major drainage.

10.3.3 Debris Flows

The Phoenix urban area has expanded out into the surrounded mountain fronts, where debris flows take place. In an initial study of the hazard to homes posed by debris flows, Dorn (2012) found that at least 89 houses are located along the pathway of former debris flows or above the debris flow chutes of the Gila Range and the Ma Ha Tuak Range of South Mountain, Camelback Mountain, Mummy Mountain, and Shaw Butte areas alone (Fig. 10.13).

Debris flows were not generally viewed as a hazard in the Sonoran Desert, until a major debris flow event occurred outside of Tucson (Youberg et al., 2008). Despite evidence to the contrary, those geoscientists living in and around Phoenix generally considered debris flows "acts of God," or extraordinarily rare geological events not worthy of study. This changed after an intense summer thunderstorm on August 12, 2014, and a hurricane that occurred on September 8, 2014, led to short and intensive precipitation events in metropolitan Phoenix, which triggered the occurrence of dozens of debris flows in one mountain range of Phoenix alone (Fig. 10.21) (Dorn, 2016).

10.3.4 Haboobs and Dust Storms

Dust and summer dust storms are part of the urban geomorphic and climate system (Brazel, 1989; Marcus and Brazel, 1992; Péwé et al., 1981). During the spring, strong dry cold fronts deflate dust from agricultural fields and abandon urban lots, producing a substantial dust hazard. During the months of July, August, and September, the Mexican monsoon's northern boundary impacts the Phoenix area and produces haboobs (Idso et al., 1972), like the one seen in Fig. 10.22, which are associated with the leading edge of cold outflow from

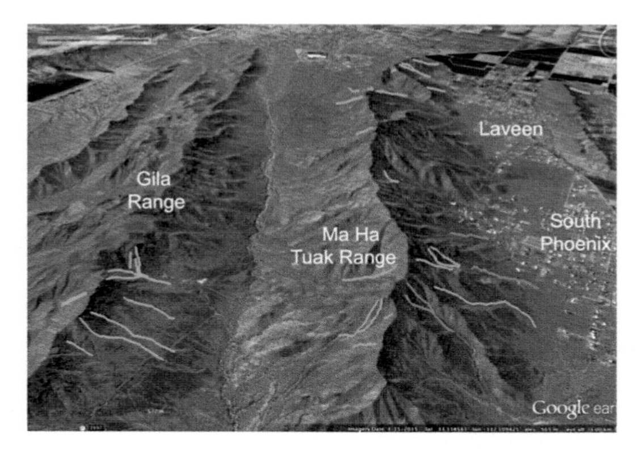

FIGURE 10.21 West-looking view of the debris-flow pathways triggered by the August 12th hurricane thunderstorm (blue) and September 8th summer monsoon event (green) at South Mountain, Phoenix in 2014 (Dorn, 2016). *Source: The base image is used following permission guidelines for Google Earth (http://www.google.com/permissions/geoguidelines.html).*

FIGURE 10.22 **A km-high, nearly 100-km wide haboob approaches Phoenix from the south at sunset.**

convective clouds. Although this is a natural phenomenon, anthropogenic activities that led to the exposure of bare ground (e.g., exposed house pads, agricultural fields, desertification) all contribute to the available surface area for dust deflation (Eagar et al., 2017).

Dust poses a regular urban hazard in terms of driving, where visibility decreases to the point where a driver is unable to see more than a few meters ahead (Baddock et al., 2013; Hyeres and Marcus, 1981). Desert dust is also associated with a number of human health issues (Goudie, 2014). Valley fever, for example, is produced by the fungus Coccidiodes that lives in soil and dust in the southwestern USA, where the Centers for Disease Control and Prevention indicated Arizona had more than 5000 reported cases each year between 2009 and 2015 (https://www.cdc.gov/fungal/diseases/coccidioidomycosis/statistics.html).

10.4 SUMMARY PERSPECTIVE ON HUMAN INFLUENCES ON THE ARID GEOMORPHIC SYSTEM IN THE URBANIZING SONORAN DESERT

From a geomorphic perspective, the Anthropocene, or the proposed new geological epoch when humans have had an overwhelming effect on the Earth system (Waters et al., 2016), requires both empirical evidence and an understanding of exactly how humans alter geomorphic processes. Accordingly, the British Society for Geomorphology maintains a Fixed Term Working Group to advise how geomorphologists should engage in scholarly analysis concerning the Anthropocene as a concept. Practical aspects include a relative magnitude problem, a boundary problem, and a spatial problem associated with "anthropogenic geomorphology" (Brown et al., 2017).

In the context of urban geomorphology (Thornbush, 2015), where human impacts result in enhanced disturbance and increased vulnerability to erosion, an arid city poses very different considerations than urban centers in wetter regions. Urban geomorphic processes in a setting like the Sonoran Desert are potentially altered by a myriad of anthropogenic influences, including: invasive species turning an ecoregion that did not naturally experience massive wildfires into an annual hazard due to invasive annual grass species; altering the armoring effects of soil crusting by widespread destruction of BSCs by periods of cattle grazing; periods of road building and home construction; and other influences, such as off-road vehicles.

An individual walking on desert landforms, before massive land-use change associated with cattle grazing and urban expansion, likely would have experienced very different surface conditions than found by the average hiker today. Extensive areas once hosted desert pavements, BSCs (Allen, 2005, 2010), and interlocking colluvium on steeper slopes that provided a net-armoring effect (Bowker et al., 2008; Granger et al., 2001; Seong et al., 2016a). Today, only patches of such armored surfaces remain, providing glimpses into the original land surfaces.

According to Brown et al. (2017), "it is clear that the relevance of the Anthropocene concept varies substantially between different branches of geomorphology", (p. 71). While Brown et al. (2017) did not consider rocky desert landscapes, such as the Phoenix metropolitan area, the basic conclusion that "the less obvious effects of humans on the geomorphic systems warrant increased research", (p. 85) certainly applies to the Phoenix metropolitan area and the surrounding Sonoran Desert. Developing a better understanding of the role of human-influenced processes at different scales will be needed to better diagnose the role of human impacts in an arid geomorphic system.

References

Abrahams, A.D., Parsons, A.J., Hirsch, P.J., 1985. Hillslope gradient-particle size relations: evidence of the formation of debris slopes by hydraulic processes in the Mojave Desert. J. Geol. 93 (3), 347–357.

Adelsberger, K.A., Smith, J.R., McPherron, S.P., Dibble, H.L., Olszewski, D.I., Schurmans, U.A., Chiotti, L., 2013. Desert pavement disturbance and artifact taphonomy: a case study from the Eastern Libyan Plateau Egypt. Geoarchaeol 28 (2), 112–130.

Allen, C., Swetnam, T., Betancourt, J., 1998. Landscape changes in the southwestern United States: techniques, long-term data sets, and trends. Sisk, T.D. (Ed.), Perspectives on the Land-Use History of North America: A Context for Understanding Our Changing Environment, 104, U.S. Geological Survey, Fort Collins, CO, USA.

Allen, C.D., 2005. Micrometeorology of a smooth and rugose biological soil crust near Coon Bluff Arizona. J. Ariz.-Nev. Acad. Sci. 38 (1), 21–28.

Allen, C.D., 2010. Biogeomorphology and biological soil crusts: a symbiotic research relationship. Geomorphologie 16 (4), 347–358.

Applegarth, M.T., 2004. Assessing the influence of mountain slope morphology on pediment form South-Central Arizona. Phys. Geogr. 25 (3), 225–236.

Baddock, M.C., Strong, C.L., Murray, P.S., McTainsh, G.H., 2013. Aeolian dust as a transport hazard. Atmospheric Environ. 71, 7–14.

Bahat, D., Grossenbacher, K., Karasaki, K., 1999. Mechanism of exfoliation joint formation in granitic rocks, Yosemite National Park. J. Struct. Geol. 21 (1), 85–96.

Belnap, J., 2003. The world at your feet: desert biological soil crusts. Front. Ecol. Environ. 1 (4), 81–189.

Belnap, J., Gillette, D.A., 1998. Vulnerability of desert biological soil crusts to wind erosion: the influences of crust development, soil texture, and disturbance. J. Arid. Environ. 39 (2), 133–142.

Belnap, J., Büdel, B., Lange, O.L., 2001. Biological soil crusts: characteristics and distribution. Ecol. Stud. 150, 3–31.

Bowker, M.A., Belnap, J., Chaudhary, V.B., Johnson, N.C., 2008. Revisiting classic water erosion models in drylands: the strong impact of biological soil crusts. Soil Biol. Biochem. 40 (9), 2309–2316.

Brazel, A.J., 1989. Dust and Climate in the American Southwest. Kluwer Academic Publishers, Dordrecht.

Brown, A.G., Tooth, S., Bullard, J.E., Thomas, D.S., Chiverrell, R.C., Plater, A.J., Murton, J., Thorndycraft, V.R., Tarolli, P., Rose, J., Wainwright, J., 2017. The geomorphology of the Anthropocene: emergence, status and implications. Earth Surf. Process. Land. 42 (1), 71–90.

Bullard, J.E., Livingstone, I., 2002. Interactions between aeolian and fluvial systems in dryland environments. Area 34 (1), 8015.

Coney, P.J., Harms, T.A., 1984. Cordilleran metamorphic core complexes: Cenozoic extensional relics of Mesozoic compression. Geology 12 (9), 550–554.

Cooke, R.U., Brunsden, D., Doornkamp, J.C., Jones, D., 1982. Urban Geomorphology in Drylands. Oxford University Press, Oxford.

Dietze, M., Dietze, E., Lomax, J., Fuchs, M., Kleber, A., Wells, S.G., 2016. Environmental history recorded in aeolian deposits under stone pavements, Mojave Desert, USA. Quat. Res. 85 (1), 4–16.

Dohrenwend, J.C., Parsons, A.J., 2009. Pediments in arid environments. In: Parsons, A.J., Abrahams, A.D. (Eds.), Geomorphology of Desert Environments. Springer, New York, pp. 377–411.

Dorn, R.I., 2011. Revisiting dirt cracking as a physical weathering process in warm deserts. Geomorphology 135 (1), 129–142.

Dorn, R.I., 2012. Do debris flows pose a hazard to mountain-front property in metropolitan Phoenix Arizona? Prof. Geogr. 64 (2), 197–210.

Dorn, R.I., 2014. Chronology of rock falls and slides in a desert mountain range: case study from the Sonoran Desert in south-central Arizona. Geomorphology 223, 81–89.

Dorn, R.I., 2016. Identification of debris-flow hazards in warm deserts through analyzing past occurrences: case study in South Mountain, Sonoran Desert, USA. Geomorphology 273, 269–279.

Dorn, R.I., Oberlander, T.M., 1982. Rock varnish. Prog. Phys. Geogr. 6 (3), 317–367.

Dorn, R.I., Gordon, M., Pagán, E.O., Bostwick, T.W., King, M., Ostapuk, P., 2012. Assessing early Spanish explorer routes through authentication of rock inscriptions. Prof. Geogr. 64 (3), 415–429.

Eagar, J.D., Herckes, P., Hartnett, H.E., 2017. The characterization of haboobs and the deposition of dust in Tempe, AZ from 2005 to 2014. Aeolian Res. 24, 81–91.

Eamer, J.B.R., Shugar, D., Walker, I.J., Lian, O., Neudorf, C., 2017. Distinguishing depositional setting for sandy deposits in coastal landscapes using grain shape. J. Sediment. Res. 87, 1–11.

Elvidge, C.D., Moore, C.B., 1980. Restoration of petroglyphs with artificial desert varnish. Stud. Conserv. 25 (3), 108–117.

Ewan, J., Ewan, R.F., Burke, J., 2004. Building ecology into the planning continuum: case study of desert land preservation in Phoenix Arizona (USA). Landsc. Urban Plan. 68 (1), 53–75.

Faist, A.M., Herrick, J.E., Belnap, J., Van Zee, J.W., Barger, N.N., 2017. Biological soil crust and disturbance controls on surface hydrology in a semi-arid ecosystem. Ecosphere 8 (3), e01691.10.1002/ecs2.1691.

Fuller, J.E., 1990. Misapplication of the FEMA alluvial fan model: a case history. In: French, R.H. (Ed.), Hydraulics/Hydrology of Arid Lands (H^2AL). American Society Civil Engineers, New York, pp. 367–372.

Fuller, J.E., 2012. Evaluation of avulsion potential on active alluvial fans in central and western Arizona. Arizona Geological Survey Contributed Report CR-12-D, 1–88.

Gilbert, G.K., 1877. Geology of the Henry Mountains. Geological and Geographical Survey, Washington, DC, U.S.

Gober, P., 2005. Metropolitan Phoenix: Place Making and Community Building in the Desert. University of Pennsylvania Press, Philadelphia.

Goudie, A.S., 2014. Desert dust and human health disorders. Environ. Int. 63, 101–113.

Graf, W.L., 1988. Fluvial Processes in Dryland Rivers. Springer-Verlag.

Granger, D.E., Riebe, C.S., Kirchner, J.W., Finkel, R.C., 2001. Modulation of erosion on steep granitic slopes by boulder armoring, as revealed by cosmogenic Al-26 and Be-10. Earth Planet. Sci. Lett. 186 (2), 269–281.

Hall, K., Thorn, C.E., Sumner, A., 2012. On the persistence of "weathering." Geomorphology 149–150, 1–10.

Harris, R.C., Pearthree, P.A., 2002. A home buyer's guide to geological hazards in Arizona. Arizona Geological Survey Down-To-Earth, 13, pp. 1–36.

Holt, W.E., Chase, C.G., Wallace, T.C., 1986. Crustal structure from 3-dimensional gravity modeling of a metamorphic core complex: a model for uplift, Santa Catalina-Rincon Mountains. Ariz. Geol. 14 (11), 927–930.

House, P.K., 2005. Using geology to improve flood hazard management on alluvial fans: an example from Laughlin Nevada. J. Am. Water Res. Assoc. 41 (6), 1431–1447.

Howard, A.D., Selby, M.J., 2009. Rock slopes. In: Parsons, A.J., Abrahams, A.D. (Eds.), Geomorphology of Desert Environments. Springer, New York, pp. 189–232.

Hyeres, A.D., Marcus, M., 1981. Land use and desert dust hazards in central Arizona. Geol. Soc. Spec. Pap. 186, 267–280.

Idso, S.B., Ingram, R.S., Pritchard, J.M., 1972. An American haboob. Bull. Am. Meteorol. Soc. 53 (10), 930–935.

Iverson, R.M., 2005. Debris-flow mechanics debris-flow hazards and related phenomena. In: Savage, W., Baum, R. (Eds.), Instability of Steep Slopes. Debris-Flow Hazards and Related Phenomena. Springer, Berlin, pp. 53–79.

Jakob, M., Hungr, O., 2005. Debris-Flow Hazards and Related Phenomena. Springer, Berlin.
Kesel, R.H., 1977. Some aspects of the geomorphology of inselbergs in central Arizona, USA. Z. Geomorph. 21, 119–146.
Larson, P.H., Dorn, R.I., Palmer, R.E., Bowles, Z., Harrison, E., Kelley, S., Schmeeckle, M.W., Douglass, J., 2014. Pediment response to drainage basin evolution in south-central Arizona. Phys. Geogr. 35 (5), 369–389.
Larson, P.H., Kelley, S.B., Dorn, R.I., Seong, Y.B., 2016. Pediment development and the pace of landscape change in the northeastern Sonoran Desert, United States. Ann. Assoc. Am. Geogr. 106, 1195–1216.
Liu, T. VML Dating Lab. http://www.vmldating.com/, 2017.
Liu, T., Broecker, W.S., 2007. Holocene rock varnish microstratigraphy and its chronometric application in drylands of western USA. Geomorphology 84 (1), 1–21.
Liu, T., Broecker, W.S., 2008. Rock varnish microlamination dating of late Quaternary geomorphic features in the drylands of the western USA. Geomorphology 93 (3), 501–523.
Liu, T., Broecker, W.S., 2013. Millennial-scale varnish microlamination dating of late Pleistocene geomorphic features in the drylands of western USA. Geomorphology 187, 38–60.
Lozano-García, M.S., Ortega-Guerrero, B., Sosa-Nájera, S., 2002. Mid-to late-Wisconsin pollen record of San Felipe Basin Baja California. Quat. Res. 58 (1), 84–92, 9.
Marcus, M.G., Brazel, A.J., 1992. Summer dust storms in the Arizona Desert. In: Janelle, D.G. (Ed.), Geographical Snapshots of North America. Guilford Press, New York, pp. 411–415.
McAuliffe, J.R., Van Devender, T.R., 1998. A 22,000-year record of vegetation change in the north-central Sonoran Desert. Palaeogeogr. Palaeoclimatol. Palaeoecol. 141 (3), 253–275.
Melton, M.A., 1965. Debris-covered hillslopes of the southern Arizona desert: consideration of their stability and sediment contribution. J. Geol. 73 (5), 715–729.
Michael, D.R., Cunningham, R.B., Lindenmayer, D.B., 2008. A forgotten habitat? Granite inselbergs conserve reptile diversity in fragmented agricultural landscapes. J. Appl. Ecol. 45 (6), 742–752.
Molnar, P., Anderson, R.S., Andersson, S.P., 2007. Tectonics, fracturing of rock, and erosion. J. Geophys. Res. Earth 112 (F3)doi: 10.1029/2005JF000433.
Nagy, M.L., Pérez, A., Garcia-Pichel, F., 2005. The prokaryotic diversity of biological soil crusts in the Sonoran Desert (Organ Pipe Cactus National Monument AZ). FEMS Microbiol. Ecol. 54 (2), 233–245.
Nations, D., Stump, E., 1981. Geology of Arizona. Kendall Hunt, Dubuque.
Oberlander, T.M., 1989. Slope and pediment systems. In: Thomas, D.S.G. (Ed.), Arid Zone Geomorphology. Belhaven Press, London, pp. 56–84.
Ollier, C.D., 1965. Dirt cracking: a type of insolation weathering. Aust. J. Sci. 27 (8), 236–237.
Page, K.J., Dare-Edwards, A.J., Owens, J.W., Frazier, P.S., Kellett, J., Price, D.M., 2001. TL chronology and stratigraphy of riverine source bordering sand dunes near Wagga Wagga, New South Wales. Austr. Quat. Int. 83, 187–193.
Parsons, A.J., Abrahams, A.D., 1984. Mountain mass denudation and piedmont formation in the Mojave and Sonoran Deserts. Am. J. Sci. 284 (3), 255–271.
Parsons, A.J., Abrahams, A.D., Howard, A.D., 2009. Rock-mantled slopes. In: Parsons, A.J., Abrahams, A.D. (Eds.), Geomorphology of Desert Environments. Springer, New York, pp. 233–263.
Pearthree, P.A., Demsey, K.A., Onken, J.A., Vincent, K.R., House, P.K., 1992. Geomorphic assessment of the flood-prone areas on the southern piedmont of the Tortilita Mountains, Pima County, Arizona. Arizona Geological Survey Open File Report 91-11, pp. 1–32.
Pelletier, J.D., 2010. How do pediments form? A numerical modeling investigation with comparison to pediments in southern Arizona USA. Bull. Geol. Soc. Am. 122 (11–12), 1815–1829.
Penck, W., 1924. Die Morphologische Analyse: Ein Kapital der physikalischen Geologie. Engelhorn, Stuttgart.
Peterson, F.F., Bell, J.W., Dorn, R.I., Ramelli, A.R., Ku, T.L., 1995. Late Quaternary geomorphology and soils in Crater Flat, Yucca Mountain area, southern Nevada. Geol. Soc. Am. Bull. 107 (4), 379–395.
Péwé, T.L., Péwé, R.H., Journaux, A., Slatt, R.M., 1981. Desert dust: characteristics and rates of deposition in central Arizona. Geol. Soc. Spec. Pap. 186, 169–190.
Reynolds, S.J., 1985. Geology of the South Mountains, Central Arizona. Arizona Bureau of Geology and Mineral Technology Bulletin, 195.
Rhoads, B.L., 1986. Flood hazard assessment for land-use planning near desert mountains. Environ. Manage. 10 (1), 97–106.
Richards, S.M., Reynolds, S.J., Spencer, J.E. Pearthree, P.A., 2000. Geologic Map of Arizona. Arizona Geological Survey Map M-35, 1 sheet, scale 1:1,000,000.

Seong, Y.B., Dorn, R.I., Yu, B.Y., 2016a. Evaluating the life expectancy of a desert pavement. Earth Sci. Rev. 162, 129–154.

Seong, Y.B., Larson, P.H., Dorn, R.I., Yu, B.Y., 2016b. Evaluating process domains in small granitic watersheds: case study of Pima Wash, South Mountains, Sonoran Desert USA. Geomorphology 255, 108–124.

Spencer, J.E., 1984. The role of tectonic denudation in warping and uplift of low-angle normal faults. Geology 12 (2), 95–98.

Stuckless, J.S., Sheridan, M.F., 1971. Tertiary volcanic stratigraphy in the Goldfield and Superstition Mountains. Ariz. Geol. Soc. Am. Bull. 82 (11), 3235–3240.

Sutfin, N.A., Shaw, J.R., Wohl, E.E., Cooper, D.J., 2014. A geomorphic classification of ephemeral channels in a mountainous, arid region, southwestern Arizona USA. Geomorphology 221, 164–175.

Thornbush, M., 2015. Geography, urban geomorphology and sustainability. Area 47 (4), 350–353.

Twidale, C.R., 1981. Granitic inselbergs: domed, block-strewn and castellated. Geogr. J. 147, 54–71.

Twidale, C.R., 1982. Granite Landforms. Elsevier, Amsterdam.

Twidale, C.R., 2002. The two-stage concept of landform and landscape development involving etching: origin, development and implications of an idea. Earth Sci. Rev. 57 (1), 37–74.

Van Devender, T.R., 1990. Late Quaternary vegetation and climate of the Sonoran Desert, United States and Mexico. In: Betancourt, J.L., Van Devender, T.R., Martin, P.S. (Eds.), Packrat Middens: The Last 40,000 Years of Biotic Change. University of Arizona Press, Tucson, pp. 134–165165.

Viles, H.A., 2008. Understanding dryland landscape dynamics: do biological crusts hold the key? Geogr. Compass 2 (3), 899–919.

Viles, H.A., 2013. Linking weathering and rock slope instability: non-linear perspectives. Earth Surf. Proc. Landf. 38 (1), 62–70.

Waters, C.N., Zalasiewicz, J., Summerhayes, C., Barnosky, A.D., Poirier, C., Gałuszka, A., Cearreta, A., Edgeworth, M., Ellis, E.C., Ellis, M., Jeandel, C., 2016. The Anthropocene is functionally and stratigraphically distinct from the Holocene. Science 351 (6269), aad2622_1–aad_1622.

Webb, R.H., Magirl, C.S., Griffiths, P.G., Boyer, D.E., 2008. Debris flows and floods in southeastern Arizona from extreme precipitation in late July 2006 Magnitude, frequency, and sediment delivery. U.S. Geological Survey Open-File Report 2008-1274, 1–95.

Youberg, A., Cline, M.L., Cook, J.P., Pearthree, P.A., Webb, R.H., 2008. Geological mapping of debris-flow deposits in the Santa Catalina Mountains, Pima County Arizona. Arizona Geological Survey Open File Report 08-06, 1–47.

Further Reading

McKenna-Neuman, C., Maxwell, C., 2002. Temporal aspects of the abrasion of microphytic crusts under grain impact. Earth Surf. Proc. Land. 27, 891–908.

CHAPTER

11

Bivouacs of the Anthropocene: Urbanization, Landforms, and Hazards in Mountainous Regions

Kevin Gamache, John R. Giardino, Panshu Zhao, Rebecca Harper Owens

Texas A&M University, College Station, TX, United States

OUTLINE

11.1 Introduction 206

11.2 Study Area 207

11.3 The New Awareness of the Critical Zone 208

11.4 Mining Town Development 210

11.4.1 Uncompahgre Mining District 211

11.4.2 Telluride (Upper San Miguel) Mining District 212

11.5 Geomorphic Processes 212

11.5.1 Glaciation 212

11.5.2 Periglacial 213

11.5.3 Fluvial 213

11.5.4 Mass Movement 214

11.5.5 Anthropogenic Building and Modifications of Landforms 217

11.6 Location, Location, Location: A Planner's Dream 221

11.7 Predicting Urban Suitability in the San Juan Mountains 223

11.8 Results 226

11.9 Conclusion 226

References 228

Urban Geomorphology. http://dx.doi.org/10.1016/B978-0-12-811951-8.00011-4

Mountain towns are changing, and whether you believe that is a good thing or a bad thing, it's a reality. ***Jen Gurecki***

11.1 INTRODUCTION

The title of this chapter probably seems a little peculiar, but is most appropriate for the topic. We specifically selected this title to reflect on urban geomorphology in a not-so-urban setting (i.e., the alpine environment). We decided to use the climbing term "*bivouacs*" to convey the idea that these mountain towns are temporary locations. While many alpine mountain towns that were developed as early mining towns have disappeared, several have made the transition to bedroom communities for recreational activities and regional service centers. Still, they all remain temporary features on the landscape.

The goal of this chapter is to discuss the location of mountain towns and the impact of built environments in the alpine landscape through an examination of interaction with various landforms and the power of various geomorphic processes on humans and their built environment. Because the focus is on the built environment of today, our attention is on the interaction between these towns and the critical zone, a concept that has become increasing important, as highlighted by the National Research Council (2001) and through various workshops sponsored by the National Science Foundation.

The interaction between humans, their built environment, and geomorphic processes can be viewed as a type of offset agreement. The settlement of the alpine environment is a partnership between the impact that the built environment has on the alpine environment and the impact that geomorphic processes have on humans and their built environment. In essence, it is a geoscience quid pro quo. The location of our study area is the largest mountain range in Colorado, namely the San Juan Mountains in the southwest part of the state (Fig. 11.1).

The mining boom of the 1800s resulted in the San Juan Mountains experiencing unparalleled growth in mining activity and numerous mining towns sprouting up throughout the San Juan

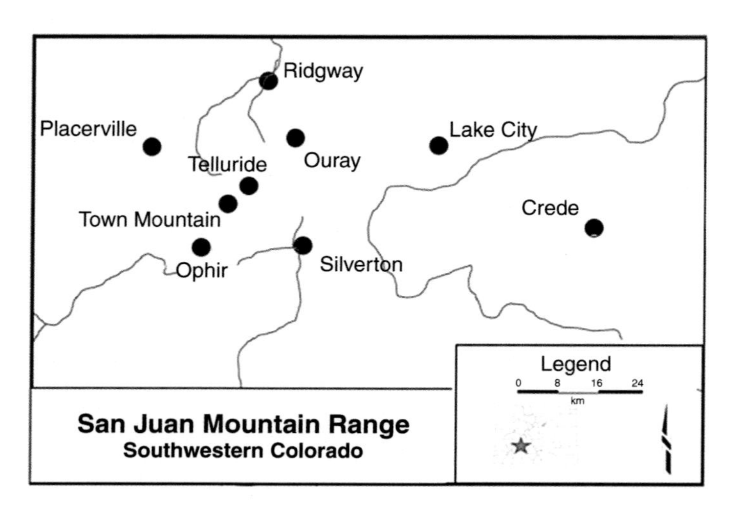

FIGURE 11.1 **Generalized map of the study area.**

Mountains. Rather than numerous associated mining camps developing during this period, only a few settlements were established because the volcanic nature of the San Juan Mountains resulted in limited placer deposits (Wyckoff, 1999). These town sites provided retail activity, limited mining, and smelting. Four towns dominated the mining/smelting scene in the San Juan Mountains: Silverton, Lake City, Ouray, and Telluride. All four still function as active towns today, although Lake City is not as well known as the others and is the least active.

Since the boom days of the 1800s, the San Juan Mountain towns have been on a roller-coaster ride both economically and demographically. Some, unfortunately, like the roller coaster that derails, have crashed and disappeared. The crash of the price of silver in the late 1800s sealed the fortunes of many of these towns. Through the end of the 19th century and into the early 20th century, these towns experienced great swings in fortune. As silver declined, the rise of gold injected new life into the towns, as did the smelting of ore. The roller-coaster ride continued from the 1920s to the present. As the importance of mining decreased throughout the area, a significant increase in both summer and winter recreational activities developed. Telluride, for example, has become a major destination for summer music festivals and winter skiing. Ouray, on the other hand, has experienced increased summer activity associated with 4 × 4 vehicles using old mining roads. Ouray has also become renowned for its winter ice climbing. These activities have left their impact on the landscape (Wyckoff, 1999). Silverton has become the mountain station stop for the Durango-Silverton Railroad. During the winter, Silverton is a cold, isolated town with one natural-state ski area for expert skiers only.

11.2 STUDY AREA

The San Juan Mountains are the largest mountain range by area in Colorado, covering 13 counties in the southwestern part of the state. Comprising volcanic summits, the San Juan Mountains feature majestic scenery, with 13 peaks rising to over 4265 m (14 000 ft) as well as many lakes, waterfalls, and streams, including the source of the Rio Grande. The San Juan Mountains are drained by the Uncompahgre, San Miguel, and Animas Rivers.

As the southernmost portion of the Rocky Mountains, the San Juan Mountains were formed as two enormous continental plates slammed into one another, folding and faulting the Earth's crust. This tectonic mountain-building process resulted in volcanic activity, which produced rich mineral veins with significant silver and gold deposits that drew miners to the region in the 1860s and 1870s.

During the Pleistocene, glaciers carved the steep mountain sides and U-shaped valleys in the range that are emblematic of the areas that eventually became home to the cities of Ouray, Silverton, and Telluride (Atwood and Mather, 1932). In addition to the spectacular landscape of the San Juan Mountains, the climate plays an important role in the geomorphology of the region. A wide variation in climatic conditions is experienced throughout the region because of its location on the flank of the mountains and because of considerable topographic relief (Burbank et al., 1969). The region is marked by short and relatively cool summers and long but seldom severe winters. Minimum temperatures several degrees below 0°C (-32°F) may occur in winter months for a short duration (Burbank et al., 1969). Maximum temperatures in the summer rarely exceed 32°C (90°F) (Burbank et al., 1969). Precipitation is typical of mountainous areas, occurring as convective thunderstorms in the summer and snow, sometimes heavy, in the winter. Afternoon thunderstorms are common in July and August and cease with the first heavy frost or snowstorm, which usually occurs before the end of September. Seasonal weather hazards include ice and snow in the winter, snow slides in early spring, and flash floods or debris flows in the summer (Burbank et al., 1969).

Weather can change rapidly in the San Juan Mountains because of the orthographic effect of air moving up and over the mountains. From November to April, cyclones transport clouds saturated with moisture from the Pacific Ocean to the Rockies, encountering the San Juan Mountains. These cyclones bring frequent, moderate snow (Fig. 11.2). March is typically the month with the greatest snowfall in the region and snow often lingers until mid-April in Telluride and Ouray. June is the driest and sunniest month before the summer monsoons begin from July until October, bringing in more adequate precipitation (NOAA, 2017).

The population density in the San Juan Mountains is low, with clusters of population being centered on the various small mountain towns located throughout the area. The total population based on the 2010 census is 14 049 (US Census Bureau, 2017). Other towns in this region, but not included in our study, are Rico, Durango, Montrose, Lake City, and Norwood.

The geology of the San Juan Mountains is dominated by volcanic processes, with a geologic history ranging from Precambrian to present. All this activity produced a highly mineralized mountain range (Blair, 2000; Moore, 2004). The San Juan Mountains are the southern extent of the Colorado Mineral Belt, which is famous for gold and silver deposits.

Volcanics are the main rock type in the area, as in the area is the San Juan Volcanic Field. Although the mountains are primarily volcanic, the spectacular landscape is the combined result of intense glaciation, fluvial erosion, and mass movement. Much of the area is federal lands administered by the United States Forest Service and the Bureau of Land Management (Moore, 2004).

11.3 THE NEW AWARENESS OF THE CRITICAL ZONE

Agreement that Earth is in a new age, the Anthropocene, continues to spread among the scientific community, although the parameters of the Anthropocene are still debated (Goudie and Viles, 2016; Mobley, 2009). Effects of human influence on the surface of Earth has been widely discussed in scientific literature for many years (Palmer et al., 2004; Wackernagel and Rees, 1998; Wilkinson and McElroy, 2007), and recently these studies have placed in new perspective the role of human influence on the near-surface critical zone. In 2001, the National Research Council defined the critical zone (Fig. 11.3) as "...the heterogeneous, near surface environment in which complex interactions involving rock, soil, water, air, and living organisms regulate the natural habitat and determine availability of life sustaining resources" (Giardino and Houser, 2015, p. 2; National Resource Council, 2001).

One of the many benefits of studying Earth science with a focus on the critical zone at specific locations is that public understanding of the link between a single garden parcel to a complete watershed to the whole planet is enhanced (Giardino and Houser, 2015). As indicated by Mobley (2009), studies of anthropogenic effects on critical zones worldwide are ongoing, but at varying rates. It is fundamental that this zone and its response to human influence be addressed as it encompasses all Earth components, from the atmosphere to groundwater. These components will be ever more needed and stressed as human population

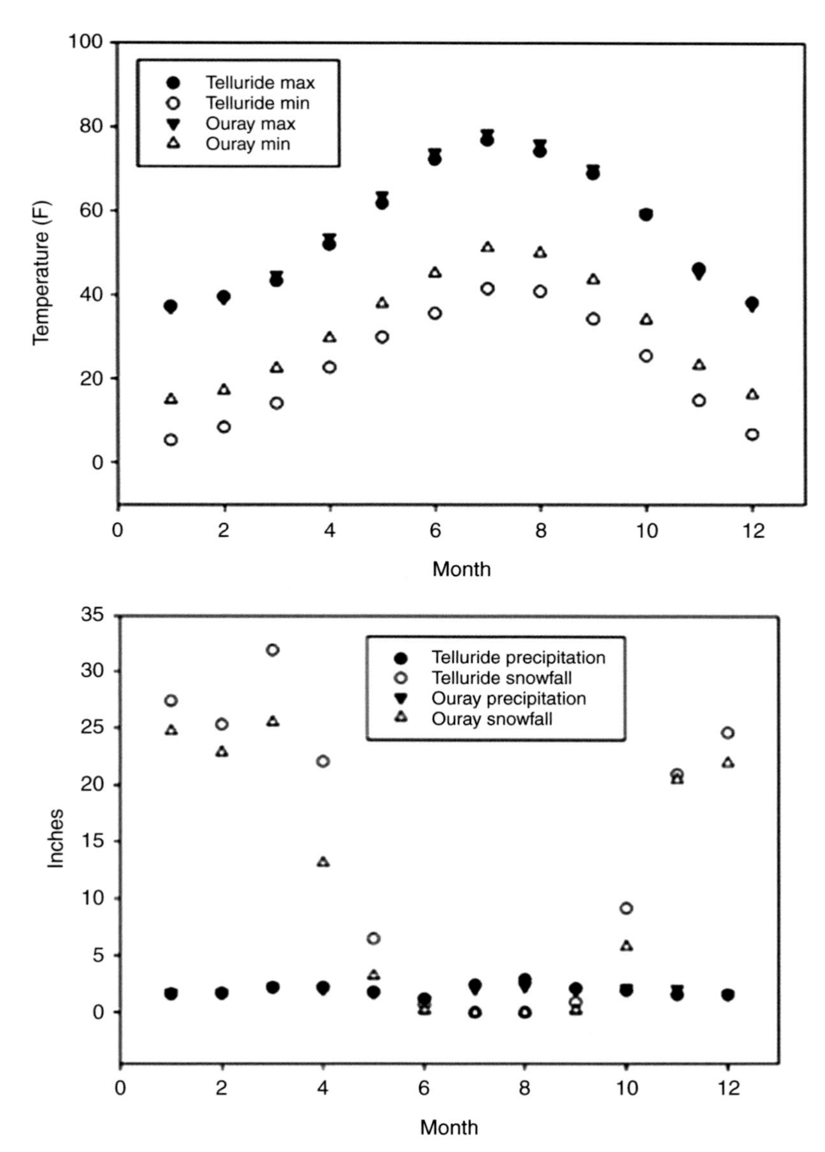

City	Elevation	Average Temperature				Temperature Extremes		Average Annual		Days with Precipitation
		July Max	July Min	Jan Max	Jan Min	High	Low	Annual Precipitation	Snow	
Silverton	6600	78	73	34	−1	96	−39	25	162	106
Ouray	7840	79	51	38	15	97	−22	23	150	98
Ridgway	7000	80	43	38	1	88	−36	19	95	68
Telluride	8800	77	42	37	6	96	−36	23	184	106

FIGURE 11.2 Climograph of Telluride and Ouray, Colorado.

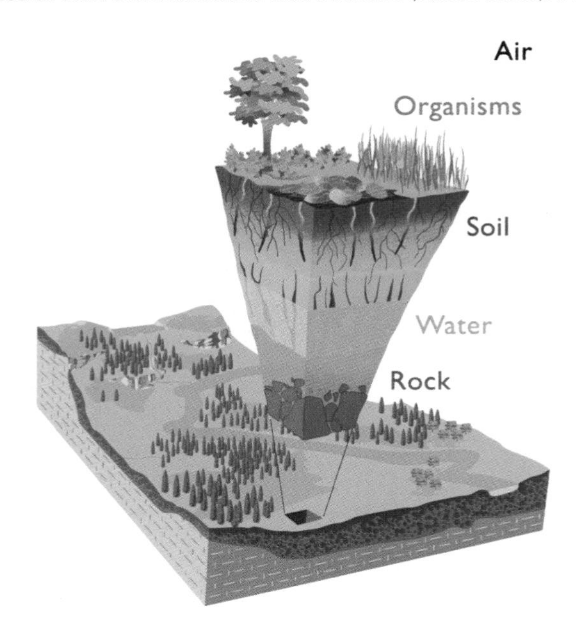

FIGURE 11.3 **Graphical depiction of the critical zone.**

increases. Thus, the critical zone in the alpine regions of the San Juan Mountains encompasses geomorphic processes ranging from natural processes of glaciation, periglacial, fluvial, and mass movement to anthropogeomorphic processes.

The traditional definition of the critical zone defines its parameters as extending from the top of the canopy to the bottom of the aquifer (Gamache et al., 2015a; NRC, 2001). In regions dominated by glaciers and associated topography, Gamache et al. (2015a, p. 363) have defined the canopy as "... extending from the top of trees, bushes, shrubs, or grasses growing on various glacial deposits." The definition of the aquifer in glaciated regions is more difficult to define, as water in the subsurface may be frozen or liquid. Even the glacier itself is a storage unit of freshwater and can be an aquifer. The lower boundary of the critical zone, thus, must be defined separately for each unique study area, considering the surface and subsurface processes in effect. Additionally, the impact of glacial processes on the critical zone is not immediately apparent, but cannot be neglected.

11.4 MINING TOWN DEVELOPMENT

The mining regions were the home of the nomadic Ute Indians for more than 500 years. These indigenous tribes hunted game in the high country during summer months and retreated to lower elevations during winter. These seasonal migration routes created paths that were later used by Spanish explorers, American miners, and eventually by road builders in the region. Whereas the nomadic people of the area utilized only what lay on top of the ground for their subsistence, the primary objective of Spanish expeditions that visited the area in the 18th century was the search for minerals. With the Utes as their guides, a wave of Spanish expeditions explored the San Juan Mountains in search of silver, but failed to find deposits of any significance.

Boasting some of Colorado's most rugged and imposing peaks, the San Juan Mountains posed a natural barrier to anyone seeking to exploit the riches of the area. The San Juan Mountains were weeks of treacherous mountain travel away from Denver, where several significant gold strikes occurred in the latter half of the 19th century (Blair, 1966). Despite the physical challenges, in 1860 Charles Baker turned rumor into reality by finding gold near where the town of Silverton now sits (Blair, 1966). In the ensuing decades, the mineral riches of the San Juan Mountains drew hordes of prospectors to the area (Blair, 1996).

The San Juan Mountains produced some of the richest mines in Colorado including the Sunnyside, Camp Bird, and Idarado (Blair, 1996). These mines produced millions of tons of ores containing gold, silver, lead, zinc, and copper between 1880 and the present (Blair, 1996). The towns of Silverton, Howardsville, and Eureka sprang up to serve the miners working in the surrounding area. In the high country, Silverton reached a population of 1000 by 1880, whereas farther north, Ouray and Telluride competed for their own share of business and prominence.

Silverton is most noted for the famous Sunnyside Mine, one of Colorado's largest gold mines containing more than 60 mi of underground workings (Blair, 1996). Originally claimed in 1873, the Sunnyside Mine was producing more than 800 tons of ore per day when it was last active in the 1990s (Blair, 1996). By the late 1870s, Silverton boasted functioning smelters but, unfortunately, the lack of finances and engineering problems delayed the arrival of the railroad to the region. By 1880, the Denver and Rio Grande Railroad was being built and in July 1882, the first train from Durango steamed into Silverton.

The arrival of rail service between Silverton and Durango initiated two decades of railroad building within the region and facilitated the long-awaited San Juan bonanza. Tons of gold and silver were transported from the mountains to the smelter in Durango. Copper, lead, and zinc were recovered from the ore as by-products and coal was also mined to fuel mining and railroad operations. While the San Juan Mountains are rich in a variety of metals, silver was the prime source of revenue. In 1891, for instance, miners in San Juan County extracted more than \$761 000 in silver compared to \$192 000 in gold (Blair, 1996).

11.4.1 Uncompahgre Mining District

Ouray, the county seat and commercial center for the Uncompahgre and nearby mining districts to the south, is nestled in a mountain park at the junction of the Uncompahgre River and Canyon Creek at an elevation of 2375 m (7800 ft) (Burbank et al., 1969). Ouray and the surrounding area have appropriately been referred to as the "Switzerland of America" because of its picturesque mountain setting (Burbank et al., 1969). The physical features of the Uncompahgre River valley, one of the major drainage systems in the western San Juan region, have controlled the commercial development of the area (Burbank et al., 1969).

Limited prospecting was conducted throughout the San Juan Mountains region, beginning about 1860. The region officially opened to prospecting and settlement in 1874 after a treaty between the Utes and the US Federal Government was signed (Burbank et al., 1969). A permanent mining camp was established in 1875, and within a very short time, the camp became a town and was platted and soon incorporated (Burbank et al., 1969). By 1880, the Town of Ouray boasted 900 inhabitants and served as an important supply center to adjacent mining operations (Burbank et al., 1969). Ouray continued to prosper and grow throughout its early years and experienced "boom and bust" periods, as do most mining communities. Today, with mining activities at an ebb, it serves as an important hub for tourism.

Gold-producing mines like the American Nettie, Jonathan, and Wanakah Mines are perched high on the cliffs overlooking Ouray and form a complex maze of tunnels and shafts (Blair, 1996; Moore, 2004). The Camp Bird Mine located south of Ouray produced high-grade gold ore from 1896 until the 1920s. It was then that the richest ore bodies were exhausted, and production switched to base metals in the 1940s (Blair, 1996). More than 2.5 million tons of ore yielded 1.4 million ounces of gold and 4.5 million ounces of silver from the Camp Bird Mine (Blair, 1996; Moore, 2004). The Revenue Virginius Mine reopened in 2010 and continues operations today.

11.4.2 Telluride (Upper San Miguel) Mining District

The principal mines of the Telluride District were the Liberty Bell Mine, Smuggler, and the Tomboy, which all together produced much of the gold and silver of the district and made the district one of the 25 top leading gold producers in the United States (Dunn, 2003). The Tomboy and the Smuggler supported their own mining camps housing hundreds of miners (Dunn, 2003). The Liberty Bell Mine closed in 1921, the Tomboy Mines in 1927, and the Smuggler-Union Mines in 1928. However, in 1940 the Tomboy and Smuggler-Union were consolidated and were worked through the 1950s (Dunn, 2003). The Idarado Mine below Red Mountain pass, which was continuously worked from the mid-1940s until it closed in 1978, contained 80 mi of interior workings that extended as far west as Telluride (Blair, 1996). The geomorphic processes that have shaped the landscape of the San Juan Mountains made mining in the area challenging.

11.5 GEOMORPHIC PROCESSES

The geomorphic processes operating in the San Juan Mountains are listed in Fig. 11.4. The figure is divided into naturally operating geomorphic processes and human-induced processes. Although many of the processes have been active in varying time frames, they are not mutually exclusive nor independent of each other. Many coexist and combine, resulting in enhanced impacts. To illustrate the important impact of the geomorphic processes on the built landscape, we present the geomorphic processes separately, but have lumped the human impacts in producing unique landforms as well as altering exiting landforms in a separate section, which addresses human interaction with the environment (Fig. 11.4).

11.5.1 Glaciation

The oldest geomorphic process that has left the biggest imprint on the present-day landscape is glaciation (Atwood and Mather, 1932; Blair, 1996). Glaciers in the San Juan Mountains directly contributed to the near-surface environment by sculpturing broad U-shaped valleys

FIGURE 11.4 **Geomorphic impacts on mountain towns.**

and depositing moraines and valley-train deposits. Other glacial landforms, such as cirques, arêtes, roche moutonnees, and so forth, are present, but are not associated directly with the various towns in the San Juan Mountains. The U-shaped valleys lined with valley deposits provided broad, flat areas for the development of the built landscape. Various moraines are situated in the Uncompahgre, Animas, and San Miguel River valleys, respectively (Gamache et al., 2015b).

11.5.2 Periglacial

Periglacial processes, which are closely associated with the glaciation of the area, can be divided into two categories: (1) those underlain by permafrost and (2) those without permafrost, but dominated by frost-action processes (Rowley et al., 2015). In the San Juan Mountains, permafrost is extremely limited today. Very few indications of active permafrost can be found. Nevertheless, there are several occurrences of stone polygons and stone stripes (Gamache et al., 2015b). In addition, many of the active rock glaciers of the region still contain internal ice. Frost action is the dominant process above the tree line. Protalus ramparts are found in all the valleys of the San Juan Mountains.

11.5.3 Fluvial

Rivers are in a constant state of adjustment to changes in the morphology and geology of the surrounding landscape. Six types of morphologies have been distinguished (Schälchli, 1991) in mountain streams, based upon increasing channel slope and particle diameters. At the lowest gradients (1.5–5%), streams exhibit uniform geometry (Schalchli 1991). These streams tend to have well-sorted sediment, with maximum diameters of 0.5–0.7 m (Sieben 1993). At slopes approaching 1.5–7% (Schalchli 1991), riffle-pool sequences form from clusters of larger sediment accumulating across the channel. Although stable under normal flow conditions, this channel sequence tends to disintegrate during floods (Sieben, 1993). At slopes of 3.5–12.5%, true step-pool sequences take form and are similar to riffle pool sequences, but with a higher distance between pools (Schalchli 1991). This larger step height allows this morphology to remain intact even at greater flows. At slopes of 9–30%, boulder step-pool sequences form (Schalchli 1991). These sequences have highly nonuniform flow over 1–3 different layers of large clasts. Finally, rounded-boulder-glide morphologies occur at slopes of 12–35%, and sharp-edged boulder glides at slopes of 29–49% (Schalchli 1991). Rounded-boulder glides are thought to be formed during high flows, whereas sharp-edged boulder glides may indicate channels in a transitional environment, with sediment not transported far from its source (Schälchli, 1991; Sieben, 1993).

Rivers at higher elevations can rapidly change in form from straight-channeled to braided as the rate of sediment deposition responds to changes in flow velocity. Rapid increase in sediment deposition creates bars in wide river valleys, quickly developing a braided river system. Sieben (1993, p. 50) defined bars as "...relatively large bed forms at the scale of the channel width." Bars can be free, developing spontaneously because of irregularities in the channel bottom or sediment supply, or they can be forced, resulting from a physical constraint imposed on the channel (Sieben, 1993).

When a confined mountain stream exits to a wide valley, such as Portland, Bear, Cascade, and Mineral Creeks, energy dissipates and much of the sediment load is deposited at the mouth of the canyon in a wide fan shape and along the upper reaches of the floodplain. Often, nonalluvial sediment tends to accumulate at the head of the fan in the wash zone. Beyond this reach, stream flow spreads laterally and is concentrated into secondary channels. The secondary channels are themselves dynamic, regularly adjusting their course in response to the deposition of newly transported or reworked alluvium (Sieben, 1993).

Step-pool sequences dominate the mountain streams and rivers in the San Juan Mountains. The step pools develop from the deposition of large clasts, logs, and debris and some are cut into bedrock, such as the sequence on Portland Creek known as the "Baby Bath Tubs" (Carter et al., 2015; Robert and Giardino, 2015), depicted in Fig. 11.5.

The three main rivers in the San Juan Mountains, the Uncompahgre, Animas, and San Miguel Rivers, have produced flat, broad floodplains, where Ouray, Silverton, and Telluride are situated (Orts et al., 2011; Zhao et al., 2016).

11.5.4 Mass Movement

The San Juan Mountains are dominated by mass wasting processes, such as downslope movement of soil, regolith, bedrock, or snow under the influence of gravity, resulting in distinct landforms. Although certain types of mass wasting can occur on very gentle slopes, the risk for hazardous mass movement events is increased by oversteepening of slopes, often by human interference, which is discussed in a section further on.

Early United States Geological Survey (USGS) geologists undertook considerable research on the landslides of the San Juan Mountains (Cross et al., 1905). In addition to the occurrence of landslides in the area are avalanches, mudflows, and debris flows. Large accumulations of talus are present in all the valleys in the San Juan Mountains.

FIGURE 11.5 Photo of Portland Creek "Baby Bath Tubs." *Source: Photo by J.R. Giardino, 2017.*

All sediment types have respective angles of repose at which the sediment can accumulate without slumping. Slump is a form of mass movement perhaps most often triggered by oversteepening of slopes. When a natural slope is altered, the natural sediment and bedrock is forced to sustain an angle of repose for which it is not suited. If not properly supported, the strength of the natural material may fail under the weight of the overlying regolith, causing slump. During slump, rupture along a curved surface occurs in the subsurface, with blocks of earth sliding along the curvature. This may be a gradual or sudden occurrence, depending on the natural material and nature of the slope alteration.

A large landslide, the Amphitheater Landslide, occurred during the Quaternary and covers moraines in the area around Ouray. The landslide is located at the southeast side of Ouray and produced considerable relief in the area. Transport of material from the landslide created a large alluvial fan in that part of Ouray. The Beaumont Hotel is built on the fan. A relatively steep gradient trends from the hotel to the rest of downtown Ouray (Fig. 11.6).

Mass-wasting events may also be triggered by the removal of vegetation or the presence of excess water. Vegetation acts as a natural protectant against mass movement, as roots serve to anchor sediment in place. This fact, coupled with sufficient water to reduce the cohesion of sediment, increases the risk of debris flows. Debris flows are chaotic mixes of supersaturated regolith that can occur suddenly, with little warning. These are most often confined to channels. The San Juan Mountains are experiencing rapid dieback of trees in the area because of a

FIGURE 11.6 Photo of Beaumont Hotel, Ouray, CO. *Source: Photo ¥tz, 2017.*

pine beetle outbreak in the area. The death of these trees is exposing large areas to increased surface runoff and accelerated erosion, resulting in increased sediment loads in the rivers (Garcia, 2017).

In the upper elevations of the San Juan Mountains, which are underlain by patches of permafrost, mass movement occurring on the gentlest of slopes is the result of solifluction processes, where water percolates through the subsurface until reaching the patches of permafrost. Unable to continue moving downward, the water migrates along the top of the frozen layer, carrying soil with it in lobes. The natural process of solifluction is complicated by increased construction of heat-generating structures on a permafrost landscape (Fig. 11.7). Figure 11.7 shows turf associated with solifluction sliding down the slope and into the road. Construction of the road altered the slope, resulting in increased movement downslope.

An additional mass movement affecting major portions of the San Juan Mountains and many of the towns in the area is avalanches. Avalanches are large quantities of snow, ice, and bedrock that move rapidly downhill. Avalanches occur because of disturbance to snowpack on a mountain slope. The upper layer of the snowpack is fractured, leading to a rapid torrent of snow, ice, and rock, as the stability of the snowpack is compromised. According to the United States National Weather Service, approximately 27 people die each year as a result of avalanches in the state of Colorado (Colorado Avalanche Information Center, www.avalanche.state.co.us).

As with any mass movement event, slope stability is key in the prevention and the risk of avalanches. The most stable slopes (i.e., least conducive to producing avalanches) will be heavily forested, allowing the rugged terrain to act as anchors for snowpack (Armstrong, 1978). It is crucial that features protruding above the ground surface (i.e., trees and boulders) be closely spaced and higher than the snowpack to be effective at preventing an avalanche. As

FIGURE 11.7 **Photo of turf solifluction near Hurricane Pass, CO.**

indicated by Armstrong (1978), an avalanche will typically not occur until the snowpack has reached heights above closely spaced surface features; therefore, surface features that are widely spaced tend to have little effectiveness in preventing avalanches. Avalanches are most common on slopes between 25 and 60° (Armstrong, 1978). Although avalanches are often triggered naturally by a sudden addition of high amounts of snow or ice, humans have also acted as triggers (Armstrong 1978). In some unfortunate cases, the additional weight of one person on a snowpack has been sufficient to trigger an avalanche, especially after heavy snowfall.

11.5.5 Anthropogenic Building and Modifications of Landforms

This section focuses on the landforms that are created by human impact as well as interaction with naturally occurring landforms. Humans have greatly modified the alpine landscape of the San Juan Mountains. Through their activities, ranging from mining, built landscapes, gravel mining, highways, roads, train tracks, water impoundment, ski trails, to river restoration, humans have brought about considerable change to the alpine environment in this area. Their modifications have also impacted landforms produced by natural geomorphic processes.

First, glaciation not only shaped the present-day landscape, but also directly contributed to the near-surface environment through the contribution of sediment accumulations in the upper levels of soil profiles. Accumulations of sediment are especially prominent along lateral, recessional, and terminal moraines. A substantial portion of the critical zone in the San Juan Mountains lies well above the surface and is capable of inhabitation by humans. Moraines, for example, are often productive sources of gravel, sand, and aggregate, which humans have exploited. The glacial deposits in the area have also provided borrow pits for sand and gravel. The terminal moraine in the San Miguel River valley has been completely removed as a borrow source as well as enhancement for the transportation entering and exiting from Telluride.

Deposits of tills in the San Juan Mountains did not produce major sources for gold, iron, copper, or zinc as placer deposits. Thus, instead of hydrologic mining, mining was restricted to underground mines, which produced large positive-relief landforms of mine tailings. The isolated locations of a few locations of permafrost are restricted to higher elevations and do not impact the mountain towns in the area. Periglacial processes present in the San Juan Mountains include protalus ramparts, stone stripes, pattern ground, rock glaciers, and other cryogenic processes.

Frost action is active in and around the mountain towns. Frost action in the San Juan Mountains is very pronounced during the fall, from late September to late November, and again from late March to early June. Many of the stonefacades on buildings in all three towns show signs of granular disintegration and spalling from daily freeze-thaw cycles.

Mining of natural resources to meet human demands is a direct disturbance of the natural workings of the critical zone. Removal of mineral matter, for example, not only reduces the quantity of solid soil material available in the critical zone, but also removes potential storage sites for soil water and habitat for near-surface organisms. In some cases, excavation may lead to over-steepening of slopes and increased danger of mass movement events (Fig. 11.8).

As previously mentioned, the major "positive" human-built landforms are the mine tailings throughout the San Juan Mountains. Figure 11.8 shows extensive mine tailings at the Camp Bird Mine.

Anthropogenic changes, such as dams and levees, induce changes upstream and downstream of engineered structures (Mueller et al., 2016; Musselman, 2006). Climate change, both natural and anthropogenic, induces channel-width adjustments to changes in base level. These processes may be considered hazards if they endanger the lives or property of people living in their midst. Ironically, sometimes it is the efforts by well-intentioned people to maintain security in a fluvial environment that can exacerbate the natural threats it poses.

Human influence on fluvial systems is not a recent development. Early human civilizations were centered on waterways and sought to reap all the benefits possible from these settings (Goudie and Viles, 2016). The modification of river systems certainly can be and often is responsibly done, but proper attention must be given to the long-lasting changes that these alterations may cause.

FIGURE 11.8 **Photo of Camp Bird Mine tailings.**

In the San Juan Mountains, humans have modified fluvial sediment transport through construction of the Ouray Hydroelectric Dam and Power Plant and the Ridgway Dam. Dam construction has effectively reduced sediment transport and deposition by trapping sediment in the reservoir, whereas accelerated soil erosion has increased sediment delivery to rivers. The Uncompahgre River channel upstream of the dam responded to a new, elevated local base level through reduced flow rates and deposition of sediment in the channel and reservoir. Water released from the dam is sediment-starved, leading to increased erosion and downcutting from Ridgway north. Dams are permanent structures that are built across river channels and engrain themselves as "positive" landforms. The associated channels are "negative" landforms, although the rate of downcutting will be modified by the presence of the dam and disruption of local base level and streamflow. Other sources of modifications to fluvial systems are channelization and river restoration projects.

Many fluvial modifications of the rivers have occurred in the study area. The Cascade Creek channel, for example, is a channelized concrete flume, which has been confined to a straight concrete channel from below the falls and through Ouray almost to its intersection with the Uncompahgre River. There is also modification of the Uncompahgre River, as the city has armored the cutbanks with large boulders. In addition, sediment is mined annually from various pointbars along the channel. The San Miguel River in Telluride has been altered by a series of stream restoration projects. The aquatic life of the stream and the adjacent riparian lands have been severely damaged by all the mining activity and built environment development in Telluride. Attempts to restore the Animas River as it flows through the town include reconfiguring the river channel, establishment of step pools, and bank stabilization (Fig. 11.9). Because the river was still being inundated with large sediment loads, upstream sediment basins were built to trap sediment and in-stream steps and hydraulic structures were reconfigured (Depke et al., 2010).

FIGURE 11.9 **Example of bank stabilization efforts on the Animas River near Silverton, Colorado.** *Source: Photo by J.R. Giardino, 2014.*

The major "positive" human-built landforms of the mine tailings seen throughout the San Juan Mountains show the extensive tailings at the Camp Bird Mine. Construction of house and condominium pads produced both a "negative" form with excavation into hillsides and a "positive" complementary form from the excavated material placed downslope to increase the size of level building pad.

The San Juan Mountains have a long history of avalanche hazards. Avalanches along US 550 and the areas surrounding Silverton and the ski runs are high activity areas. Each year avalanche debris consisting of trees and earth materials are transported downslope. Various steps have been taken to minimize avalanche hazards. For example, Fig. 11.10 shows the impact of human development on landforms in the area. The photograph shows an avalanche shed that has been constructed at the location of the greatest number of avalanches above US 550. Avalanche debris is shunted over the highway and deposited in the canyon of the Uncompahgre River, resulting in increased sediment load. Anthropogenic influence in mountainous regions may increase the risk of avalanches by alteration of natural slopes, deforestation, and implementation of smooth, impervious surfaces.

As a result of the rapid growth of tourism and year-round settlement, new construction of houses and condominiums has encroached up the valley walls. In some cases, excavation may have led to oversteepening of slopes and increased danger of mass movement events. In Ouray, house pads excavated into the sandstone along the west wall of the valley have resulted in some houses being inundated with rockfalls and debris flows.

Humans alter the angle of slopes to produce flat surfaces for construction projects, typically by carving into hillsides and replacing the natural slope angle with a horizontal surface adjacent to a newly oversteepened slope. Because of pad excavating, shear stresses

FIGURE 11.10 Photo of avalanche shed on US Hwy 550 south of Ouray, CO. *Source: Photo by J.R. Giardino, 2017.*

have been increased by loads on slopes resulting from the built environment. Additionally, growth of these mountain towns has modified the drainage such that, when subdivisions are built, many minor geomorphic features are destroyed, increasing the likelihood of potentially hazardous geomorphic processes. Specifically, steeper slopes are produced to create flat surfaces for building. In addition, the covering of pervious surfaces with house footprints, concrete, roads, and parking lot surfaces has greatly decreased infiltration and increased surface runoff.

Ski runs, mountain roads, bike roads, and hiking paths cause impacts ranging from minor to significant, but as humans acquire more leisure time, mountain sports are growing dramatically and the impact will be significant. The introduction of nontraditional "mountain sports" also alters the landscape. In Telluride, for example, a vast area has been used for creation of a golf course and the landscape has been completely altered through the removal of vegetation and flattening of slopes.

Moreover, city infrastructure plays a role in altering the environment. Early mountain roads, although less frequently used, are still present and provide narrow regions of slope modification and impervious surfaces. The present road in the region with high amounts of mass movement (US 550) increases these risk factors. The high traffic flow on this corridor also contributes external heat and vibrations to the soil, compromising slope stability. Early train routes remain, as well as a current narrow-gauge train route that terminates in Silverton. These railroads introduce the same potentially hazardous factors as do the highways for vehicular traffic. Additionally, recreational trails contribute to increased mass movement. Off-road vehicles create channels and then compact the materials, allowing water to flow freely down newly created channels. New "positive" landforms, such as landfills, have been established in the area, creating increased threats of slope failure and pollution.

11.6 LOCATION, LOCATION, LOCATION: A PLANNER'S DREAM

In real estate, the three most important factors are location, location, and location. But with location comes a consideration for safety. Up to this point, we have discussed the importance of geographic processes on the built environment and vice versa. The question then arises, "How can the impact of these natural processes and the impact of the built environment be avoided, or at least minimized?"

Urban geomorphology focuses on solving various problems regarding geomorphologic interactions associated with urban development, such as examining the suitability of different landforms for specific urban planning, evaluating geomorphic impacts during urban construction, analyzing the geomorphologic consequences from human-introduced landforms, and hazard control in both urban and suburban areas (Coates, 1976; Cooke, 1984). Each aspect of urban geomorphology requires not only a solid understanding of Earth science, but also sufficient skill-sets for conducting the relevant evaluation, planning, and implementation. Traditional geomorphological techniques such as field surveys, have merit because of their long-established tradition (Goudie, 2003). However, the rapid progress of urban development in mountain regions has not been isolated from the flourishing information age. Thus, computational efficiency is an urgent requirement for rapid, accurate, and planning. The development of geomorphometry fulfills such a task (Hengl and Reuter, 2009).

Geomorphometry is the study of quantitative land-surface analysis that integrates mathematics, geoscience, engineering, and computer science (Pike, 1995). Qualitative depiction of landform evolution has been augmented by numerical analysis approaches since the mid-20th century. Improved remote-sensing platforms, coupled with geographic information systems (GIS) offer high-quality acquisition of Earth surface's information modeling and management, driven by the design and production of efficient algorithms for solving geoscience problems (Church, 2010).

Specifically, geomorphometry can aid in two areas that are important in urban geomorphology: revealing the nature of geomorphology-related processes and facilitating an understanding of human-environment interactions. Data alone do not necessarily convey meaning to the geomorphologist or planner. In most cases, they must employ the correct geomorphometric techniques to extract information from raw data. The extracted information can be further coupled by its own domain knowledge to understand the underlying processes, such as geological, hydrological, climatological processes, and so forth. Thus, by understanding the nature of processes revealed from the data, the extracted information can be further studied to identify and explain human-environment problems (Evans, 2012; Miller and Laflamme, 1958). For example, geomorphometry can be used to predict urban-development suitability, as is shown at the end of this section. This will be demonstrated in a case study.

Even though geomorphometry is based on mathematical and statistical principles, it still obeys the basic principles of process geomorphology. For instance, to characterize the surface of Earth, six classic factors of topographic form must be considered: elevation, terrain-surface shape, topographic position, topographic context, spatial scale, and landform type (Deng, 2007). With geomorphometry, the first four parameters can be accurately calculated and the last two parameters can be effectively analyzed.

The combination of GIS and remote sensing has significantly enhanced the computational capacity to solve Earth science-related issues (Goodchild, 1992). Various geomorphometry software use specific analytical functionality; and the calculation of a simple geomorphological factor, such as profile curvature of a slope, can have different results. Thus, it is fundamental to have a somewhat a priori "feeling" for the outcome.

Geomorphometry has been used to map glacier termini, monitor crop health conditions, extract river channels, optimize urban transportation systems, assess terrain for military support, and model solar radiation. By coupling GIS with remote sensing, the computational capacity of geomorphometry is significantly enhanced.

Sequential acquisition of data by remote sensing is beneficial because it can access extremely rough terrain that humans cannot. Remote sensing also provides a chronological record that is paramount in the study of urban geomorphology, which is so dynamic. Synthetic aperture radar and light detection and ranging are two additional remote-sensing technologies that are used in geomorphometry. Synthetic aperture radar is good at penetrating dry materials, such as sand or dry snow. Light detection and ranging is excellent for delineating land-surface details (Jensen, 1996).

11.7 PREDICTING URBAN SUITABILITY IN THE SAN JUAN MOUNTAINS

Artificial neural networks (ANNs) have been widely used in environmental studies, including land-surface classification, water-quality forecasting, rainfall-runoff modeling, and geomorphology mapping (Haykin, 2004; Palani et al., 2008). The ANN technique is a machine-learning technique, which is good at learning relationships between specified input and output variables. An ANN constitute an information-processing model that stores empirical knowledge through a learning process and, subsequently, makes the stored knowledge available for future use. An ANN can mimic a human brain in acquiring knowledge from the environment through a learning process. A neuron is the fundamental processing unit in all ANNs (Jain et al., 1966). These neurons consist of connecting links, with each assigned a specific weight (Fig. 11.11). Input is passed from one end of the links, multiplied by the connection weight, and transmitted to the summing junction of the neuron (Hsu et al., 1995).

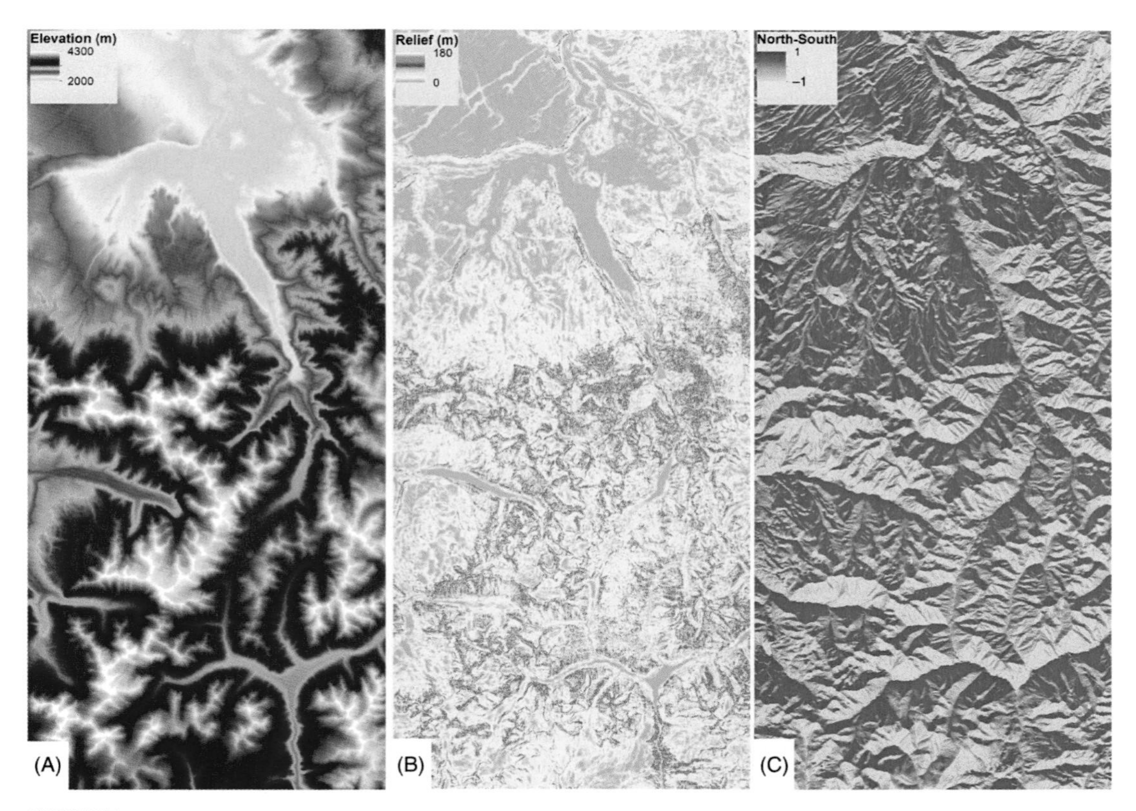

FIGURE 11.11 **Topographic Input layers into ANN. (A) DEM of study area; (B) Local relief; (C) North-South orientation.**

The application of ANNs in geomorphometry requires reasonable input layers. The main purpose of this case study illustration is to demonstrate urban suitability based on six different towns in the San Juan Mountains. In this case study, we first developed seven indices as ANN input layers. Then we utilized the ANN to predict the urban suitability of the whole study area in the San Juan Mountains. Seven parameters were selected as input layers: elevation, relief, north-south orientation, winter solar radiation, summer solar radiation, topographic shielding, and distance to a river. These seven parameters can be reduced to three categories: topographic (Fig. 11.11), topo-climate (Fig. 11.12), and human-environmental (see Table 11.1 for ANN overview).

For the topographic category, elevation was derived from the USGS digital elevation map (DEM) for the area (Fig. 11.11A). The spatial resolution of the DEM is 10 m. The local-relief layer was calculated by subtracting minimum elevation from maximum elevation within a 3×3 moving window (Fig. 11.11B). Local relief can be used to approximate surface roughness and as an alternative layer for surface slope. The north-south orientation factor was calculated from slope aspect (Fig. 11.11C). To contrast north-facing slopes from south-facing slopes, a cosine function was applied. Thus, south-facing slopes appear darker than north-facing slopes.

For the topo-climate interaction category, winter solar radiation (Fig. 11.12A) and summer solar radiation (Fig. 11.12B) were modeled separately. We assumed that solar radiation has an impact on urban development because humans prefer relatively warmer conditions in winter and relatively cooler conditions in summer. Thus, topographic shielding, also known as topographic exposure, was produced (Fig. 11.12C). One human-environment interaction variable was produced: that was distance to the nearest river (Fig. 11.12D). To examine the differences among six towns in the San Juan Mountains (Mountain Village, Ophir, Ouray, Ridgeway, Telluride, and Silverton), we used GIS to calculate the average statistics for all seven parameters within town limits.

Thus, the six towns served as training sites for urban development suitability. In addition, we selected a portion of image that contained the highest elevations and longest distances to a river as the seventh training site and it served as an unsuitable site. Our ANN was then implemented based on all the training sites and input layers as discussed previously.

TABLE 11.1 Artificial Neural Network (ANN) Parameters and Categories

Input layer	Category
Elevation	Topographic
Relief	
North-south	
Winter radiation	Topo-climate
Summer radiation	
Topographic shielding	
Distance to river	Human-environment

Input layers for ANNs.

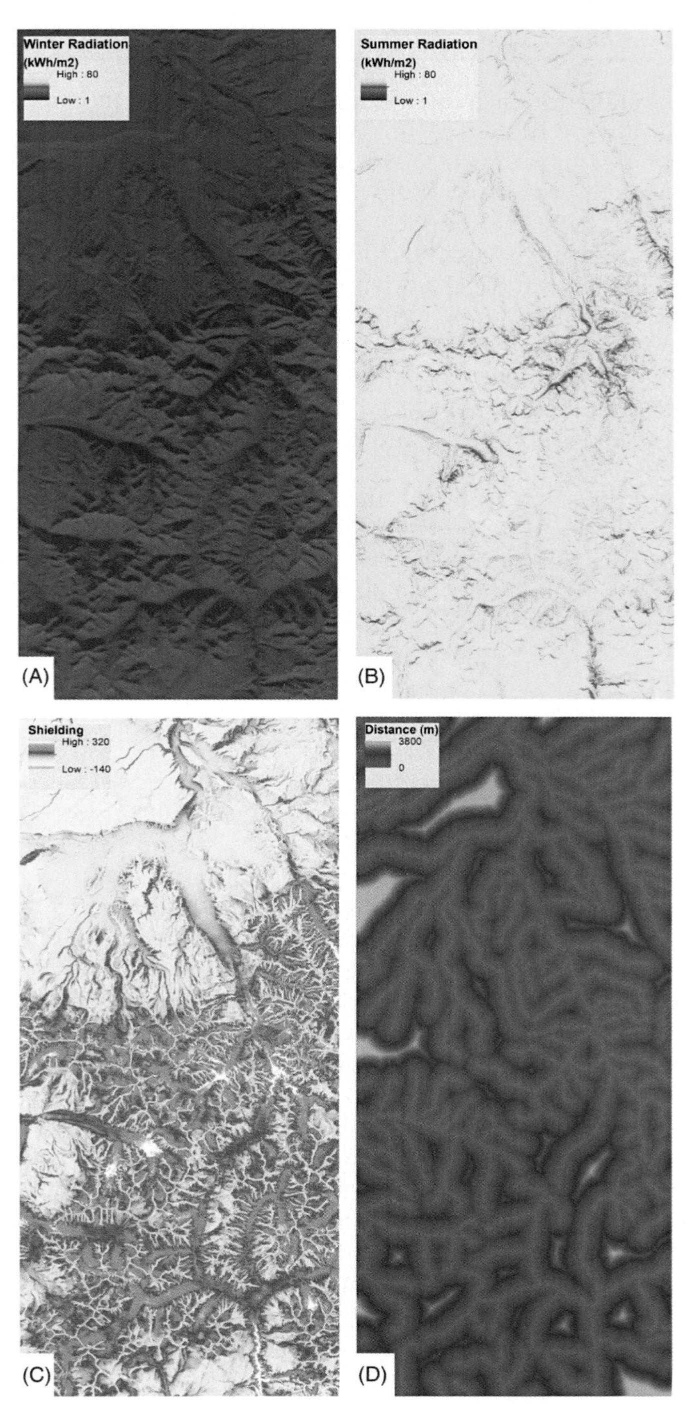

FIGURE 11.12 **Topo-climate and human-environment input layers into ANN. (A) Winter radiation; (B) Summer radiation; (C) Topographic shielding; (D) Distance to rivers.**

11.8 RESULTS

According to the statistics, each town exhibits its own characteristics based on the different perspectives. Ridgway and Ouray are the two towns at the lowest elevations, whereas Ophir and Mountain Village are the two towns at the highest elevations (Fig. 11.13A). Regarding local relief, Silverton and Ridgway have the smoothest surfaces, whereas Ouray and Mountain Village have the roughest surfaces (Fig. 11.13B). Silverton and Ophir are predominantly south-facing, whereas Mountain Village faces moderately north and Telluride faces moderately south. The orientations for Ridgway and Ouray are not evident (Fig. 11.13C). It appears that Mountain Village and Ridgway are not strongly shielded by the neighboring topography, whereas the other four towns are all well-protected (Fig. 11.13D). Ouray receives the least amount of solar radiation, both during winter and summer, whereas Ophir receives the most solar radiation during the two seasons (Fig. 11.13E and F). Ouray, Telluride, and Ophir are all much closer to nearby rivers compared to the other three towns (Fig. 11.13G).

The ANN result (Fig. 11.14) shows the urban suitability for the western San Juan Mountains. Thus, it appears that most of the southern part of the San Juan Mountains is unsuitable for urban development, except for Mountain Village. It is interesting to note that the northern part of the San Juan Mountains is more suitable for development as demonstrated by Ridgway.

11.9 CONCLUSION

Since the beginning of human occupation, the face of the planet has been changed by the encroachment of human activities and their built environments and infrastructures on the natural landscape. The expansion of human society has required more natural resources and more land, resulting in dramatic modification of the natural landscape. Humans have developed ways, techniques, and technology to extract natural resources, capture and store water, stabilize slopes, modify river floodplains and habitats, create built environments, and facilitate recreational activities. The effects of these endeavors have been exacerbated by increased human population, increased disposable incomes, and increased leisure time, resulting in encroachment into pristine, sometimes vulnerable environments. Through all of these efforts, landforms have been altered or removed as the built environment advances into the alpine mountain regions.

With all of these modifications, geomorphology plays a pragmatic role. Unfortunately, many times this role is undervalued, not understood, or completely ignored, with resulting consequences of loss of life and loss or damage of infrastructure. Thus, it is the responsibility of geomorphologists to use a variety of tools to exact understanding and positive change. As increased leisure time drives more mountain-town expansion and new development in the future, ANNs are a very important tool that can be used to preselect appropriate locations for future development.

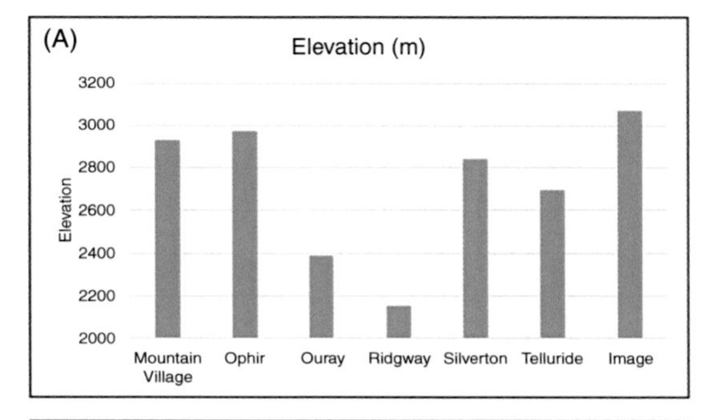

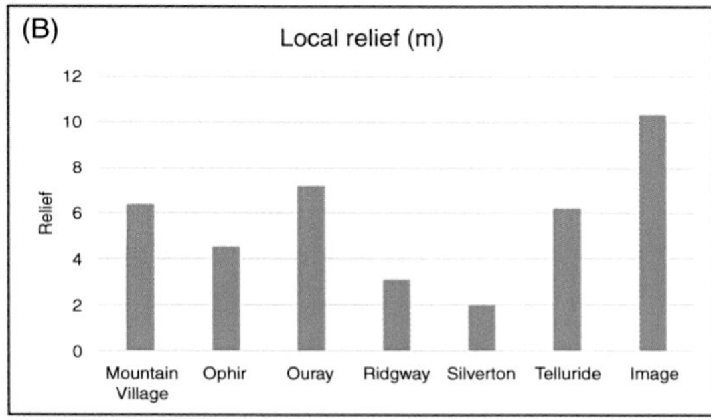

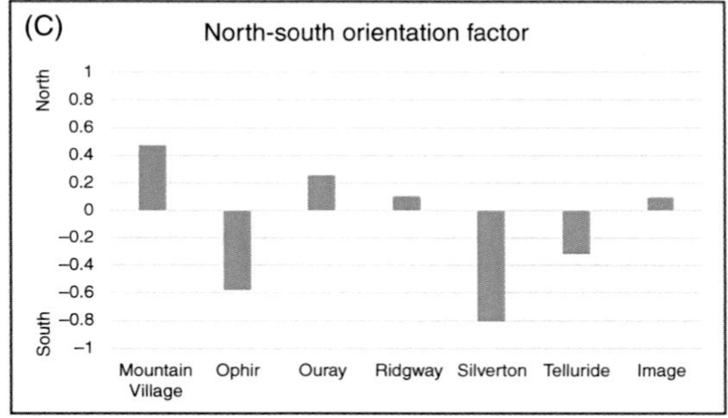

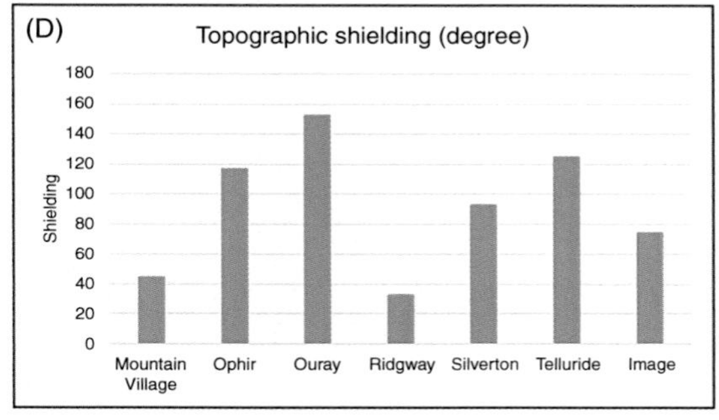

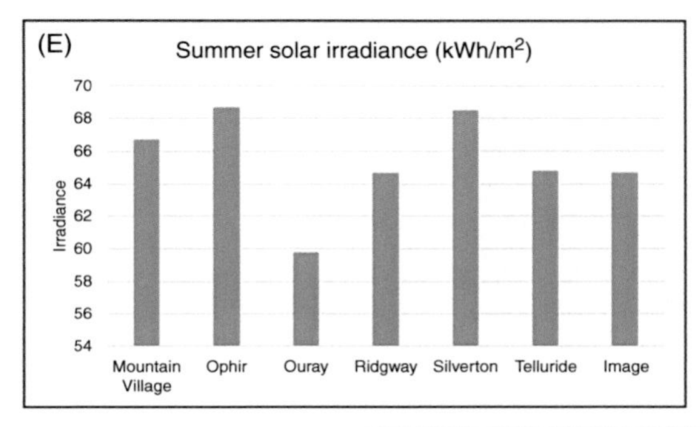

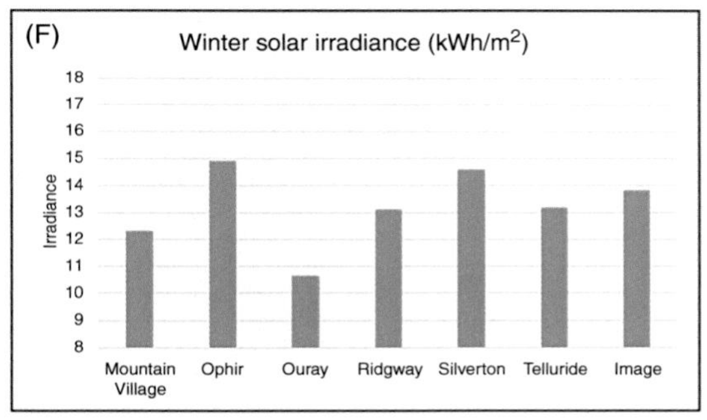

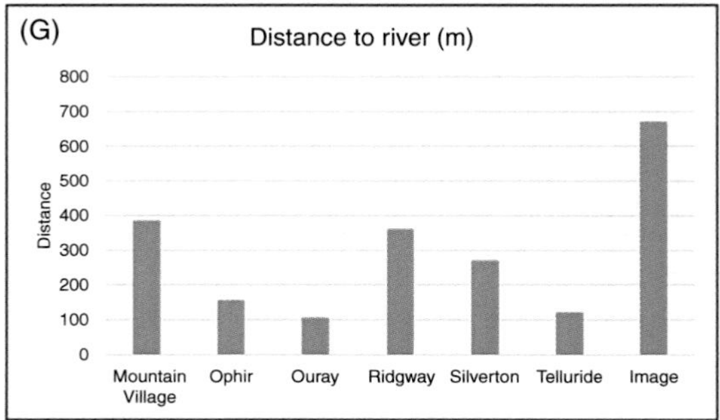

FIGURE 11.13 (A) Elevation of study area towns; (B) local relief; (C) north-south orientation of study area towns; (D) topographic shielding of study area towns; (E) summer solar radiation of study area towns; (F) winter solar radiation of study area towns; and (G) distance to river.

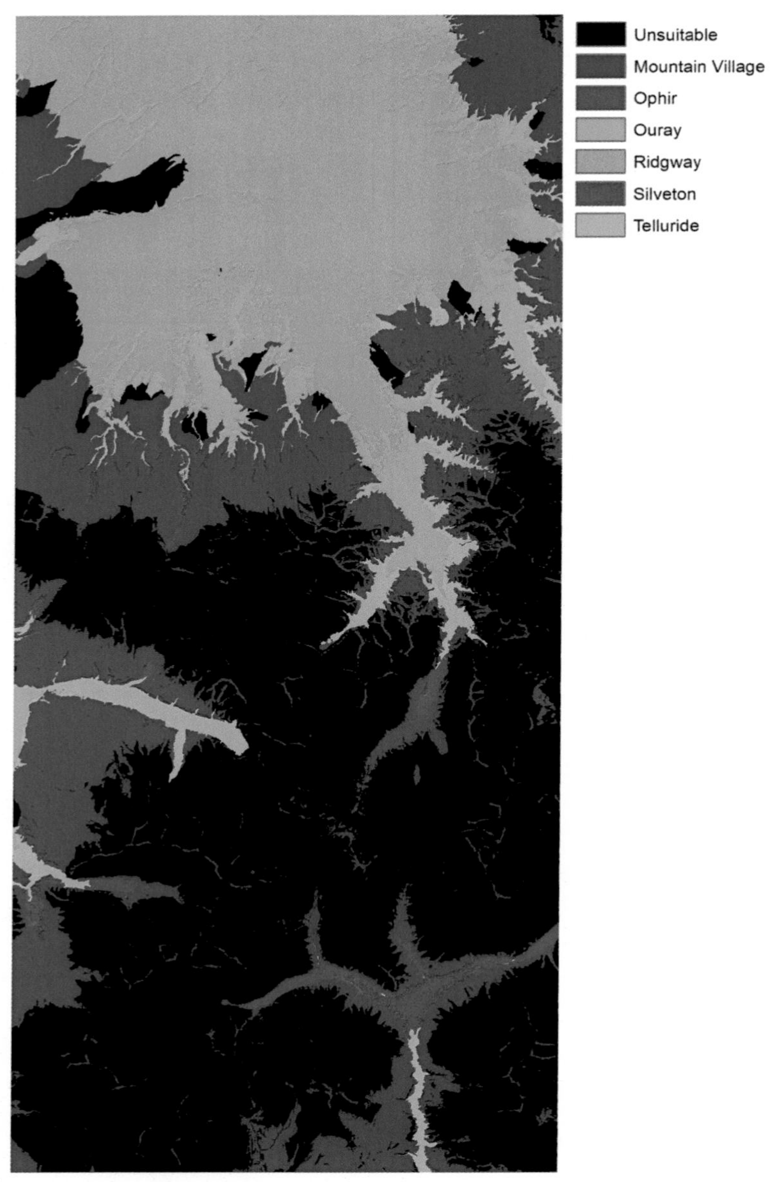

FIGURE 11.14 **ANN prediction of urban suitability.**

References

Armstrong, B., 1978. A History of Avalanche Hazard in San Juan and Ouray Counties. National Research Council, Colorado, Associate Committee on Geotechnical Research, Technical Memorandum, 120, pp. 199–218.

Atwood, W.W., Mather, K.F., 1932. Physiography and Quaternary Geology of the San Juan Mountains, Colorado. US Government Printing Office, Washington, DC, p. 176.

Blair, R., 1996. The Western San Juan Mountains: Their Geology, Ecology, and Human History. University Press of Colorado, p. 406.

Blair, R., 2000, Personal Communication, July 2000, Ouray, CO.

Burbank, W.S., Luedke, R.G., Colorado State Mining Industrial Development Board, 1969. Geology and ore deposits of the Eureka and adjoining districts, San Juan mountains, Colorado. US Government Printing Office, Washington, DC, Geological Survey Professional Paper 535, iv, p. 73.

Carter, C., Fedotova, A., Giardino, J.R., Price, A.E., 2015. Physical and water chemistry characteristics of mountain stream step pools, San Juan mountains, CO. Geolog. Soc. Am. Abstracts Prog. 47 (7), 727.

Church, M., 2010. The trajectory of geomorphology. Prog. Phys. Geogr. 34 (3), 265–286.

Coates, D.R., 1976. Urban Geomorphology. Geological Society of America, p. 167.

Cooke, R.U., 1984. Geomorphological hazards in Los Angeles. Lond. Res. Series Geogr. 7, 43–44.

Cross, W., Howe, E., Ransome, F., 1905. Description of the Silverton quadrangle. US Geolog. Surv. Atlas 120, 34.

Deng, Y., 2007. New trends in digital terrain analysis: landform definition, representation, and classification. Prog. Phys. Geogr. 31 (4), 405–419.

Depke, T.J., Giardino, J.R., Vitek, J.D., 2010. Terrestrial photogrammetry using consumer-grade cameras to study short-term temporal change: Uncompahgre river, CO, USA. Geol. Society Am. Abstr. Progr. 42 (5), 313.

Dunn, L.G., 2003. Colorado Mining Districts: A Reference. Colorado School of Mines, Arthur Lakes Library, Golden, CO, p. 364.

Evans, I.S., 2012. Geomorphometry and landform mapping: what is a landform? Geomorphology 137 (1), 94–106.

Gamache, K.R., Giardino, J.R., Regmi, N.R., Vitek, J.D., 2015a. The impact of glacial geomorphology on critical zone processes. Principles and Dynamics of the Critical Zone. Elsevier, Amsterdam, The Netherlands, vol. 19, Chapter 12, p. 363.

Gamache, K., Giardino, J.R., Allen, H., Rowley, T., 2015b. Glacial and periglacial geology of Yankee Boy basin, CO: a mapping pilot project using an unmanned aerial vehicle. Geolog. Soc. Am. Abstracts Prog. 47 (7), 788.

Garcia, C.J., 2017. Influence of Bark-Beetle-Induced Tree Mortality on Slope Erosion in the San Juan Mountains of Ouray, CO, USA.

Giardino, J.R., Houser, C., 2015. Principles and Dynamics of the Critical Zone. Elsevier, Amsterdam, The Netherlands, p. 674.

Goodchild, M.F., 1992. Geographical information science. Int. J. Geogr. Inform. Syst. 6 (1), 31–45.

Goudie, A., 2003. Geomorphological Techniques. Routledge, London, p. 592.

Goudie, A.S., Viles, H.A., 2016. Geomorphology in the Anthropocene. Cambridge University Press, Cambridge, p. 323.

Haykin, S, Network, N, 2004. Neural Networks: A Comprehensive Foundation, second ed. Prentice Hall, p. 823.

Hengl, T., Reuter, H.I., 2009. Geomorphometry: Concepts, Software, Applications. Elsevier, Amsterdam, The Netherlands, p. 765.

Hsu, K.l., Gupta, H.V., Sorooshian, S., 1995. Artificial neural network modeling of the rainfall-runoff process. Water Resources Res. 31 (10), 2517–2530.

Jain, A.K., Mao, J., Mohiuddin, K.M., 1966. Artificial neural networks: a tutorial. Computer 29 (3), 31–44.

Jensen, J.R., 1996. Introductory Digital Image Processing: A Remote Sensing Perspective, second ed. Prentice-Hall, p. 526.

Miller, C.L., Laflamme, R.A., 1958. The Digital Terrain Model: Theory and Application. MIT Photogrammetry Laboratory, p. 20.

Mobley, M.L., 2009. Monitoring Earth's critical zone. Science 326 (5956), 1067–1068.

Moore, G.E., 2004. Mines, mountain roads, and rocks: geologic road logs of the Ouray area. Ouray Hist. Soc. 1, 250.

Mueller, E.R., Schmidt, J.C., Topping, D.J., Shafroth, P.B., Rodríguez-Burgueño, J.E., Ramírez-Hernández, J, Grams, P.E, 2016. Geomorphic change and sediment transport during a small artificial flood in a transformed post-dam delta: The Colorado River delta, United States and Mexico. Ecol. Eng. 106, In Press, Online Access September 2016.

Musselman, Z.A., 2006. Tributary Response to the Lake Livingston Impoundment—Lower Trinity River, Texas.

NOAA (National Oceanic and Atmospheric Administration), 2017. Climate of New Mexico. https://www.ncdc.noaa.gov/climatenormals/clim60/states/Clim_NM_01.pdf. Retrieved 16 January 2018.

National Resource Council, 2001. Basic Research Opportunities in Earth Science. National Academies Press.

Orts, A., Giardino, J.R., Vitek, J.D., Gamache, K., Gamache, G., Van Winkle, R.S., 2011. Restoration of the Animas river through Silverton, Colorado: an applied engineering geomorphology class project. Geolog. Soc. Am. Abstracts Prog. 43 (5), 416.

Palani, S., Liong, S.-Y., Tkalich, P., 2008. An ANN application for water quality forecasting. Marine Pollut. Bull. 56 (9), 1586–1597.

Palmer, M., Bernhardt, E., Chornesky, E., Collins, S., Dobson, A., Duke, C., Gold, B., Jacobson, R., Kingsland, S., Kranz, R., 2004. Ecology for a crowded planet. Science 304 (5675), 1251–1252.

Pike, R., 1995. Geomorphometry: process, practice, and prospect. Z. Geomorphol. Suppl. 101, 221–238.
Robert, J.H., Giardino, J.R., 2015. Lithological impacts on step pools in alpine streams in the San Juan mountains, Colorado: a preliminary study. Geolog. Soc. Am. Abstracts Prog. 47 (7), 739.
Rowley, T., Giardino, J.R., Granados-Aguilar, R., Vitek, J.D., 2015. Periglacial processes and landforms in the critical zone. Princip. Dynam. Crit. Zone 19, 397.
Schälchli, U., 1991. Morphologie und Strömungsverhältnisse in Gebirgsdbächen. Ein Verfahren zur Festlegung von Restwasserabflüssen: Mitteilungen der Versuchsanstalt fur Wasserbau. Hydrologie und Glaziologie an der Eidgenossischen Technischen Hochschule Zurich, vol. 113,p. 7–112.
Sieben, J., 1993. Hydraulics and morphology of mountain rivers: literature survey. Commun. Hydr. Geotech. Eng. 93 (4), 157.
US Census Bureau, 2017. Washington, DC. Available from: https://www.census.gov/.
Wackernagel, M., Rees, W., 1998. Our Ecological Footprint: Reducing Human Impact on the Earth. New Society Publishers, vol. 9, p. 160.
Wilkinson, B.H., McElroy, B.J., 2007. The impact of humans on continental erosion and sedimentation. Geol. Soc. Am. Bull. 119 (1–2), 140–156.
Wyckoff, W., 1999. Creating Colorado: The Making of a Western American Landscape 1860–1940. Yale University Press, p. 336.
Zhao, P., Giardino, J.R., Kelkar, K., 2016. Characterization of river basins in the San Juan mountains, Colorado: an advanced geospatial mapping technique. Geolog. Soc. Am. Abstracts Prog. 48 (7), 59.

Further Readings

Ahlstrom, A.K., Gamache, K., Giardino, J.R., Vitek, J.R., 2011. The Ouray amphitheater landslide: a preliminary analysis. Geolog. Soc. Am. Abstr. Prog. 43 (5), 445.
Burbank, W.S., Luedke, R.G., US Geological Survey, n.d., Geology and Ore Deposits of the Uncompahgre (Ouray) Mining District, Southwestern Colorado. Geological Survey Professional Paper 1753, p. 119.
Colorado Encyclopedia, 2017. Available from: https://coloradoencyclopedia.org.
Kelkar, K., Giardino, J.R., 2015. Mass movement vulnerability in the San Juan mountains, Colorado: a pilot 3-D mapping approach. Geolog. Soc. Am. Abstracts Prog. 47 (7), 386.
Kelkar, L., Giardino, J.R., Zhao, P., 2016. The susceptibility of mass movement in the western San Juan mountains, Colorado: a 3-D mapping approach. Geolog. Soc. Am. Abstr. Prog. 48 (7), 315.
Reed, J.C., Giardino, J.R., 2015. Dendrogeomorphic assessment of the Ouray, Colorado Amphitheater landslide. Geolog. Soc. Am. Abstr. Prog. 47 (7), 735.
Richter, D.D., Mobley, M., 2009. Monitoring Earth's critical zone. Science 326, 1067.
Western Regional Climate Center (WRCC), 2017. Climate Summaries, v. 2017. WRCC, Reno, NV, Available from: http://www.wrcc.dri.edu/.

CHAPTER

12

Pokhara (Central Nepal): A Dramatic Yet Geomorphologically Active Environment Versus a Dynamic, Rapidly Developing City

*Monique Fort**, *Basanta R. Adhikari***, *Bhawat Rimal*[†]

*Département de Géographie, Université Paris-Diderot-SPC, Paris Cedex 13, France;
**Civil Engineering Department, Institute of Engineering, Tribhuvan University, Kirtipur, Nepal;
[†]Institute of Remote Sensing and Digital, Earth (RADI), CAS, Beijing, China

OUTLINE

12.1 Introduction 232

12.2 Pokhara City in Its Valley: A Long, Dramatic, and Complex History 232
- 12.2.1 *An Active Mountain in a Subtropical Environment* 233
- 12.2.2 *A Catastrophic, Geomorphic Evolution* 234
- 12.2.3 *The Recent Birth of a Major City in Nepal* 238

12.3 A Tourist City With Major Attractions Related to Its Geomorphology 241
- 12.3.1 *Exceptional Viewpoints* 243
- 12.3.2 *The Lakes: Legacies of Catastrophic Events* 245
- 12.3.3 *Karstic Features: Gorges and Caves Related to the Old Ghachok Conglomerates* 246
- 12.3.4 *Bhim Kali Dhunga* 246
- 12.3.5 *Hot Springs* 246

12.4 Potential Threats: Natural Hazards and Risks 248
- 12.4.1 *Flood Hazards* 248
- 12.4.2 *Earthquake Hazards* 250
- 12.4.3 *Sinkholes and Subsidence* 250
- 12.4.4 *An Example of Anthropogenic-induced Hazards: Sand Mining* 253

Urban Geomorphology. http://dx.doi.org/10.1016/B978-0-12-811951-8.00012-6

12.4.5 *Environmental Impacts of Urbanization Versus Geoheritage Preservation* 255
12.5 Conclusions 255
References 257

12.1 INTRODUCTION

Pokhara (28°13′ N, 83°59′ E, 870 m a.s.l.), the second largest city in Nepal (population circa 265 000 in 2015), lies in an intermontane basin located in the midland hills of the Lesser Himalayas and is characterized by a seasonally contrasting monsoon climate. Pokhara is built on an exceptionally broad, flat plain located at the foot of the Annapurna Range (>8000 m). It is one of the most popular tourist destinations in Nepal because of its natural beauty, with magnificent views of glaciated peaks, including the prominent Machhapuchhre Peak (Nepalese for "Fishtail Peak") standing close by, only 28 km north of the city. As well as being a central node at the crossing of N–S and E–W trade routes and an important administrative center, Pokhara offers a wide variety of tourist attractions (lakes, caves, gorges, temples) related to its geomorphological evolution and occupation history. The Himalayan collisional tectonics, on the one hand, and the monsoon rains enhanced by the steepest gradients in the world, on the other hand, are controlling factors of the complex evolution of the Pokhara valley and make the city and its surroundings a potentially disaster-prone area, with increasing vulnerability due to its rapid growth during recent decades (its population has tripled in the past 25 years, according to Rimal et al., 2015).

In this chapter, we first describe the main geomorphological characteristics of the Pokhara valley and explain how they are the result of several catastrophic debris flows, most of them triggered by large, Medieval earthquakes. We then present a brief history of Pokhara, its gradual urban expansion in connection with trading and tourism development based on its exceptional geomorphic sites. Finally, on the basis of some recent events like the slurry flood of the Seti Khola (May 2012) and the 7.8 Mw Gorkha earthquake (April–May 2015), we show how natural hazards (including landslides, river bank erosion, sinkhole collapse, and land subsidence) must be seriously considered by urban planners, particularly because the city is expanding very rapidly and will soon be provided with an international airport, scheduled for completion in 2021.

12.2 POKHARA CITY IN ITS VALLEY: A LONG, DRAMATIC, AND COMPLEX HISTORY

Administratively, the Pokhara valley lies in parts of the Kaski, Syangja, and Tanahu Districts of the Gandaki zone, western Nepal (Fig. 12.1A and B). The valley includes the very flat Pokhara plain, in sharp contrast to its surrounding hills (Sarangkot, Begnas, Kalika, Phoksing) (Gurung, 1965) and is drained by the Seti Khola (Nepalese for the "white river") and its small tributaries (Phurse, Bijayapur, Yangdi, Mardi, Tal, Saraudi Kholas). Although dissected, the Pokhara plain extends in the form of large alluvial fans, from Bharabhari (Kaski) at the foot of the Annapurna Range to Bhimad (Tanahu) to the south (Fig. 12.1C).

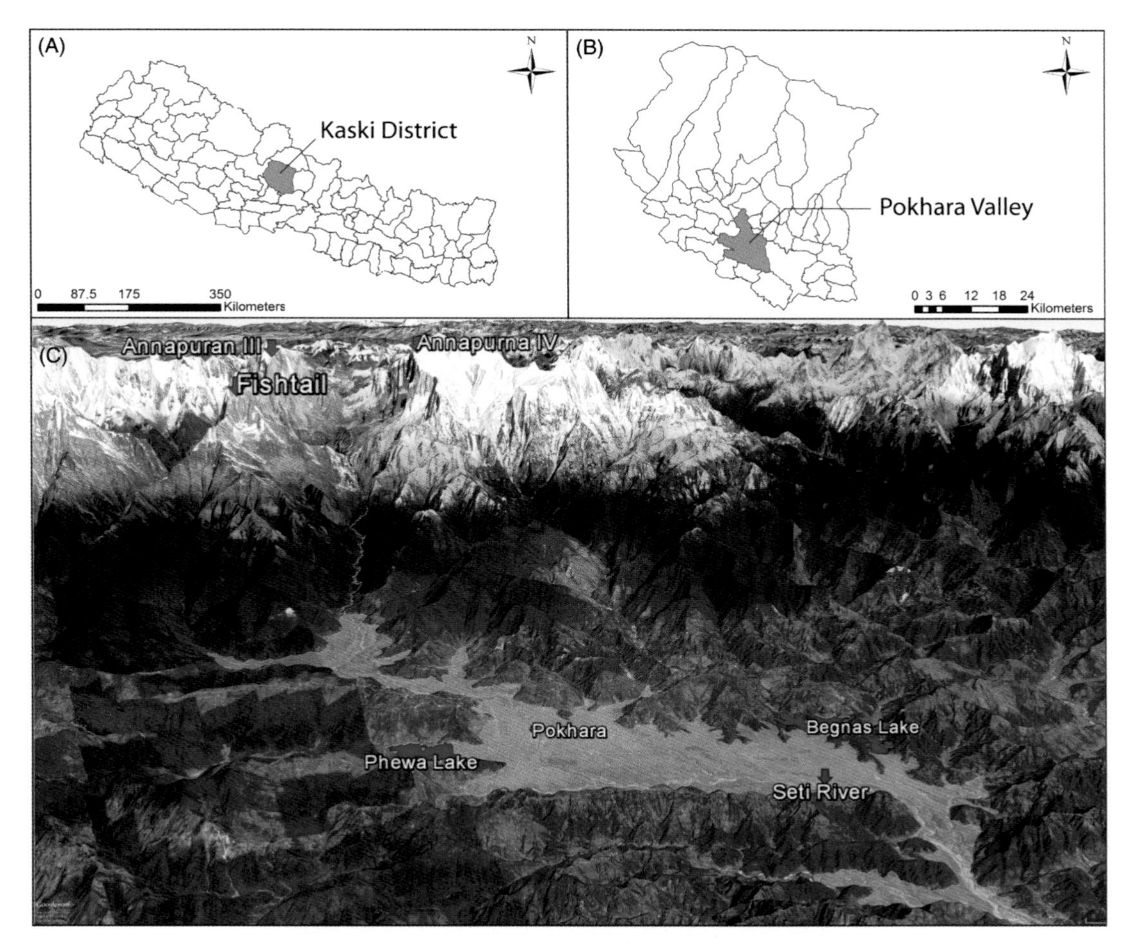

FIGURE 12.1 **Location map of the Pokhara valley.** (A) Administrative context of the Pokhara valley. (B) Location of the Kaski District Pokhara within Nepal. (C) The physiographic context of the Pokhara valley, with snow-capped Annapurna in the background, the Seti River (the major drain of the valley), and some lakes. The *gray /yellowish area* represents the Pokhara valley plain.

12.2.1 An Active Mountain in a Subtropical Environment

Physiographically, the Pokhara valley belongs to the Pahad, a relatively depressed area between the Mahabharat (Lesser Himalayas) and the Higher Himalayas, which are two elongated, more or less parallel ridges that were formed in response to the collision between the Asian and Indian plates (Fig. 12.2). Originating from the overhanging, glaciated Sabche cirque (>4500 m) carved into the sedimentary rocks of the "Tibetan series," the Seti Khola profile is deeply entrenched in the Higher Himalayan crystalline, gneissic bedrock and then becomes gentler when entering the Pokhara valley, where the Seti Khola flows within terraced alluvial deposits, whereas the surrounding hills are shaped across the Lesser Himalayan metasedimentary rocks (mostly schists, quartzites, dolomitic limestones) (Fig. 12.3).

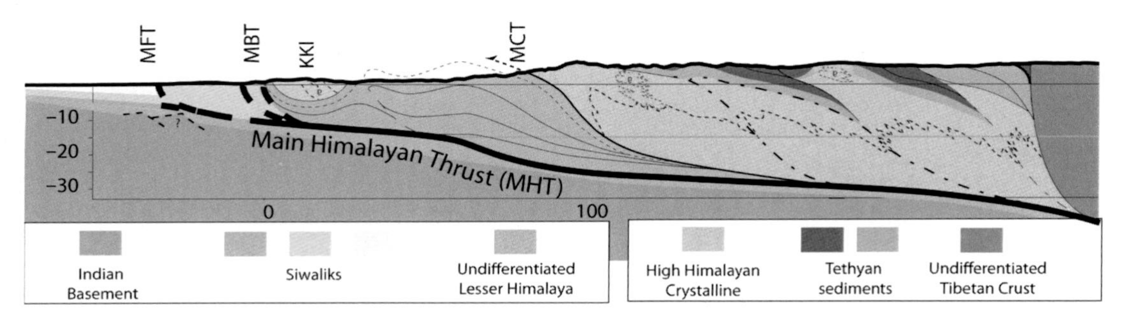

FIGURE 12.2 Cross-section of the Himalayas. The Himalayan Arc results from the collision between the Asian and Indian plates, about 50–45 Ma. The compressional motion between the two plates has been, and continues to be, accommodated by slip on a suite of major thrust faults, connected at depth along a major detachment plane, the Main Himalayan Thrust. The Higher Himalayas develop above the Main Central Thrust (MCT), whereas the Pahad and Lesser Himalayas extend between the MCT and the Main Boundary Thrust (MBT) (©Bollinger). *Source: From Bollinger, L., Avouac, J.P., Cattin, R., Pandey, M.R., 2004. Stress build up in the Himalaya. J. Geophys Res. 109, B11405, doi:10.1029/2003JB002911.*

The climate is tropical to subtropical. The temperature varies between 30° and 32°C in summer and 6° and 8°C in winter (Gurung, 1965). Daily temperature ranges are highest during the premonsoon period (April–May) and lowest during July and August, when high-humidity and cloudiness decrease radiation (Gurung, 1965). The rainfall pattern is dominantly monsoonal, with 80% of the total rainfall occurring from June to September, quite often preceded by convective storms in late spring. The average annual precipitation in Pokhara is 3951 mm, with amounts exceeding 5000 mm north of the valley (elevations circa 3000 m) due to orographic amplification, which makes the south Annapurna front zone the area with the highest annual precipitation in the country. In winter, the circulation of westerlies prevails and the sky remains clear, although the atmosphere is cold and rather foggy in the morning.

The Pokhara valley and surrounding hills are covered with luxuriant vegetation, whose species vary depending on the local edaphic conditions. The plain itself is rather dry, being constituted of gravelly material that favors savanna-type vegetation, such as xerophytic plants—whereas, the traditional villages on the plain are surrounded by hedges planted with *Euphorbia*, *Acacia*, and *Ficus* trees (to name but a few). In contrast, the hills are naturally covered by moist deciduous forests, gradually evolving to wetter, more permanent trees above 2300 m; (Gurung, 1965) nucleated settlements are located on the hilltops, overlooking the terraced fields developed on slopes that are generally <20°.

Nowhere in the Himalayan Range is this bioclimatic gradient greater than north of Pokhara. This reflects the specific nature of the geomorphic context, where the mountain front is steepest, where it is the most impressive yet potentially the most dangerous (Fig. 12.3).

12.2.2 A Catastrophic, Geomorphic Evolution

A few distinctive landforms characterize the Pokhara valley: its flat morphology, the dramatic set of the alluvial terraces, and the localized canyons of the Seti Khola and its tributaries.

FIGURE 12.3 **Pokhara in its mountainous setting.** Panoramic view of the northern part of the Pokhara valley from the Sarangkot ridge: the High Himalayan front is dominated by Macchapuchare (6997 m), Gangapurna (7454 m), Annapurna III (7555 m), Annapurna IV (7524 m), Annapurna II (7937 m), and Lamjung Himal (6983 m), from left to right. The Seti Khola, in the lower foreground, originates from deep gorges stemming from the glaciated cirque of Sabche between the Macchapuchare and Annapurna IV Peaks. The flat Pokhara "plain" in fact corresponds to a major terrace of the Seti Khola that is nowadays subject to intense urbanization: Lamachaur and the Kali Khola are visible in the middle ground (*central and right part*), whereas the large Tibetan refugee camp of Tollo Hemja appears in the foreground on the right bank of the Seti Khola. The hills are traditionally inhabited by Gurung people. *Source: ©Fort, 2013.*

The flat morphology of the Pokhara plain is striking. The plain slopes gently to the south, with a general longitudinal gradient varying from 32‰ upstream to 9‰ downstream (Fort, 1987). It also slopes laterally from a central axis, a feature that has caused diversions of tributary streams (Gurung, 1965). Locally, the surface of the plain displays a braided-channel pattern (Fig. 12.3; see also Fig. 12.8A). All of these are characteristic of a large alluvial outwash fan, upon which the recent urbanization of Pokhara has taken place.

The terraces developed by the meandering Seti Khola and its tributaries are quite spectacular, especially north of Pokhara where several sets can be identified (Fig. 12.4A). Their elevation relative to the Seti Khola bed can reach >80 m (Fort, 1987; Yamanaka et al., 1982). Although particularly suitable for agricultural use, they have been gradually abandoned by villagers immigrating to the city.

The canyons of the Seti Khola and its tributaries are among the most intriguing features of the valley (Fort, 2010). Interrupting the long sections of wide floodplain confined by flights of terraces, they occur along limited reaches a few hundred meters to 1 km long and are deeply entrenched (up to 50 m) in a material made of gravel and boulders, cemented together to form a hard, conglomeratic bedrock (the "Ghachok Formation," Fort, 1987), with a lime-rich matrix that is known locally as "*gaunda*" (Hormann, 1974). These gorges can be very narrow (<1 m wide) and the stream has sometimes disappeared into underground tunnels (Fig. 12.4B). In fact, these landforms are closely related to the geomorphic history of the Pokhara valley, characterized by a series of catastrophic debris-flow events followed by fluvial readjustments, all processes that developed in relation to the proximity of the steep Main Central Thrust (MCT) front (Fort, 1987, 1988).

Several quaternary formations have been identified in the Pokhara valley (Fort, 1987; Hormann, 1974; Yamanaka et al., 1982). To the north of the valley are the oldest perched slope deposits (Gyarjati limestone breccia) (Fort, 1987); the old, deeply weathered alluvium and the Ghachok formations are stepped above the most recent Pokhara gravel filling; whereas in the center of the valley, the Pokhara gravel is superposed on the Gaunda-Ghachok conglomerates (Fig. 12.5). This particular setting clearly indicates the rise of the High-Himalayan front relative to the valley center. Here, we focus on the Pokhara gravel and its significance.

Several centuries ago, mountain collapses of exceptional magnitude affected the west face of Annapurna IV, upheld by Nilgiri limestones (Fig. 12.3). These collapses mobilized a volume of debris of $\sim$4–5 km^3, the products of which were transported as a megadebris flow (the "Pokhara gravel," Gurung, 1965), which buried the lower part of the Pokhara valley and its former differentiated topography (Fort, 1987, 2010); hence, explaining the varying thickness in space of this gravel. The sudden input of such a large volume of debris drastically changed the landforms of the Pokhara valley, hiding the initial topography (rocky spurs, conglomeratic alluvial terraces, deep river bed) and transforming the valley into a deadly, coarse debris field, in the form of a wide alluvial fan, with convex-up topography. It was suggested that such catastrophic events were triggered by high-magnitude earthquakes that generated mountain collapse and subsequent megafloods (Fort, 1987). Recent studies have confirmed this hypothesis and have further demonstrated that several giant debris-flow events occurred during the Medieval period (Schwanghart et al., 2016). These authors showed that at least three major pulses of generalized debris aggradation occurred in the Pokhara valley, whose dates (radiocarbon on buried woods and charcoals) match very well with the identified nearby M >8 past earthquakes, that is, $\sim$1100, 1255, and 1344 AD, respectively (Bollinger et al., 2014, 2016; Lavé et al., 2005; Mugnier et al., 2013; Pant, 2002). Ongoing studies have shown that the sediment pulses had different magnitudes and extents, with coarser debris deposited along the main axis of the Pokhara valley, whereas finer material flowed down to the margins of the deposits (Stolle et al., 2017).

One major consequence of this huge input of debris was the damming of the former tributary valleys of the Seti Khola and the subsequent development of lakes, the shape and size of which reflect the type of hydrographic network where they were formed (Gurung, 1965). After the last catastrophic event, which probably carried and dropped the large, gneissic boulders like the Bhim Kali Dhunga (Fig. 12.6) on top of the plain, the rates of geomorphological adjustments were exceptionally high, for example, with an average dissection rate of the Seti Khola of 10–20 cm y^{-1} and a sediment yield of 22 860 m^3 y^{-1} km^2 calculated for the upper-Seti catchment and averaged over the last 500 years (Fort, 1987; Fort and Peulvast, 1995). Such

FIGURE 12.4 Distinctive landforms of the Pokhara valley. (A) View of the Seti Khola terraces from Milanchowk, north of Pokhara: the upper terrace, outlined by a steep, pink cliff, corresponds to the Ghachok, conglomeratic formation. The intermediate, flat terrace corresponds to the top of the Pokhara gravel, which was dissected several times by the Seti Khola and buried again, at the origin of the lower terraces. Extensively cultivated four decades ago, these lower terraces have been gradually abandoned, as shown by tree regrowth. Note that during monsoon time, the mountain front (in the back ground) is hardly visible (©Fort, 2016). (B) Gorges of the Seti Khola at Ramghat, in the heart of Pokhara city: upstream, the river spreads in a large amphitheater (Fig. 12.14D), before entering downstream the 1–4 m narrow gorge, cut through the hardened Ghachok conglomerates. In some places, the canyon is so narrow and so deep (70–80 m) that only the sound of water can be heard (©Fort, 2016).

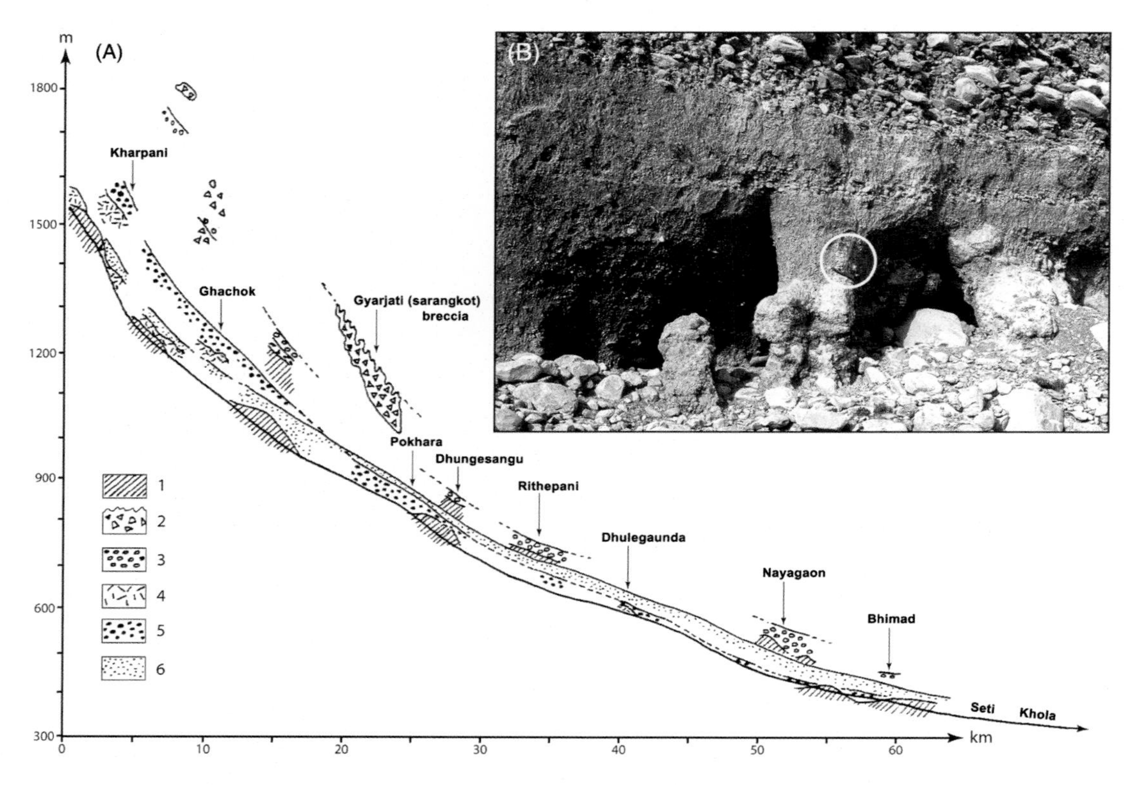

FIGURE 12.5 **Quaternary deposits of the Pokhara valley.** (A) Longitudinal profile of the Seti Khola and altitudinal distribution of the various quaternary formations deposited in the Pokhara valley. (1) Bedrock; (2) oldest perched slope deposits (limestone breccia); (3) old deeply weathered alluvium; (4) pre-Ghachok, calcareous, till-like breccia; (5) thousands-year old Ghachok (Gaunda), limestone conglomerates; and (6) centuries-old Pokhara limestone gravel. Note that north of the valley, the Ghachok formation is stepped above the Pokhara gravel filling, whereas in the center of the valley, the Pokhara gravel have buried the Ghachok conglomerates. Pokhara city and its suburbs are built on areas where the two units are superposed, hence, explaining the formation of the Seti Khola gorges (Modified from Fort, M., 1987. Sporadic morphogenesis in a continental subduction setting: an example from the Annapurna Range, Nepal Himalaya. Z. Geomorph. N. F., Suppl.-Bd. 63, 9–36.). (B) Karstified Ghachok, limestones conglomerates, as displayed by the *pillar shapes* (*lower part*; hammer for scale). They are fossilized under the Pokhara gravel, whose sedimentary facies are rich in a limy matrix, characteristic of a debris flow mode of deposition (©Fort, 1981).

large-magnitude hazards create a permanent, low-recurrence, but significant threat for the growing population living in this intermontane basin, as exemplified by the 2012 flood.

12.2.3 The Recent Birth of a Major City in Nepal

The hills around the Pokhara valley were most probably inhabited by prehistoric people, as attested by Neolithic tools found at sites like Kaskikot, NW of the present city (Adhikari and Seddon, 2002; Pandey, 1987). Then, the Gurungs and the Magars, two closely related ethnic groups who most certainly migrated from the northern parts of the Himalayan Range,

FIGURE 12.6 **Bhim Kali Dhunga.** This giant boulder was carried to the Pokhara area by the last catastrophic megaflood several centuries ago. The late Harka Bahadur Gurung, a pioneer in studying Pokhara valley geography and geomorphology, is providing the scale (©Fort, 1980).

settled in the higher hills, living off agriculture and pastoralism close to the glaciated mountains (Gurung, 1965). Next, groups of Indian origin came from the westward expansion during the 12–16th centuries, settling down and establishing Hindu-type principalities. As rulers of Gorkha (one of these states), the Shah Kings conquered Kathmandu valley in 1769 and built up present-day Nepal. Newars from the Kathmandu valley came to Pokhara to trade and settled specifically in the plain area of the valley (Udas, 2013). While the inhabited valley bottom was used as grazing land, the Batulechaur terrace (north of Pokhara) became a winter capital and religious gatherings were held there. At that time, Pokhara was essentially a strategically located marketplace at the crossing between the E–W route within the Gorkha state, and the S–N trans Himalayan salt-trade route from Tibet (von Fürer-Haimendorf, 1975), with good connections to the Terai plains (Himalayan piedmont) and India. Pokhara became a crucial location to secure the new Gorkha state and it rapidly developed as a military and administrative center (Caplan, 1975), while the Newar influence aided the development of its urban architecture.

The real change occurred during the second half of the 20th century under the influence of several factors (Fig. 12.7) that gradually changed the Pokhara area's predominantly rural landscape into a small town and then into a rapidly growing city (Adhikari and Seddon, 2002; Poudel, 2002). First, Pokhara became the winter capital of the King. Second, the migration of Thakkhali and Gurung people was very significant. Thakkhali trader migration was linked to the closure of the Tibetan border in 1950 and encouraged them to move their business to Pokhara, where they had warehouses. Several thousand Tibetan exiles came to Pokhara, where refugee camps were installed (Fig. 12.3). Many Gurungs that were enrolled in Gurkha regiments of the British and Indian armies preferred to settle in the Pokhara valley rather than return to their remote villages in the hills; and their remittances allowed them to open shops, selling inexpensive products initially produced in the villages (clothes, cooking utensils, tools, crafts) (Gurung, 1965; Macfarlane, 1976). Both Thakkhali and Gurung groups and their relative wealth initiated what would become a major marketplace as it is today and indirectly encouraged further migration of people from the hills down to the Pokhara plain.

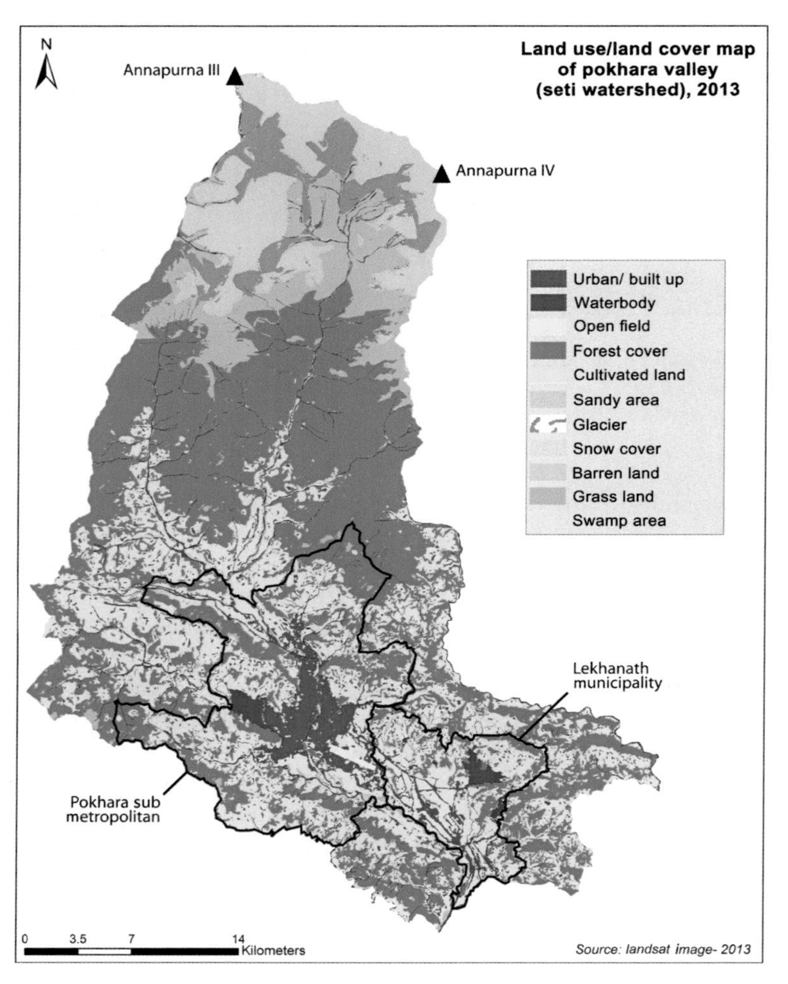

FIGURE 12.7 Land use/land cover map of the Pokhara valley (Seti watershed), 2013. The map shows the sharp contrast between the High Himalayan Mountain front to the north, with forest, grassland, barren rocks, and glaciated mountains, and the Lesser Himalayas where urban areas are progressively encroaching cultivated land. *Source: Adapted from Rimal, B., Baral, H., Stork, N., Paudyal, K., Rijal, S., 2015. Growing city and rapid land use transition: assessing multiple hazards and risks in the Pokhara Valley, Nepal. Land 4(4), 957–978, doi:10.3390/land4040957.*

Third, this migration was accelerated by the eradication of malaria in the late 1950s, which enabled the extension of irrigated perimeters in the southern part of the Pokhara plain and the development of health services (Military Hospital, Leprosy Hospital) and education facilities (Mierow, 1997). A fourth determining factor to boost the economy was the construction of major transportation infrastructure in the late 1960s and early 1970s: the Siddhartha Highway from Pokhara to India via Butwal and Bhairawa in the Himalayan foothills, and the Prithivi Highway from Pokhara to Kathmandu, respectively. Urbanization progressed, along these major axes and within the new Pokhara, leading to improvements in the transport network (e.g., the building of the Mahendra Bridge across the Seti Khola gorges), which in turn encouraged the construction of new buildings and the densification of the city to the detriment of cultivated areas (mostly rice fields) (Fig. 12.8; Rimal, 2011).

Nowadays, the Pokhara metropolitan city and its outskirts extend outward along the major roads (Fig. 12.9), whereas in 1959 the eight core localities of Pokhara had a total population of 3295 (Gurung, 1965). Between 1959 and 2011, the population increased dramatically, with a 10.7% increase from 1959 to 1970 (reaching 20 611 inhabitants by 1971), to 255 465 inhabitants in 2011 (CBS, 2011). However, an accurate comparison is difficult because population growth and related urbanization was accompanied by the annexation of 10 adjacent Village Development Committees to form the Pokhara metropolitan city area, with a population of 313 841 in 2011 (Rimal et al., 2015). If the population of Lekhanath municipality is added, then the joint population of both municipalities reaches 390 657 (CBS, 2011) and corresponds to a continuous, densified built-up area. Such a dramatic increase in population is mostly explained by migration from rural villages: the Pokhara metropolitan city offers better services (schools, hospitals, access to higher living standards) and much greater economic opportunities, among which tourism seems the most appealing.

12.3 A TOURIST CITY WITH MAJOR ATTRACTIONS RELATED TO ITS GEOMORPHOLOGY

In parallel with its urbanization, Pokhara gradually became a major place for tourism (Shakya, 1995). As early as 1899, Ekai Kawaguchi, a Japanese Buddhist monk and one of the first foreigners to visit the Pokhara valley, was noted as saying, "In all my travels in the Himalayas, I saw no scenery so enchanting as that which enraptured me in Pokhara" (Kawaguchi, 1909: pp. 42–43). The global media coverage of the first ascent of the Annapurna Peak (8091 m) by the Frenchman Herzog in 1950, followed by hippies settling along the Phewa lakeside in the early 1960s, made Pokhara an attractive city for its unique scenery of glaciated peaks and peaceful environment. In addition, being an ancient settlement, the Pokhara valley is culturally very rich with a combination of different religions, diverse ethnic communities, and their traditional festivals and religious sites. The first tourist information center was established in 1961, as tourism was rapidly perceived as a significant sector for socioeconomic development (Upreti et al., 2013). The Annapurna Conservation Area Project was launched in 1986 to preserve the exceptional natural environment (in the districts of Kaski, Myagdi, Lamjung, Mustang, and Manang). Progressively, Pokhara

FIGURE 12.8 **Landscape change between 1980 (aerial photo) and 2017.** (A) Detail of the Pokhara plain surface in the late 1970s between the Pokhara airstrip (*upper left corner*) and Amintara village (*lower right corner*). Note the Seti Khola gorge section to the north (aerial photo). (B) The same area in 2017 (Google Earth image) showing the large development of urban areas close to the airport. The International Mountain Museum (*large white building*) is prominent. (C) Future housing estate, Lekhanath Municipality (aerial view, ©M. Fort, 2009). (D) Progression of urbanization in Chorepatan: new buildings constructed to the detriment of rice fields (©Fort, 2009).

became a successful destination, either as the starting point for popular or more adventurous trekking routes in the mountains (Annapurna Circuit, Annapurna South Base Camp, etc.) or as a place to recover and relax: a place of peace and harmony, a place to discover

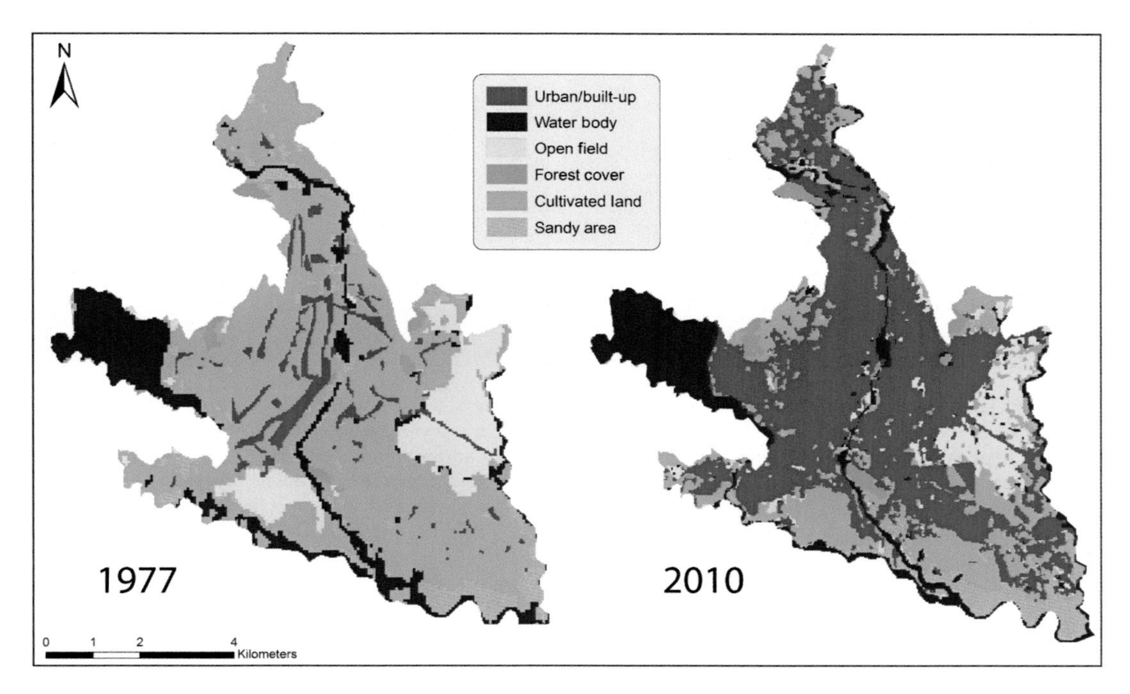

FIGURE 12.9 **Urbanization growth between 1977 and 2010 in Pokhara metropolitan city.** These two land-use maps show the considerable increase in urban area to the detriment of cultivated land. The built-up area increased from 3.50 km^2 in 1977 to 212.44 km^2 in 2010, which represents 6.33 and 51.42% of the total area, respectively. *Source: Adapted from Rimal, B. 2011. Urban growth and land use/land cover change of Pokhara sub metropolitan city. Nepal. J. Theor. Appl. Inf. Techn. 26(2), 118–129.*

both environmental and cultural heritage and diversity (having lakes, caves, gorges, waterfalls, and temples). The large, flat Pokhara valley, which is quite unusual in the Himalayan Mountains, appeared to be an ideal place for developing infrastructure (hotels, museums, transport network) and providing facilities and services (travel, trekking agencies) for tourism (Fig. 12.10A and B), in addition to those of a henceforward big city with health, education, industrial, commercial, and real estate opportunities.

All of these assets and attractions for urban and tourism development are directly related to the geomorphic formation of the Pokhara plain. Here, we briefly present some sites of geomorphic interest that are attractive to all visitors, including foreigners on their way back from mountain adventures or just spending some vacation time (Fig. 12.11).

12.3.1 Exceptional Viewpoints

In the early morning before sunrise, tourist buses rush to two exceptional viewpoints: the World Peace Stupa (Fig. 12.12A), which is a holy Buddhist shrine standing at 1020 m on the southern side of Phewa Lake, and Sarangkot (1592 m) on the north side of the valley (Fig. 12.3). Both offer a panoramic view of the Higher Himalayan Range, from Dhaulagiri

FIGURE 12.10 Importance of tourism in the Pokhara economy. (A) The International Mountain Museum, located south of the Pokhara Airport strip (Fig. 12.8B). Subdivided into several sections, it provides information about Himalayan geology, culture and environment, geography, ecology, and mountaineering history. Recent sections deal with climate change and natural hazards (©Fort, 2013). (B) Different generations of hotel construction in Lakeside, Baidam area, west of Pokhara city. Note the increase in the size and number of floors and the category upgrade from older to more recent buildings (from right to left), showing how prosperous this area is in terms of tourism business (©Fort, 2016).

FIGURE 12.11 Main places of tourism interest.

(8167 m) and Annapurna I (8091 m) to Manaslu (8163 m), with the light changing as fast as the sun rises in the sky. From the Peace Stupa, the lake is visible in the foreground, whereas Sarangkot offers an overview of the complex terrace system of the Pokhara valley.

12.3.2 The Lakes: Legacies of Catastrophic Events

Representing remnants of the Medieval megafloods, which transported thick Pokhara gravel that plugged the mouths and lower reaches of the Seti Khola tributaries (Fort, 2010; Gurung, 2002; Schwanghart et al., 2016; Stolle et al., 2017), the Pokhara valley's seven lakes (Phewa, Begnas, Rupa, Maidi, Khaste, Gunde, Dipang) enhance the region's reputation as a natural wonderland. In fact, Phewa Lake, which is considered to be the second largest lake in Nepal, is certainly the most appealing, being the closest to airport and bus stations. It provides opportunities for boating, with the nearby snow-clad mountains reflecting in its still waters. One popular boat trip visits a small island, a paleotopographic relict of the drowned Pardi Khola valley, upon which the two-storied pagoda Tal Barahi temple devoted to the

FIGURE 12.12 **Major attractions of the Pokhara valley.** (A) World Peace Buddhist Stupa located on Rani Ban hilltop, south of the Phewa Lake: from there, visitors enjoy breathtaking views of the Himalayan Range (©Fort, 2016). (B) Tal Barahi Temple (©Fort, 2013). (C) The newly built Baidam zone viewed from the Peace Stupa ridge. The height of some buildings is noteworthy (©Fort, 2016). (D) Paragliding flight above the lake, with the glaciated mountains in the background (©Fort, 2013).

Hindu goddess Durga has been built partly hidden within the grove rich in "*Pipal*" (*Ficus religiosa*) sacred trees (Fig. 12.12B).

The flat Baidam area, east of the lake, which was grazing land until 30 years ago, has experienced the most rapid development in terms of hotels, guesthouses, restaurants, and trekking shops (Fig. 12.12C). More recently, activities like paragliding enable visitors to enjoy both the beauty of the mountains and the traditional landscapes with paddy fields and hamlets, newly urbanized areas, and the Phewa Lake below (Fig. 12.12D).

12.3.3 Karstic Features: Gorges and Caves Related to the Old Ghachok Conglomerates

Gorges and caves add more value to Pokhara's natural beauty. The Seti gorges, cut into the Ghachok conglomerates, traditionally provided easy ways to cross the Seti Khola due to their narrowness. Nowadays, crossed by road bridges, they may be observed at several places, such as north of Pokhara (K.I. Singh Bridge), north of Ramghat hollow (Mahendra Pul, "*pul*" means bridge in Nepalese) and south of Ramghat (Prithivi Highway Bridge), and south of the domestic airport (Fig. 12.13C). The caves are more special: they also formed in the Ghachok, limestone-rich conglomerates that are found below the Pokhara gravel. Mahendra cave (Fig. 12.13B), and Bat's cave are sited north of Pokhara, close to the confluence with the Kali Khola, whereas the holy Gupteshwar Mahadev cave was formed downstream of Davi's falls (Fig. 12.13A), a site where the Pardi Khola waters, draining Phewa Lake, abruptly disappear along a narrow crack into the calcareous bedrock. All of these features of karstic origin are still developing underground and may be the cause of serious hazards, such as sinkholes.

12.3.4 Bhim Kali Dhunga

Preserved on the Prithivi Narayan Campus, Tribhuvan University, weighing 3000 t and 10 m in diameter, the famous Bhim Kali Dhunga ("*dhunga*" means stone in Nepalese), a giant gneissic boulder, is quite unique (Fig. 12.6). A local legend tells that this rock was thrown down from Machapucchare Peak by the powerful hero Bhim (a brave god in Hindu mythology) and is devoted to the deity Kali. In fact, Bhim Kali Dhunga represents the final stage of Pokhara gravel deposition and its date coincides with the timing of a large earthquake in 1681 AD (Schwanghart et al., 2016). This boulder and many other smaller ones, such as Chiple Dhunga and Mandre Dhunga, were brought by a highly competent, highly muddy flow that was nourished by coarse debris detached from the lower part of the High-Himalayan front. Most of these boulders have now disappeared after being quarried for building purposes.

12.3.5 Hot Springs

The hot springs ("*tatopani*" in Nepalese) of Kharpani are another site of interest, located in the very north of the Pokhara valley (Fig. 12.19 for the location). As in many other places in Nepal, the Kharpani tatopani are located close to the MCT, the major tectonic discontinuity between the Lesser and Higher Himalayas. At depth, groundwater in contact with heated

FIGURE 12.13 **Major geosites of the Pokhara valley.** (A) A very popular, tourist place: Davi's falls along the Pardi Khola canyon, a few hundred meters north of its junction with Phusre Khola (©Fort, 2009). (B) Mahendra cave (Batulechaur, northeast of Pokhara) is named after the name of the late King Mahendra Bir Bikram Shah Dev. The cave follows the coarse stratification of the Ghachok conglomerates (note coarse *blocks on the left*) and contains stalagmites and stalactites (the latter hanging from the roof) (©Fort, 2016). (C) The gorges of the Pardi Khola can be observed from the Siddhartha Highway, south of Davi's falls and east of the Gupteshwar Mahadev cave (©Fort, 2009). (D) Old Gurung man taking a therapeutic bath in the "*tatopani*" (hot springs) of Kharpani, located closed to the Main Central Thrust of the Higher Himalayas (©Fort, 2016).

rocks (geothermal energy) rises up along fault cracks before springing out at the surface, close to river level (Girault et al., 2014). Due to their high-mineral content, the springs are traditionally attended by local people for therapeutic purposes (healing effects, blood purification, muscle relaxation, rheumatism treatment) (Fig. 12.13D). The Kharpani hot springs are a very busy place, especially on Saturdays when young people come for picnics and bathing.

12.4 POTENTIAL THREATS: NATURAL HAZARDS AND RISKS

The expansion of Pokhara and its suburbs was rapid and largely unplanned yet, in such an area, natural hazards are quite frequent (floods, earthquakes, slope, sinkhole collapses). These, together with social inequalities and inadequate land use by local people, have contributed to increasing the vulnerability of urban populations (Rimal et al., 2015). We present here a few natural hazard events that occurred recently and, as expressed by Gurung et al. (2015, p. 102), we should stress that "the event however complex it may seem, is a natural process, which went on to become a disaster due to lack of preparedness." This will be a real challenge for the authorities in charge of Pokhara submetropolitan city and Lekhnath municipality.

12.4.1 Flood Hazards

On May 5, 2012, a devastating hyperconcentrated flood occurred in the upper part of the Seti Khola valley, seriously affecting infrastructure, settlements, and tourism spots located along the river (Fig. 12.14A). It was unpredicted and the consequences were fatal, with more than 70 deaths and missing people reported, mostly concentrated near the Kharpani hot springs, and nearby settlements of the upper Seti Khola valley (Bhandary et al., 2012; Oi et al., 2014).

The nature of this event is very complex and unique: a rocky wall collapse of about 22 million m^3 occurred on the western flank of Annapurna IV (Fig. 12.14B), east of the Sabche cirque (the source of the Seti Khola). The fall of a rock and ice section of nearly 1600 m caused the two materials to shatter and liquefy, creating a debris flow downstream of the Seti Khola gorge (Kargel et al., 2013). The flood developed in different waves and propagated along the inhabited valley at an average speed of 13 m s^{-1}, with a peak flow volume of about 12 300 m^3 s^{-1}, reaching a +15 m elevation relative to the river bed in some places (Oi et al., 2014), which has flood characteristics that explain the heavy human losses. The sandy flow affected many informal settlements on the lower Seti Khola terraces and ultimately passed through the heart of Pokhara, at Ramghat, where a pond formed and threatened the life of people ashore in Pokhara (Fig. 12.14D). Riverbank erosion was very destructive, and damaged several road sections north of Pokhara, while a major water pipe supplying Pokhara metropolitan city was cut along several hundred meters near Puranchaur (Fig. 12.14C; Gurung et al., 2015). Overall, this event resulted in about NR 49,25 million of property losses (Gurung et al., 2015, p. 106).

The cause of this flood has been widely debated, particularly because it may happen again, with even more serious consequences across the urbanized Pokhara valley (Bhandary et al., 2012). The flood has been successively interpreted as the result of (1) a glacial lake collapse; (2) a landslide outburst flood; or (3) related to a sudden release of subterranean or impounded waters across the 2-km deep Seti Khola gorge (Kargel et al., 2013), this last option being the most likely. The fine sediment sampled in the slurry flow had similar sedimentological and compositional characteristics as the Seti Khola floodplain deposits (including both recent sediment and ancient terraces underlain by Pokhara gravel) that were rich in limestone and calcschists belonging to the Nilgiri formation (Colchen et al., 1981), indicating that they have all come from the same source, as from the west face of Annapurna IV (Kargel et al., 2013).

FIGURE 12.14 **Impacts of the May 15, 2012, Seti Khola flood.** (A) Kharpani Bridge, *upper part* of the Pokhara plain (upstream view). The muddy flow overpassed the bridge, destroying houses settled on the opposite flat terrace. Many people were killed, including those bathing in the hot springs (Nep. "*tatopani*") visible in the *lower left corner* near the river (©Fort, 2013). (B) The effect of the 2012 Seti flood, which destroyed the tatopani Bazaar (hot spring), killing 72 people. Annapurna IV in the background from where the rock collapse took place (upstream view) (©Adhikari, 2013). (C) Near Puranchaur (upstream view), the terrace collapse (*left bank*) triggered by the higher, muddy flows of the Seti Khola damaged the water pipe feeding the Pokhara metropolitan city. Repaired, but set along the same design (i.e., very close to the river level), this water pipe might be affected again by another flood (©Fort, 2016). (D) The 2012 Seti Khola flood reached Ramghat, an open, wide space between two gorge sections of the Seti Khola. A large amount of the debris brought by the flood was mined soon after the flood (downstream view) (©Fort, 2016).

In fact, as sudden and destructive as the 2012 event was, it appears as a "weak" reminder of past Medieval events (1100, 1255, and 1344 AD, Schwanghart et al., 2016; Stolle et al., 2017) that built up the "plain" of the Pokhara valley. This means that an even greater disaster could happen at any time, as there are several possible modes of catastrophic discharge of water and sediment into the Seti River that could wipe all settlements and infrastructure entirely off the map. One crucial point is the potential trigger of such an event. The 2012 rockwall collapse was not triggered by rainfall or an earthquake, in contrast to the aforementioned Medieval events, but the volume of the collapse was large enough to be recorded by the seismometer network (Dwidedi and Neupane, 2013). Among the possible explanations, either a classic decohesion process of high-mountain cliffs and/or the impact of climate warming (melting of permafrost ice present within the rock joints and cracks), are reasonable options.

12.4.2 Earthquake Hazards

The Pokhara valley should be considered as potentially subject to infrequent, very large seismic ruptures, as its recent history indicates. In fact, it is located right at the foot of the Greater Himalayas near the ramp of the detachment fault, where earthquakes are most frequent. According to recent studies (Bollinger et al., 2014, 2016; Lavé et al., 2005; Mugnier et al., 2013; Pant, 2002; Rajendran et al., 2015), the return periods of very large earthquakes (Mw >8) have been assessed on the basis of historical events, such as Pokhara's. These studies have identified along the Himalayan Range those segments that have not been severely affected for the past circa 500–600 years, that is, those zones that have not accommodated the continuous convergence (5 cm y^{-1}) of the Indian and Asian plates by a slipping movement. This accumulation of energy (or slip deficit) makes these areas the most susceptible to be next affected by a high-magnitude (>8) earthquake. The segment from western Nepal to the east of Pokhara appears to be the most important seismic gap in Nepal (Bollinger et al., 2016; Fig. 12.15); all the more so because the recent April 25, 2015, Gorkha earthquake, only 70 km away from Pokhara, did not impact the Pokhara valley; instead, it propagated eastward, severely affecting Kathmandu and adjacent areas (Bollinger et al., 2016). This means that along the area extending from Gorkha to the western Nepal border, another large earthquake is already due and may occur at any time.

Although earthquakes are the least predictable natural hazards, assessing the seismic risk and the vulnerability of such a city as Pokhara has become an important issue. A detailed study, prepared by the United Nations Development Programme/UNDP on Earthquake Risk Reduction and Recovery Preparedness/ERRRP (2009), has clearly assessed potential hazards, stressing the elements at risk, their vulnerability, the costs of potential damages, and, more importantly, preparedness issues and evacuation plans. This study provides an Earthquake Disaster Preparedness and Response Framework Report for the Pokhara metropolitan city. Although many aspects are carefully considered (evacuation sites, open spaces, emergency planning, accessibility for rescue), it is not certain that the scale used is fully relevant: the focus is more on buildings and infrastructure and does not include indirect hazards at the catchment scale (mountain collapse, landslides, subsequent megafloods) that could be induced by such a giant earthquake similar to those caused by the Medieval ones that shaped the Pokhara valley. In particular, the newly developed built-up areas on the lower terraces of the northern Pokhara valley make them particularly vulnerable to such events of low frequency, but exceptional magnitude.

12.4.3 Sinkholes and Subsidence

Karst topography is very common in the Pokhara valley because of the large extent of carbonate-cemented terraces of Ghachok conglomerates. The most common karst features are caves (e.g., the aforementioned tourism, religious caves) as well as sinkholes caused by dissolution, which progress either on or below the surface. This hidden, subterranean process may lead to the collapse of the surface layers without warning, such as in Chorepatan (southwest of the domestic airport). There are also areas susceptible to subsidence due to soft clay, peat, and lime mud at the entrance of tributary valleys (Koirala et al., 1996).

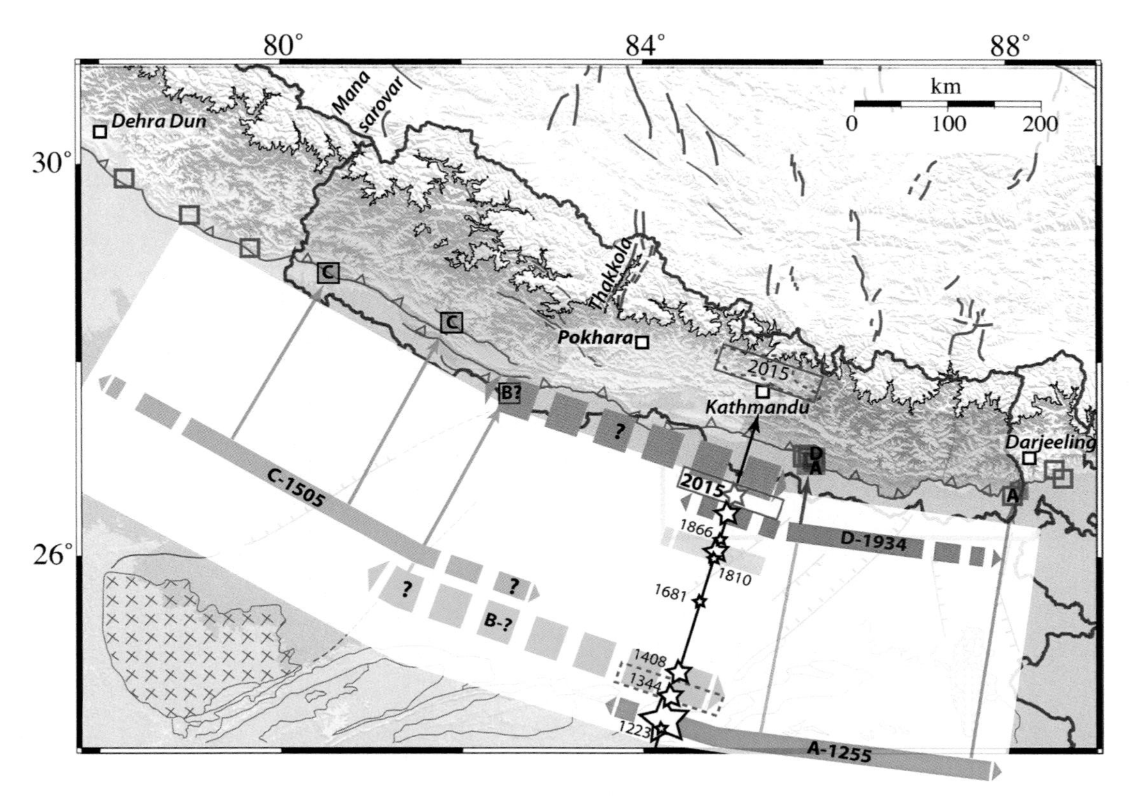

FIGURE 12.15 **Palaeoseismicity of Nepal.** Rupture lengths and return times of great Himalayan earthquakes in Nepal since 1223 AD, as deduced from limited macroseismic historical evidence and available paleoseismological/ morphotectonic data (*squares*: investigated sites). Surface ruptures are noted as A (1255); B (possible 1344 or 1408 AD rupture trace, assuming that one of these events was a great earthquake rupturing the front all the way from the east of Kathmandu to west of Pokhara); C (1505); and D (1934). The position in time of the *gray dotted line* closest to the Himalayan front, within the gap between Kathmandu and Pokhara, is model-dependent. The line drawn from Kathmandu to the south corresponds to the extent of blind Mw 7+ events similar to that in 2015. Note the question marks referring to high uncertainty. The probability of an earthquake occurring in the area is high due to the long gap following the next surface rupturing event. *Source: Modified from Bollinger, L., Sapkota, S.N., Tapponnier, P., Klinger, Y. 2016. Slip deficit in central Nepal: Omen for a repeat of the 1344 AD earthquake? Earth, Planets Space. 68, 12, doi:10.1186/ s40623-016-0389-1.*

In the context of the dramatic urban growth of the Pokhara valley, the development of both natural and anthropogenic sinkholes is progressing and now poses a serious threat to buildings, infrastructure, and farmland depending on the different parts of the valley (Rimal et al., 2015; Fig. 12.16). More specifically, there is little control of superficial water fluxes (runoff, irrigation waters), so that subsurface flow, solution cavities, and sinkholes are widely developed in some areas of the Pokhara valley (Gautam et al., 2000). Dhital and Giri (1993) clearly illustrated the effect of karst-related processes by showing the local collapse of the Seti Khola

Bridge. The wide area of Ramghat, open between two very narrow sections of gorges, in which the Seti Khola is not visible, is probably where large subterranean karstic pipes did collapse.

Another place where sinkholes are of great concern is the Jamire area of Armala VDC, located on the northeast side of the valley in the Kali Khola catchment. This is a suburban area, where urbanization is progressing in the middle of irrigated rice fields. The soils are composed of fine, silty clay sediments (the legacy of backwash water sediments deposited during and after one Medieval event) (Fig. 12.17A), overtopped by recent alluvial gravel. In November 2013, many sinkholes appeared, spherical in shape (Fig. 12.17B) and generally of small diameter (<10 m) yet with a depth close to 7 m, posing a serious threat to the already settled population and their built structures as well as agricultural farmland (Khatiwada, 2015; Pokhrel et al., 2015). It seems that poor control of irrigated water, combined with a lowering of the water level in the Kali Khola, generated these sinkholes, which, although back-filled

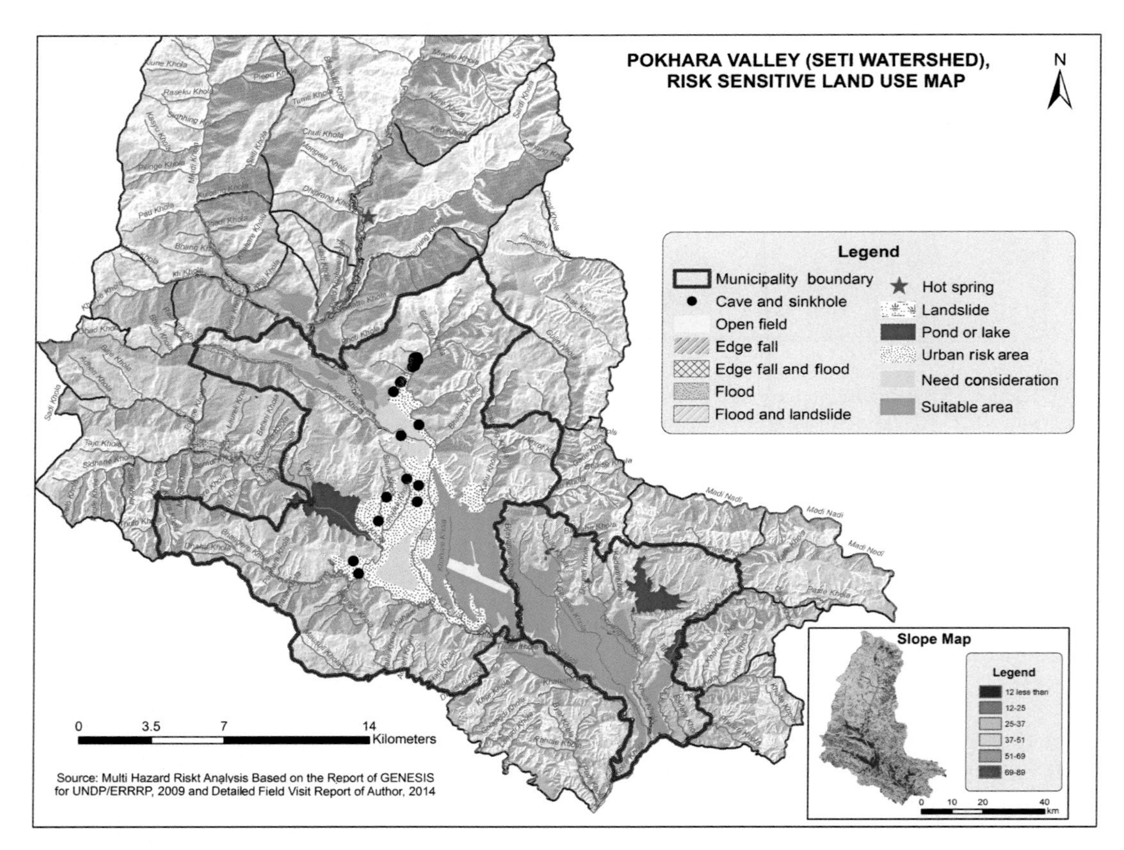

FIGURE 12.16 **Risk sensitivity map of the Pokhara submetropolitan city and its surroundings.** Most of the risks are close to the vicinity of the Seti Khola valley and its tributaries and, in the city itself, consist mostly of caves and sinkhole hazards. The northernmost sinkholes correspond to the Armala area in the Kali Khola valley. *Source: Adapted from Rimal, B., Baral, H., Stork, N., Paudyal, K., Rijal, S. 2015. Growing city and rapid land use transition: assessing multiple hazards and risks in the Pokhara Valley, Nepal. Land 4(4), 957–978, doi:10.3390/land4040957.*

with gravel by local people, continue to progress. In fact, human encroachment in the Duhuni Khola, a right-bank tributary of the Kali Khola, increases subsurface waterflow and, hence, indirectly helping the development of a piping mechanism in the silty limey layer. This, in turn, favors the progression of underground erosion and sinkhole formation and poses a serious threat to future, planned urbanization.

All of these geohazards have affected the growth and development of Pokhara, particularly as a result of varying land prices and housing development (Adhikari and Seddon, 2002). This is why larger and taller buildings are generally avoided in some places due to soil instability that is inadequate for deep foundations. Currently, the old *bazaar* (market) and the main shopping zone around the Mahendra Pul area appear to be the safer places (Fig. 12.11). Yet, the city's growth and the need for new homes are so high that they may lead to contradictory behavior by the population.

12.4.4 An Example of Anthropogenic-induced Hazards: Sand Mining

In recent decades, the rapid development of the city has dramatically increased the demand for construction materials. Many sand and gravel pits have been opened along the Seti Khola, particularly after the 2012 Seti Khola flood, because much debris was trapped in the upstream part of the sites where the flood current was constrained, such as Ramghat, or where the river could spread (meander loops) (Fig. 12.18A and B). Although sand and gravel mining is allowing some poor families to make a living, such practices are also the cause of environmental deterioration (disappearance of aquatic life and habitat) and natural hazards. More specifically, they may cause bank erosion (undercutting, shallow landslides, collapses), thus, increasing the risk for adjacent settlements, especially when the river widens its active channel during the monsoon season. Mining also makes the land surface more unstable and may cause it to collapse. This process may be further accentuated by a deficit in the solid load of the river, leading to river scouring and incision, which may

FIGURE 12.17 **Land subsidence and sinkholes.** (A) Cross-section of the Armala valley, showing the backwash limy layers (©Fort, 2016). (B) Sinkhole formation in the Armala area (northeast of Pokhara) due to anthropogenic factors: sand and gravel mining in the river lowered the bed level and groundwater eroded all of the subsurface silt and clay material (©Adhikari, 2013).

FIGURE 12.18 **Sand mining in the Seti River floodplain.** (A) North of Pokhara, as seen from the Upallo Dip area (*left bank* of the Seti Khola, downstream from Lamachaur), people are carrying out sand mining for their livelihoods. They have constructed groins to trap sand particles carried by the river during the monsoon (©Fort, 2016). (B) Dhulegaunda gorges are cut into the Ghachok conglomerates, so that the Seti Khola flow is slowed down, forcing the aggradation of gravel and sand upstream. This provides an unlimited, renewable resource for sand mining (©Fort, 2015).

locally endanger hydraulic structures like dams, culverts, and bridges and, more generally, induce changes in river morphology and eventually a lowering of the groundwater table. In turn, the drop in the watertable may locally favor the development of sinkholes, such as those observed nearby Armala. This has been observed elsewhere, for example, in Florida, where several sinkholes formed following overabstraction of water (Thornbush, 2017).

12.4.5 Environmental Impacts of Urbanization Versus Geoheritage Preservation

The previous examples illustrate the recent progression of construction in naturally hazardous areas and the indirect impacts of the search for building materials. Other negative trends are also observed.

First, the high-demand for building land has resulted in the densification of habitat and its extension to more dangerous places, such as closer to the insecure edges of the Seti Khola (Fig. 12.19A), while favoring multistoried buildings that hide the mountain view of the inhabitants living in smaller, older buildings.

Second, constructions are observed in scenic places like along the Pardi canyon and the Davi's falls area. This will progressively reduce the attractiveness of these areas and destroy the aesthetics of such geoheritage.

Third, pollution has significant impacts. Air pollution is everywhere, with the increase in traffic and, hence, spread of smoke clouds above the valley, hiding the mountain view from the city, especially during the premonsoon and dry season. Water pollution has also become a problem: whereas Phewa Lake was the main source of drinkable water in the early days, it is now largely polluted by the septic tanks of adjacent hotels and restaurants as well as chemicals, pesticides, and fertilizers. Similarly, the Seti Khola and its tributaries are polluted by domestic sewerage pipes (Fig. 12.19B) and/or rubbish directly thrown into river gorges (Fig. 12.13C). Undoubtedly, haphazard urbanization and poor urban planning are destroying the natural beauty of Pokhara.

There is clearly a need for strict environmental regulations without which the tourism potential will be spoiled and the most attractive geosites hidden within a totally urbanized valley. There are maps of hazardous/risky places (Rimal et al., 2015), but they certainly need continuous updating to avoid construction in threatened areas (Fig. 12.16). The preservation of geoheritage sites is another priority because tourism is the most important and lucrative industry in the Pokhara valley; and because most visitors will soon be able to land directly at the new Pokhara International Airport and enjoy the proximity of the Higher Himalayan snowy peaks.

12.5 CONCLUSIONS

The Pokhara valley has a very unique development history compared to other areas of the Himalayan Range. Over the last centuries, the Pokhara valley has been shaped by catastrophic events triggered by earthquakes and amplified by the glaciated environment and steep topographic gradient. Already a key administrative and commercial center, Pokhara has now become a major tourist attraction in Nepal. Tourists flow there every year because of its many natural geoheritage sites, with magnificent views of the Annapurna Range, and its position as

FIGURE 12.19 **Anthropogenic risks.** (A) Lack of control of wastewaters (sewerage pipes), which flow directly into the Seti Khola canyon, left bank, south of Mahendra Pul (©Fort, 2009). (B) Densification of urban areas leads to building in hazardous places like this along the right bank of the Seti Khola canyon, south of Mahendra Pul, with a risk of edge collapse (©Fort, 2009).

the starting point of mountain treks. However, natural hazards and other anthropogenic factors are degrading the environmental appeal of the valley. The rapid growth in population has induced haphazard urbanization, increasing the vulnerability and potential risks for people day after day. The 2012 Seti Khola flood, for example, is a warning of what could happen again in the near future—and, furthermore, seismic risk cannot be excluded as well. Therefore, to protect the Pokhara valley residents from future hazards recommendations, such as the implementation of detailed hazard and risk zoning maps, building codes, geotechnical tests prior to construction, and informing the population (which includes many migrants from rural areas), would raise awareness collectively and help the Pokhara valley to achieve a sustainable and harmonious development in one of the most dramatic landscapes in the world.

Acknowledgments

Our contribution is a tribute to the late Dr. Harka Bahadur Gurung (1939–2006), a Nepalese geographer, experienced, respected politician, and one of the most brilliant intellectuals of his country. His seminal papers on Pokhara valley and Nepal were of great inspiration to us. We are very grateful to Rémi de Matos-Machado (Univ. Paris-Diderot) for his help in finalizing illustrations. We are also deeply indebted to Laurent Bollinger (CEA-DASE, Bruyères-le-Chatel, France), who provided personal data and gave authorization to reproduce them. Finally, we acknowledge with warm thanks the fruitful discussions we had in the field with W. Schwanghart, A. Stolle and O. Korup (Institute of Earth and Environmental Sciences, University of Potsdam, Germany), S.N. Sapkota (Department of Mines and Geology, Kathmandu), and N. Gurung (Kadoorie Agricultural Aid Association, Pokhara).

References

Adhikari, J., Seddon, D., 2002. Pokhara, Biography of a Town. Mandala Book Point, Kathmandup. pp. 273.

Bhandary, N., Dahal, R.K., Okamura, M., 2012. Preliminary understanding of the 1255 Seti River Debris-flood in Pokhara, Nepal. Nepal Eng. Ass. JC Newsletter 6 (1), 29–38.

Bollinger, L., Sapkota, S.N., Tapponnier, P., Klinger, Y., 2016. Slip deficit in central Nepal: omen for a repeat of the 1344 AD earthquake? Earth Planets Space 68, 12. doi: 10.1186/s40623-016-0389-1.

Bollinger, L., Sapkota, S.N., Tapponnier, P., Klinger, Y., Rizza, M., Van Der Woerd, J., Tiwari, D.R., Pandey, R., Bitri, A., Bes de Berc, S., 2014. Estimating the return times of great Himalayan earthquakes in eastern Nepal: evidence from the Patu and Bardibas strands of the Main Frontal Thrust. J. Geophys. Res. 119, 7123–7163.

Caplan, L., 1975. Administration and Politics in a Nepalese Town. Oxford University Press, Londonp. pp. 266.

CBS (Central Bureau of Statistics), 2011. Government of Nepal.

Colchen, M., Le Fort, P., Pêcher, A., 1981. Geological map of Annapurnas–Manaslu–Ganesh Himalaya of Nepal. In: Gupta, H.K., Delany, F.M. (Eds.), Zagros-Hindu Kush, Himalaya: Geodynamic Evolution. American Geophysical Union, Washington, DC, (scale 1:200,000).

Dhital, M.R., Giri, S., 1993. Engineering-geological investigations at collapsed the Seti Bridge site, Pokhara. Bull. Depart.Geol., Tribhuvan University 3 (1), 119–141.

Dwidedi, S., Neupane, Y., 2013. Cause and mechanism of the Seti River flood, 5th May 2012, western Nepal. J. Nepal Geol. Soc. 46, 11–18.

Fort, M., 1987. Sporadic morphogenesis in a continental subduction setting: an example from the Annapurna Range, Nepal Himalaya. Z. Geomorph. N. F. Suppl.-Bd. 63, 9–36.

Fort, M., 1988. Catastrophic sedimentation and morphogenesis along the High Himalayan front: implications for palaeoenvironmental reconstructions. In: Whyte, P. (Ed.), The Palaeoenvironments of East Asia From the Mid-Tertiary. Centre of Asian Studies, Hong Kong, pp. 171–194.

Fort, M., 2010. The Pokhara Valley: a product of a natural catastrophe. In: Migon, P. (Ed.), Geomorphological Landscapes of the World. Springer, Dordrecht, The Netherlands, pp. 265–274.

Fort, M., Peulvast, J.P., 1995. Catastrophic mass-movement and morphogenesis in the Peri-Tibetan Ranges, examples from West Kunlun, East Pamir and Ladakh. In: Slaymaker, O. (Ed.), Steeplands. Wiley and Sons, pp. 171–198.

Gautam, P., Pant, S.R., Ando, H., 2000. Mapping of subsurface karst structure with gamma ray and electrical resistivity profiles: a case study from Pokhara valley, central Nepal. J. Appl. Geophys. 45 (2000), 97–110.

Girault, F., Bollinger, L., Bhattarai, M., Koirala, B.P., France-Lanord, C., Rajaure, S., Gaillardet, J., Fort, M., Sapkota, S.N., Perrier, F., 2014. Large-scale organization of carbon dioxide discharge in the Nepal Himalayas. Geophys. Res. Lett. 9 (2014). doi: 10.1002/2014GL060873.

Gurung, D.R., Maharjan S.B., Khanal N.R., Hoshi G., Murthy M.S.R., 2015. Seti flash flood: technical analysis and DDR interventions. In: Nepal Disaster Report 2015, MoHA and DpNet-Nepal, 4.4, 101–110.

Gurung, H.B., 1965. Pokhara Valley, Nepal Himalaya: A Field Study in Regional Geography. Unpublished (Ph.D. thesis), University of Edinburgh, pp. 246.

Gurung, H.B., 2002. The Pokhara Valley: A Geographical Survey. Nepal Geographical Society c/o Central Department of Geography Tribhuvan University, Kirtipur, Kathmandu.

Hormann, K., 1974. Die Terrassen an der Seti Khola: Ein Beitrag zur Quartären Morphogenese in Zentral Nepal. Erdkunde 28 (3), 161–176.

Kargel, J.S., Paudel, L., Leonard, G., Regmi, D., Joshi, S., Poudel, K., Thapa, B., Watanabe, T., Fort, M., 2013. Causes and human impacts of the Seti River (Nepal) disaster of 2012. In: Proceedings of Glacial Flooding and Disaster Risk Management Knowledge Exchange and Field Training (July 11–24, 2013). Huaraz, Peru., HighMountains.org/workshop/peru-2013.

Kawaguchi, E., 1909. Three Years in Tibet. The Theosophical Office, Madras pp. 42-43.

Khatiwada, B., 2015. Hydrogeological Study in Thulibeshi Phat of Armala valley, Kaski District With Special Reference to Recent Occurrence of Sinkholes (MSc. dissertation), Tribhuvan University, Central Department of Geology, Kirtipur, Kathmandu.

Koirala, A., Rimal, L.N., Sakrikar, S.M., Pradhananga, U.B., Pradhan, P.M., Hanisch, J., Jäger, S., Kerntke, M., 1996. Engineering and Environmental Geological Map of the Pokhara Valley (1:50000). Department of Mines and Geology, in cooperation with BGR, Kathmandu and Hannover.

Lavé, J., Yule, D., Sapkota, S., Basant, K., Madden, C., Attal, M., Pandey, R., 2005. Evidence for a great medieval earthquake (~1100 A.D.) in the Central Himalayas Nepal. Science 307, 1302–1305.

Macfarlane, A., 1976. Resources and Population: A Study of the Gurungs of Nepal. Cambridge University Press, London, pp. 364.

Mierow, D., 1997. Thirty Years in Pokhara, first ed. Pilgrims Book House, Varanasi, pp. 126.

Mugnier, J.L., Gajurel, A., Huyghe, P., Jayangondaperumal, R., Jouanne, F., Upreti, B., 2013. Structural interpretation of the great earthquakes of the last millennium in the central Himalaya. Earth-Sci. Rev. 127, 30–47.

Oi, H., Higaki, D., Yagi, H., Usuki, N., Yoshino, K., 2014. Report of the investigation of the flood disaster that occurred on May 5, 2012 along the Seti River in Nepal. Int. J. Eros. Control Eng. 7 (4), 111–117.

Pandey, R.N., 1987. Palaeo-environment and prehistory of Nepal. Contr. Nepal Stud. CNAS-TU, Kathmandu 14 (2), 111–124.

Pant, M.R., 2002. A step toward a historical seismicity of Nepal. Adarsa 2, 29–60.

Pokhrel, R.M., Kiyota, T., Kuwano, R., Chiaro, G., Katagiri, T., Arai, I., 2015. Preliminary field assessment of sinkhole damage in Pokhara, Nepal. Int. J. Geoeng. Case Hist. 3 (2), 113–125. doi: 10.4417/IJGCH-03-02-04, http://casehistories.geoengineer.org.

Poudel, K.R., 2002. Causes of Land Use Change and Their Effect in Environment in Pokhara Valley, Mountain Hills of Nepal. HPE Forum, Tribhuvan University, P.N. Campus, Pokhara, 2(2), pp. 31–41.

Rajendran, C.P., John, B., Rajendran, K., 2015. Medieval pulse of great earthquakes in the central Himalaya: viewing past activities on the frontal thrust. J. Geophys. Res. Solid Earth 120 (3), 1623–1641. doi: 10.1002/2014JB011015.

Rimal, B., 2011. Urban growth and land use/land cover change of Pokhara sub-metropolitan city. Nepal. J. Theor. Appl. Inf. Technol. 26 (2), 118–129.

Rimal, B., Baral, H., Stork, N., Paudyal, K., Rijal, S., 2015. Growing city and rapid land use transition: assessing multiple hazards and risks in the Pokhara Valley, Nepal. Land 4 (4), 957–978. doi: 10.3390/land4040957.

Schwanghart, W., Bernhardt, A., Stolle, A., Hoelzmann, P., Adhikari, B.R., Andermann, C., Tofelde, S., Merchel, S., Rugel, G., Fort, M., Korup, O., 2016. Repeated catastrophic valley infill following medieval earthquakes in the Nepal Himalaya. Science (New York, N.Y.) 351 (6269), 147–150. doi: 10.1126/science.aac9865.

Shakya, S.R., 1995. A Glimpse of Pokhara. Asta K. Shakya, Pokhara.

Stolle, A., Bernhardt, A., Schwanghart, W., Hoelzmann, P., Adhikari, B.R., Fort, M., Korup, O., 2017. Catastrophic valley fills records large Himalayan earthquakes, Pokhara, Nepal. Quat. Sci. Rev. 177, 88–103.

Thornbush, M.J., 2017. Part 2: Spatial-temporal occurrences of sinkholes as a complex geohazard in Florida, USA. J. Geol. Geophys. 6 (3), 286–291.

Udas, G.M., 2013. Historical review of natural and cultural heritage of paradise Pokhara. In: Upreti, B.R., Upadhayaya, P.K., Sapkota, T. (Eds.), Tourism in Pokhara Issues, Trends and Future Prospects for Peace and Prosperity. Pokhara Tourism Council (PTC), South Asia Regional Coordination Office of the Swiss National Centre of Competence in Research (NCCR North- South) and Nepal Center for Contemporary Research (NCCR), Kathmandu, pp. 27–69.

UNDP/ERRRP (United Nations Development Programme/Earthquake Risk Reduction and Recovery Preparedness Programme for Nepal), 2009. In: Joshi, A. (Ed.), Earthquake Vulnerability Profile and Preparedness Plan of Pokhara Sub-Metropolitan City. United Nations Development Programme/Earthquake Risk Reduction and Recovery Preparedness Programme for Nepal, Kathmandu, pp. 100.

Upreti, B.R., Upadhayaya, P.K., Sapkota, T. (Eds.), 2013. Tourism in Pokhara Issues, Trends and Future Prospects for Peace and Prosperity. Pokhara Tourism Council (PTC), South Asia Regional Coordination Office of the Swiss National Centre of Competence in Research (NCCR North- South) and Nepal Center for Contemporary Research (NCCR), Kathmandu, pp. 356.

Von Fürer-Haimendorf, C, 1975. Himalayan Traders. J. Murray, London, pp. 315.

Yamanaka, H., Yoshida, M., Arita, K., 1982. Terrace landform and quaternary deposit around Pokhara Valley, Central Nepal. J. Nepal Geol. Soc. 4, 101–120, Special issue.

SECTION V

URBAN STONE DECAY: CULTURAL STONE AND ITS SUSTAINABILITY IN THE BUILT ENVIRONMENT

13 *Urban Stone Decay and Sustainable Built Environment in the Niger River Basin* 261

14 *A Geologic Assessment of Historic Saint Elizabeth of Hungary Church Using the Cultural Stone Stability Index, Denver, Colorado* 277

15 *Photographic Technique Used in a Photometric Approach to Assess the Weathering of Pavement Slabs in Toronto (Ontario, Canada)* 303

CHAPTER

13

Urban Stone Decay and Sustainable Built Environment in the Niger River Basin

*Olumide Onafeso**, *Adeyemi Olusola***
*Olabisi Onabanjo University, Ago Iwoye, Ogun State, Nigeria;
**University of Ibadan, Ibadan, Nigeria

OUTLINE

13.1 Introduction 261
13.2 Decay of Clay Sandstones and Mudstones 262
13.3 Evidence of Rock Decay Consequent to Urban Stone Decay 264
13.3.1 Geology, Materials, and Deep Rock Decay 264
13.3.2 Chemical Rock Decay and Sediment Production 266
13.4 Warm Wet Climates of the River Niger Basin Region 269
13.5 Prevailing Atmospheric Pollution of the Urban-built Environment 270
13.6 Conclusion 273
References 274

13.1 INTRODUCTION

Stone-built urban cultural heritage in many places has left rich impressions about historic settlements and its people. Research around physical and chemical rock decay (i.e., weathering) processes leading to urban stone decay have been of significant influence to the understanding of the events and peoples of prehistoric times. However, such studies have only come to the fore in recent times and, as such, little is yet known of many regions of the world. Although the processes behind stone decay are to be understood to avert the

Urban Geomorphology. http://dx.doi.org/10.1016/B978-0-12-811951-8.00013-8

further decadence of urban-built environments, the challenge of rapid urbanization today is rather that of planning instead of sustainable development.

Although the processes of rock decay abound in most of West Africa, extensive studies in the area are quite limited. While wind drives the predominant decay activities to the north of the region, the erosional surfaces in the central to southern flanks are marked by the action of running water, as virtually all of the landscape is dissected by rivers and streams of varying characteristics. Gbadegesin and Onafeso (2010) posited that the region of West Africa belongs largely to the great continental shield that always acted as a rigid block from the late Precambrian era.

This chapter, following attempts to sum up the artifacts left behind from past civilizations as indicators of the vintage urban-built environment, provides a theoretical review of the grounds for the ruins and the plausible future challenges for contemporary urban architecture. Although lacking in detailed empirical analysis, for want of time and funding, this chapter relies chiefly on history, geology, and archaeology to describe the urban geomorphological importance of the West African region. Emphasizing the effects of rock decay on the widespread landscapes underlain by hardpan crustal original blocks of igneous extrusions, which are either exposed or covered by different sedimentation materials, the structural framework of urban-built environment architecture is also described from ancient to contemporary times.

13.2 DECAY OF CLAY SANDSTONES AND MUDSTONES

The entire heritage of humanity is situated within the built environment, such that the stones used for its construction hold some sort of significance for its sustainability. For this reason, building conservation strategies have become very important in both architectural and constructional sciences to engender the maintenance of these relics. According to Alves and Sanjurjo-Sánchez (2011), materials are crucial fragments of human culture and traditions indicated in the diverse surfaces of the built environment, from extensive walls to kitchen utensils. All of these materials, regardless of their structural and chemical composition, experience transformations after emplacement over time. This is because decay, in general, but especially with rock masses either exposed on slopes or affected in situ by the effects and intermittent actions of precipitation and temperature, significantly alter the geochemical properties of rocks within engineering timescales and, thus, induce rapid changes of rock materials from initial rock-like properties to eventual soil-like properties (i.e., pedogenesis).

Although Meybeck (1987) suggested that sandstone also occupies a similar proportion of the Earth's surface as granite or limestone, sandstone remains the most important architectural stone material in West Africa, especially during prehistoric to early modern periods. Remarking on the possible reasons for the paucity of research on the decay of sandstones, Young and Young (1992) noted the widespread belief that sandstone decay is a rather direct case of physical breakdown that is controlled by simple lithological and structural conditions. The key modification arising from the chemical composition, due to the decay of primary minerals into secondary ones, often results from an increase in water content. Fig. 13.1A and B illustrate comparative examples of a typical tropical lateritic soil and deep rock decay around the Ikorodu area of Lagos State in Nigeria.

FIGURE 13.1 **Typical tropical lateritic soils.**

The alteration of oxygen atoms or ions into hydroxyls, which is the constitution of water or lattice water, and the introduction of water molecules hydrating inner absorbent surfaces, are established prevailing mechanisms in clay genesis or synthesis processes. As rocks of differing mineralogical compositions get mechanically reduced into smaller sized particles, surface area per unit weight progressively increases allowing higher quantities of adsorbed water to interface with rock constituents. Simultaneously, however, these rock constituents are exposed to more or less marked temperature effects and oxidation processes. Albeit, the general transformation process into secondary minerals would be difficult to understand without considering the significant properties of water, and especially of water molecules in the adsorbed state, as well as the ambiguous behavior of alumina and dynamics of the silica phase (Fripiat and Herbillon, 1971).

Turkington and Paradise (2005) conducted a review of sandstone decay research in the past 100 years, exposing the course of research from the early explanations and taxonomy of landscapes to the development of process-based explanations and the diminishing scales of inquiry in the disparity concerning understanding of process(es) and descriptions of the genesis of sandstone decay features. Sandstones, which are also sometimes called arenites or wacke, are a clastic form of sedimentary rocks composed largely of sand-sized minerals or rock grains and mostly containing combinations of quartz or feldspar, which are the most common minerals in the Earth's crust Turkington and Paradise (2005).

Sandstones are lithified accumulations of grains <2 mm in size and the second most abundant sedimentary rock types after shale and are very important indicators of erosional and depositional processes. In terms of their texture, sandstones consists of interstitial volumes between grains, which may either be empty or filled with a chemical cement of silica or calcium carbonate. As such, both texture and mineralogical properties are used in classifying sandstone. While the grains of modern sandstones of later age are usually empty with high levels of porosity, the interstices of ancient ones are usually filled up with mineral materials depending on the source area and the rate of deposition Turkington and Paradise (2005).

Rock-decay studies have been an essential part of geomorphological research because rock decay is thought of as the beginning of many dynamic systems and, as such, exerts a somewhat dominant force in the evolution of various landscapes Gbadegesin and Onafeso (2010). Prior to the shift toward reductionism and quantification in physical geography in the past 50 years or so, inquiries into sandstone decay have concentrated on the responses of landforms and landscapes Fripiat and Herbillon, 1971. However, contemporary research into sandstone decay seemed to trail the general shift in geomorphology toward process studies. This is because the two traditional views of (1) structural theory that emphasizes that regional geology dictates landform development and (2) climatic theory, which implies that mesoscale climatic variability is the major control on geographical variability in rock decay, have been agreed to be oversimplifications. As such, rock decay seems to have now been acknowledged as the consequence of the actions of a wide range of processes, operating either sequentially or simultaneously. An emphasis on critical and intensive study of relationships between process(es), decay form(s), rock properties, and environmental conditions has now emerged (Turkington and Paradise, 2005).

Increasingly sophisticated assessment techniques, which allow investigation of morphology, rock structural, chemical and mineralogical properties, and environmental conditions at finer, more detailed, resolutions have been advanced as supplementary to the traditional methodologies of field observations, measurement, and characterization of rock decay features and products. The assessment of the prolonged existence of worked stones is a multidisciplinary study, which is not only peculiar to geomorphology. As such, in order to analyze the probable future instability of the architecture of the urban-built environment, it is important to establish scientifically valid and replicable techniques of classifying rock decay and, consequently, the relation of rock decay processes to future erosion. There are several methods of measuring the chemical decay of rocks, most of which have been established for several decades (Campbell, 1991; Dorn, 1995).

13.3 EVIDENCE OF ROCK DECAY CONSEQUENT TO URBAN STONE DECAY

Rock decay can be summed up as the process wherein the texture and composition of rocks and minerals are altered as a result of exposure to natural Earth surface agents, such as water, oxygen, organic, and inorganic acids as well as temperature fluctuations (Pope, 2013). Such changes in processes, which may be either chemical or mechanical, or even both, usually occur in situ, continuing during and after transportation Faniran, 1975. Erosion, which is the removal of decayed materials from where they were formed, is a natural process that has been occurring throughout geological time to create landforms such as rivers, valleys, caves, and coastal platforms Pope, 2013. Erosional rates are, however, often accelerated by human interference. This is because agricultural activities like cropping, grazing, and the construction of buildings and roads as well as other urban infrastructures have sometimes resulted in large area soil erosion, landslides, and even desertification due to biodepletion.

13.3.1 Geology, Materials, and Deep Rock Decay

Gbadegesin and Onafeso (2010) suggested that the basement pan of the region has remained a relatively rigid block despite fractures due to widespread subsequent denudations and erosional washings. Because much of the region is composed of an ancient crystalline basement rock complex, as well as older sedimentary rocks that are mainly sandstones, a number of uplands and scarps are found dotting the landscape of the region. Examples of such uplands includes the Fouta Djallon highlands in Guinea, which form the source of the great River Niger, the Banfora and Hombori Mountains in Mali, the Gamkaza and Manpong uplands in Ghana, as well as the Adamawa, Jos plateau, Atakora and Yoruba hills, among others.

The basement complex of West Africa was affected by the 600 Ma Pan-African orogeny, which occupies the reactivated region that results from plate collision between the passive continental margin of the West African craton and the active Pharusian continental margin (Burke and Dewey, 1972; Dada, 2006). The basement rocks are believed to have resulted from at least four major orogenic cycles of deformation, metamorphism, and remobilization, corresponding to the Liberian (2700 Ma), the Eburean (2000 Ma), the Kibaran (1100 Ma), and the Pan-African cycles (600 Ma). The first three cycles were characterized by intense deformation and isoclinals folding accompanied by regional metamorphism, followed by extensive migmatization (Obaje, 2009). The Pan-African deformation was accompanied by a regional metamorphism, migmatization, and extensive granitization and gneissification that produced syntectonic granites and homogeneous gneisses (Abba, 1983).

The end of the orogeny was marked by faulting and fracturing (Gandu et al., 1986; Olayinka, 1992). This basement complex shows great variations in grain size and mineral composition. The rocks are quartz, gneiss, and schist, consisting essentially of quartz with small amounts of white micaceous minerals Olayinka, 1992. The remaining sections are composed of crystalline rocks of the basement complex, consisting mainly of folded gneiss, schist, and quartzite complexes, which belong to the older intrusive series (deSwardt et al., 1965). In general, the landforms of West Africa south of the Sahara, are dominated by erosion surfaces cutting across rocks of the basement complex.

In the Volta basin of Ghana, lower Paleozoic rocks are present, but Mesozoic and young strata occur only in coastal areas and downfaulted troughs. In a number of places, large upwarping brings the basement complexes to a sufficient elevation to enable some of the older erosional surfaces of the continent to be preserved. Unlike the arid Sahara and major physiographic divisions to the north, West Africa has a tropical or equatorial climate regime, which provides sufficient moisture and warmth for rapid rock decay, with sufficient water for permanent or at least seasonal streams (Bridges, 1990).

These basement complexes are more than 750 Ma old and have all been subjected to tectonic and exogenic change. From an environmental perspective, however, it is their rock decay and erosional properties that are most important. Specifically, the granites and quartzites are the predominant rock type in Africa and basement complexes, respectively. Under humid conditions, granites can experience decay deep underground, as opposed to drier regions where the decayed material is often easily eroded. The often crumbly material derived from humid climate granitic decay covers much of the surface material over wide areas of the African continent (Lewis and Berry, 2012). Quartzites are resistant both in humid and semiarid conditions and tend to form topographic highs, such as ridges. Schists and gneisses break down relatively fast in both humid and dry areas. Their products are not generally as thick as those derived from granites. Basic igneous rocks decay rapidly in the humid tropics and, together with schists and gneisses, generally form low-lying areas, where they are identified, resulting in a low-relief landscape (Lewis and Berry, 2012).

Similarly, Mabbutt (1952) described the relief forms and destructional processes occurring in a group of granite hills in Damaraland, southwest Africa. He suggested that the hills are attributed to the dissection of a former erosion surface by rejuvenated streams and that drainage incision has followed longitudinal and diagonal joint lines, which control the orientation and subdivision of individual groups of granite domes. He further noted that the destruction of granite masses by plating, or the unloading of thick shells, produces the dome outline, whereas the development of vertical jointing leads to granite tors. He concluded that domes may be cleaved along transverse joints to form secondary domelets or split longitudinally to form narrow whalebacks. Thus, the idea that highlands are lowered by plating to form flat domes that are reduced by marginal fracturing was termed "mural weathering" by Mabbutt (1952), who also suggested that further wastage occurs by slow granular disintegration on low granite outcrops.

13.3.2 Chemical Rock Decay and Sediment Production

Bayon et al. (2012) suggested that changes in chemical rock decay intensity on continents are driven primarily by natural factors, such as physical rock decay rates, vegetation, rainfall, and temperature (Gaillardet et al., 1999; White and Blum, 1995). They also advocated that intensive land use and accelerated soil denudation, by increasing the surface area of minerals and exposed rocks, can also dramatically lead to much higher rates of chemical alteration (Raymond and Cole, 2003). According to Bayon et al. (2012), the degree of chemical decay of fine-grained sediments can be inferred from the ratio of aluminum to potassium (Al/K). This they argued on the premise that potassium is highly mobile during chemical decay and typically depleted in soils, whereas aluminum is one of the most immobile elements incorporated into secondary clay minerals, such as kaolinite.

As such, Schneider et al. (1997) considered high Al/K in Congo fan sediments to be indicative of periods of intense chemical rock decay in the Congo basin. Bayon et al. (2012) further suggested that downcore variations of bulk chemical composition can also reveal changes in sediment source. Such a conclusion follows the Bayon et al. (2009) measurement of neodymium (Nd) and hafnium (Hf) isotopic ratios to discriminate between both rock decay and provenance signals in our sediment record. The study found that the Nd isotopic signature of terrigenous sediments is retained during continental rock decay and subsequent transport, thereby providing direct information on the geographical provenance of sediments just as Goldstein et al. (1984) had earlier suggested. The two studies (Bayon et al., 2006, 2009) then concluded that Hf isotopes exhibit globally similar behaviors, but are also prone to substantial fractionation during chemical weathering because incongruent dissolution of silicate rocks leads to products of erosion having very distinctive but systematic Hf isotopic signatures. As similar studies have not been conducted in parts of West Africa, the conclusions may yet hold true as historical conditions have shown significant similarities in virtually all parts of Africa.

Citing Nahon (1986), Beauvais and Colin (1993) affirmed that lateritic weathering of the Earth's crustal mantle has been found to cover a third of the emerged continental areas of the world. More importantly in the tropics, Nahon observed that substantial portions of this mantle consist of some 10-m thick iron duricrust. Duricrusts are materials found on the surface or near surface of the Earth consisting of a hardened accumulation of silica (SiO_2), alumina

(Al_2O_3), and iron oxide (Fe_2O_3) in varying proportions, with admixtures of other substances that may be enriched with oxides of manganese or titanium within restricted areas Nahon (1986). It is then inferred that siliceous, ferruginous, and aluminous crusts constitute duricrusts proper (Faniran, 1969–1971, 1975). Moreover, the encrusted layers of calcium carbonate, gypsum, and salt are often also considered as forms of duricrust Faniran, 1972. As such, laterites, bauxites, and quartzites are examples of duricrust layers representing the chemical alteration of the upper parts of plains and other features of low relief in West Africa (Faniran, 1972).

Beauvais and Colin (1993) reviewed several studies on lateritic systems carried out in Africa and suggested they have focused on the iron duricrust formed either under tropical contrasted climate or under humid equatorial climate conditions. Beauvais and Colin (1993) further suggested that investigations of iron duricrust have been from the perspectives of geomorphological distribution at a large scale, or in terms of petrological differentiation at the scale of profiles and minerals, and recently in terms of geochemical pathways at the scale of the landscape. As such, all of these studies have shown that secondary ferruginization processes involve the development of successive layers from the bottom to the top of profiles for each system, such as a mottled clay layer, indurated mottled clay layer, soft nodular layer, and iron duricrust. These layers are subjected to hematitization and goethitization processes depending on morphopedoclimatic changes.

Goethite (FeO·OH), which is a form of hydrated iron oxide formation, affects iron duricrusts as well as nodules of the soft nodular layers or indurated mottled clay layers which, in turn, allows goethite content to increase at the expense of hematite (Formenti et al., 2014). This may have been the condition in the region that encouraged the abundance of brass and bronze materials widely used in most of West Africa for the casting of artifacts and sculptures. However, it is not clear whether such materials were used in building construction, although evidence suggests they clearly form major parts of the aesthetics in the urban-built environment and, as such, are predominant in urban architecture from the old Bini Kingdom (now Benin-City in Nigeria) to Timbuktu in Mali.

The mineralogical transformation of goethite can be accompanied by gibbsite crystallization in the nodules, which is often derived either from in situ kaolinite hydrolysis or from inheritance of past more humid climatic conditions Gbadegesin and Onafeso (2010). Similarly, tin-doped goethite may have been the material behind the developments of the Nok civilization around the areas surrounding the present city of Jos in Plateau State, Nigeria as well as the much talked about Bantu people civilizations of the Iron Age spread around the ancient Barombi Mbo regions of present-day eastern parts of Nigeria and western Cameroon. These are considered to be the origin of the third-millennium BP Bantu farmers, who spread both southward (across Atlantic equatorial Africa) and eastward (through the Congo watershed), reaching Angola and the African Great Lakes region by ~2500 years BP, respectively, as well as the subsequent migration waves toward southern Africa, Central African Republic, and Democratic Republic of the Congo Shaw (1981).

Koita et al. (2013) suggested that various models of rock decay profiles have been proposed in granitic rocks, but never for the hard rocks of West Africa. They posited that there is no description of the rock decay profile in volcano-sedimentary rocks in the literature and, as such, proposed three models to describe the weathering profiles in granites, metasediments, and volcanic rocks for hard rock formations located in West Africa. Studying the Dimbokro catchment in the Ivory Coast, they described vertical layered rock decay profiles based on the various decay and erosion cycles specific to West Africa. They found that the geological formations of the Dimbokro catchment originated from the Eocene to the recent Quaternary period and, thereby, based the characterization of rock decay profiles on bedrocks and rock decay profile observations at outcrops as well as on the interpretation and synthesis of geophysical data and lithologs from different boreholes.

Koita et al. (2013) also suggested that the related rock decay profile for each of the geological formations (i.e., granites, metasediments, and volcanic rocks) after modeling comprises four separate layers from top to bottom. The suggested that layers include alloterite, isalterite, fissured layer, and fractured fresh basement. They concluded that these rock decay profiles are systematically covered by a soil layer and that, although granites, metasediments, and volcanic rocks of the Dimbokro catchment experienced the same decay and erosion cycles during palaeoclimatic fluctuations from the Eocene to recent Quaternary period, they exhibit differences in thickness. They found the rock decay profile of granites to be relatively thin due to the absence of iron crust that protects decayed products against further breakdown, while metasediments and volcanic rocks develop iron crusts better than granites, indicating that alterite is more resistant to decay.

The chemical decay of silicate rocks serves as a key sink for atmospheric carbon dioxide, particularly on continental scales, and this has played a significant role in the evolution of the Earth's climate (Bastian et al., 2017). Although, the nature and size of the interaction of rock decay with climate is yet to be understood, most especially the timescale over which chemical rock decay acts in response to climate change. However, erosion on continents include both mechanical and chemical processes, as these act together to shape the Earth's surface and contribute to the extensive capture of atmospheric carbon through the export of organically rich clay fractions and metamorphism of silicate minerals.

Whereas, the link between climate and physical erosion rates are recorded in river chemistry data and sedimentary records, the response of chemical rock decay to climate change at continental scales is not well understood and, as such, requires further investigation (Burbank et al., 2003; Millot et al., 2002). Considering a continent's size, the comparative significance of physical erosion, temperature, rainfall, vegetation, and lithology on chemical rock decay over both long and short periods of time is yet to be conclusively agreed upon. Several studies that have attempted to establish links between climate and silica-based rock decay have been based on the analysis of dissolved phases in modern river basins or on the reconstruction of past ocean chemistry during the Cenozoic era, from marine carbonates or deep-sea ferromanganese deposits (Gaillardet et al., 1999).

Similarly, only a few studies have investigated historical variations in silicate decay over short timescales, and these have yielded contradictory results (Bayon et al., 2012; Beaulieu et al., 2012; Dosseto et al., 2015; Li and West, 2014; von Blanckenburg et al., 2015). Yet, this type of evidence is significant for predicting and improving knowledge on the progression of the short-term carbon cycle and its impact on the global climate system. Sediments from rock decay profiles transported by large rivers and deposited along channel margins are able to provide important evidence of the short-term evolution of rock decay at the subcontinental scale. Thus, in a bid to proffer controls on the links between hydroclimate and rock decay, Bastian et al. (2017) in a study in the Nile basin, reconstructed the Late Quaternary evolution of chemical rock decay by employing a marine sediment record recovered from the Nile deep-sea fan off the coast of Egypt.

The tropical region of subSaharan Africa experienced major hydrological fluctuations during the Quaternary period (Shanahan et al., 2015). These changes have radically exaggerated fluvial discharge and particle delivery to the surrounding ocean margins (Skonieczny et al., 2015). It is, therefore, safe to conclude that most material unearthed at the earthwork of Sungbo Eredoat Oke Erinear Ijebu-Ode in Nigeria were sourced from secondary locations, especially southward toward the coast. It is also interesting to note that these materials include pottery materials, quartz, quartzites, and gneiss Lasisi and Aremu, 2016. However, the pottery materials excavated from the site at 1–1.9 m have been washed, the charcoal already in a crumbled state, and the character of some of the stones were indiscernible (Lasisi and Aremu, 2016).

Although the charcoal retrieved from 2.47 m is small, it is significant as it correlates with the same period as Late Stone Age materials. The cultural manifestations in the form of pottery are not an end in themselves, but a means to an end in that these materials represent a culture, an idea, a people, a great culture in antiquity (Lasisi and Aremu, 2016). Also, the gneiss boulder found within the mound has decayed mostly to the consistency of clay, as the breakdown of gneiss is characterized by mineral dissolution, with the formation of clays and ferruginous products occurring as the replacements for feldspar and biotite and, thus, filling the interstices between grains (Baynes and Dearman, 1978). It is difficult to calculate how long it would take gneiss from the trench to decay into kaolinite because environmental conditions differ greatly from place to place, the decay of gneiss lasts for a protracted period of time and, as such, constitutes a pre-Holocene relict product of paleorock decay found during warmer and more humid climatic conditions (Critelli et al., 1991; Mongelli et al., 1998). Therefore, the decomposed gneiss found within the upper level actually came from the ditch and formed the reversed stratigraphy (Lasisi and Aremu, 2016).

13.4 WARM WET CLIMATES OF THE RIVER NIGER BASIN REGION

Some 25 000 years ago, the wide-ranging climate evolution in the region around the River Niger basin indicates similar trends from the Atlantic coast to the Red Sea. This condition in the southern Sahara and the Sahel regions reflects the end of a humid epoch of the upper Pleistocene and the commencement of an arid epoch. Servant (1973) studied lacustrine deposits in the Chad basin and showed that the link between precipitation and evaporation was adequate to permit widespread lakes to endure. However, according to him, the arid zone extended throughout the subsequent 8 millennia about 400 km farther south than its contemporary bounds.

The modification from a lacustrine episode to an arid epoch has also been observed in the deposits of the Afar lakes. Gasse (1975) showed the existence of three lacustrine epochs in the upper Pleistocene, when the lacustrine environment depreciated and the dried-up beds of Lake Abay covered by Gramineae. Shaw (1981) suggested that the lakes in the Saharan region expanded significantly from the Atlantic coast to the Red Sea, due to the extreme dryness of the period extending from 16 000 to 14 000 years BP and from 12 000 years BP onward. Such conclusions were drawn from the fact that lacustrine deposits consisting of diatoms were observed in virtually all of the low-lying regions adjacent to the River Niger basin (Gasse, 1975).

Furthermore, the works of Servant (1973) in Niger and Chad suggested a continuous curve of the precipitation/evaporation ratio from studies of several types of lake, considering their sources of supply as well as their hydrogeological and geomorphological situations. These significant oscillations are indicative of a climatic curve with a seemingly general character, which is suggestive of great expansions of the lakes around 8500 years BP and two peaks of attenuations around 4000 and minor fluctuations after 3000 years BP. As such, Gasse (1975) suggested that these major oscillations affected several lakes in Afar, although with some minor differences due to their sources of supply, thereby insinuating a definite correlation between the Chad curve and the humidity curve for the Siberian continental zone.

The rainforests are the great green heart of Africa and present a unique combination of ecological, climatic, and human interactions. Malhi et al. (2013) reviewed the past and present state of changes in African rainforests and explored the challenges and opportunities for maintaining a viable future for these biomes. Essentially, these climatic drift episodes have had significant effects, especially in terms of algal growth due largely to the Neolithic palaeoenvironmental deterioration of archaeological sites. This trend is unlikely to abate, as recent studies using downscaled HadCM3 rainfall predictions under the A1–A2 and B1 SRES emission scenarios have shown significant plausible alterations in future rainfall-runoff relationships over the lower course of the River Niger (Onafeso, 2012).

13.5 PREVAILING ATMOSPHERIC POLLUTION OF THE URBAN-BUILT ENVIRONMENT

In West Africa generally, microenvironments and lithological sequences are indicative of climate change scenarios. The Sahara Desert, for example, has shown overwhelming evidence that it has in the recent past experienced wetter climates. Such indications are arrayed from the distribution pattern of flora and fauna to sedimentary features, which are unexplainable without an assumption of a damper climate in the past (Jeje, 1980).

According to Shaw (1981), a couple of animals native to the central African region may have lived in the desert, according to fossil records, thereby indicating that such animals must have migrated there through corridors of vegetation or water. For example, crocodile species of the central parts of Africa have been found in waterholes in the deep ravines of both Ahaggar and Tibesti massifs Shaw (1981). Similarly, African mudfish have also been found in the extreme north around the oasis of Biskra in southern Tunisia Shaw (1981).

Several studies have indicated that the drainage features of the Sahara Desert indicate previous greater rainfalls (Hansen et al., 1998). For example, to the west of the Ahaggar, a huge plain extends a few hundred kilometers from the Atlantic, sloping gently from the margins of the El Juf depression. This clearly must have formed the evaporation basin of an extensive system of streams in the past. Similarly, considerable research attention should be allotted to the drainage lines leading from the southern slopes of the Atlas southward, of which WadiSaoura has been traced for more than 500 km Diester-Hass (1976). This is because conclusions have been drawn that it is a valley that in the past carried sufficient volumes of water to remove the aeolian sands that today choke its middle reaches (Burke and Durotoye, 1971; Diester-Hass, 1976; Grove and Warren, 1968).

Varied pollutants intermingle in the atmosphere and, in turn, interface with facades in the surfaces of the urban-built environment, thereby causing unwarranted modifications and leaving marks of these interactions. This has been the incentive behind long-term modifications of urban architectural materials in most parts of the world, especially in the presence of moisture Onafeso, 2012. Extrinsic agents have been known to affect the surfaces of materials, thereby resulting in the changing of their original characteristics, most especially physical and chemical. These alterations, which are either based on changes in the initial compositions of the material or mechanical destruction of the material often results in modification of the material until a balance is achieved between the agent, the material, and the environment.

Although there has been a significant shift in the types and composition of building materials in most of the urban built environment in West Africa, the transition is perhaps not a progressive one, as most of the old clay sandstone and mudstone structures have now been replaced with concrete limestone ones. This transition is evident in virtually all towns and cities of the region, even in the centers of some of the ancient towns, where old the mudstone bricks of ancient mud houses are being replaced with limestone-cemented layers of mud bricks. An example observed in Ijebu-Ode, Ogun State in Nigeria is shown in Fig. 13.2. Furthermore, marble represents the building material of choice in the modern-day West African urban-built environment, a development that has attracted little to no research attention. While it is important to note that the decay processes of clay sandstones and mudstones differ significantly from those of limestones, especially with respect to the environment in which the process takes place, the other issue is the sustainability of the building material.

It has been established that fine-grained rocks generally decay faster than more coarse-grained rocks because they have a larger accessible surface area (Bridges, 1990). Similarly, rocks with increased numbers of joints and fractures decay more quickly than a solid mass of rock with the same dimensions Bridges, 1990. These gaps provide pathways for rock decay agents to enter a rock mass and speed up the decay process. Meanwhile, carbonation involves the reaction between minerals and carbonate or bicarbonate ions Lea, 1970. Increased atmospheric carbon dioxide has been noted to generate increases in the acidity of rainwater, thus, increasing the carbonic acid content Townsend (2002). This condition is more plausible in West Africa due to rainfall formation processes and dissipation, which is not in the present scope of discourse.

However, in urban-built environments, small amounts of acid in the rain may be considered as being capable of slowly dissolving buildings, particularly those made from limestone

FIGURE 13.2 **Clay mudstone bricks sandwiched by limestone cement.**

and concrete. A similar process may occur where acid in water moves through the soil. Carbonic acid creates hydrogen (H^+) ions, which can be substituted for other ions within minerals, thus, altering their composition and eventually causing them to break down. Acid rain is one of the prospective devastating problems expected as part of anthropogenic climate change, especially for those regions where high sulfur coal burning; as well, carbon dioxide emission are nearly unabated and the result is most likely increased sulfuric acid content of the rains (Hansen et al., 2013).

A different aspect associated to the modification of building materials is the release of pollutants. This is because building materials can act as pollution sources affecting the surrounding environment, especially in relation to "cementitious" materials like limestone. In the same vein, the corrosion of metallic materials has also been linked to pollution of the atmosphere through various emissions due largely to chemical interactions of building materials and atmospheric chemical components (Townsend, 2002; van der Sloot, 2000). The release of substances from building materials can also affect those materials or other nearby materials. Pollutants from building materials can arise from their pore content, such as soluble salts on natural rocks; and in the case of cements in the initial stages of setting (Odler, 1998).

However, in the case of artificial materials prepared with water, as, for example, mortars and pavements, water has also been identified as a potential source of pollutants (Netterberg and Bennet, 2004). In corollary, Alves and Sanjurjo-Sánchez (2011) suggested that the context of epidemiological studies of material decay should involve the study of pollutant release due to the modification of the components of the materials. This is especially relevant in relation to the decay of natural stone constituents and particularly those used in building construction. Special reference in this light has, therefore, been made about calcite (Cooper et al., 1991), dolomite (Rodríguez-Navarro et al., 1997), silicates (Smith et al., 2002), iron sulfides (Dreesen et al., 2007; Honeyborne, 1998), organic components (Hartog and McKenzie, 2004; Lea, 1970), and metal corrosion (Winkler, 1994). Alves and Sanjurjo-Sánchez (2011) concluded that these aspects of the roles in material deterioration and atmospheric pollution in the built environment may be considered in relation to the formulation of building materials as well as the control of chemical composition of different constituents, especially concerning also the way that they are applied in the built environment in the form of architectural options. This is because aggregates and cement ratios are important as they are used to arrive at optimized workable and cohesive materials for building construction.

Meanwhile, increased vehicular traffic and an associated rise in atmospheric sulfur and nitrogen oxides have continued to pose a threat to built heritage around the world. The potential for stone surface change is shown by field trials in which small limestone tablets affixed to buildings in exposed and sheltered locations exhibited a range of features arising from the action of different rock decay processes (Hartog and McKenzie, 2004). Furthermore, limestone is a widely used building stone primarily because of its appearance and quality. However, it is particularly susceptible to deterioration, principally through the effects of chemical dissolution. Even unpolluted rain contains carbon dioxide, creating a weak carbonic acid that is able to dissolve calcite (Thornbush and Viles, 2007), the main mineral component of limestone. This natural acidity of rain is further increased by reactions with other atmospheric pollutants, such as sulfur and nitrogen oxides, with a resultant increase in the rate of limestone dissolution. It is, therefore, interesting to note the need to focus research attention on the gradual decay of materials both in construction and in situ deposits, as these would improve knowledge of the weakening of the shear strength of urban-built environment rocks in West Africa.

13.6 CONCLUSION

The history of society in West Africa is one of a city-based civilization, where most human settlements are congested into significant urban-built environments that are surrounded by several rural settlements, most of which are agroallied. It is nearly commonplace to have most of these city-states with walls for security and boundary demarcations. Although an undifferentiated basement complex underlies the entire area of West Africa, the developments of urban centers in the region have been traced along the Niger River largely due to the availability of water to sustain human life and establish trade routes.

However, the continued decay of rocks and erosion of weakened materials, which undergo subsequent deposition, remains critical to the sedimentation of silts in the form of clay sandstones and mudstones. This process, aided by intermittent wet-dry climatic fluctuations in the region, has fostered the chemical alterations of rocks that have been readily employed as construction materials in the building of some of the most advanced West African civilizations.

Whereas clay sandstones and mudstones have been the age-old constructional materials in West Africa, contemporary times have shifted to limestones for its "cementability" and availability. While clay sandstones and mudstones have not proven to be durable materials, as most of the relics of past civilizations have shown, it is also very unlikely that the current reliance on limestones and marble finishing can be sustainable. This follows from limestone's dissolutional (i.e., dissolvability) tendencies, as well as atmospheric pollution potentials due to chemical decay processes enhanced with the presence of water found abundantly around equatorial West Africa.

Essentially, it is established that erosion shapes landscape evolution and the connections between climate and tectonics in West Africa, as it does in many humid tropical parts of the world. It is, therefore, important to understand the dynamics of erosion rates through time, as this could be of utmost significance for decrypting Earth's topographic evolution and appraising the processes and relationships between climate and tectonics.

Even though we have traced the evolution of urban stone decay using proxy historical records as well as limited some archaeological and geochemical surveys, it is evident that we have a limited understanding of the landform processes involved in these near losses of priceless (ancient) cultural heritage. There is, therefore, the dire need to increase research efforts toward improving knowledge into the operative conditions that drive urban geomorphology in the region adjacent to the great Niger River.

Along the geological timescale, it is pertinent to note that magnitudes of erosional rhythms and durations of erosional intervals expectedly possess certain crucial connections that are yet to be well-researched in the region of the Niger River basin. The landscapes of this region, where erosion is dominated by several river incisions due largely to drainage density, are expected to have erosion rates positively correlated with the rates of tectonic uplifts.

Similarly, the fluctuations and high variability of precipitation rates as well as contemporary anthropogenic climate change are yet to be adequately connected to the characteristic recurrence time driving processes of landscape erosion. It is, therefore, not clear how the contemporary constructional materials of the urban-built environment are likely to respond to climate-tectonic adjustments in time.

References

Abba, S.I., 1983. The structure and petrography of alkaline rocks of the Mada Younger Granite complex, Nigeria. J. Afr. Earth Sci. 3, 107–113.

Alves, C., Sanjurjo-Sánchez, J., 2011. Geoscience of the built environment: pollutants and materials surfaces. Geosciences 1, 26–43.

Bastian, L., Revel, M., Bayon, G., Dufour, A., Vigier, N., 2017. Abrupt response of chemical weathering to Late Quaternary hydroclimate changes in northeast Africa. Sci. Rep. 7 (44231). doi: 10.1038/srep44231.

Baynes, F.J., Dearman, W.R., 1978. The relationship between the microfabric and the engineering properties of weathered granite. Bull. Int. Assoc. Engr. Geol. 18 (191). doi: 10.1007/BF02635370.

Bayon, G., Vigier, N., Burton, K.W., Brenot, A., Carignan, J., Etoubleau, J., Chu, N.-C., 2006. The control of weathering processes on riverine and seawater hafnium isotope ratios. Geology 34 (6), 433–436.

Bayon, G., Burton, K.W., Soulet, G., Vigier, N., Dennielou, B., Etoubleau, J., Ponzevera, E., German, C.R., Nesbitt, R.W., 2009. Hf and Nd isotopes in marine sediments: constraints on global silicate weathering. Earth Planet. Sci. Lett. 277 (3–4), 318–326.

Bayon, G., Dennielou, B., Etoubleau, J., Ponzevera, E., Toucanne, S., Bermell, S., 2012. Intensifying weathering and land use in Iron Age Central Africa. Science 335 (6073), 1219–1222.

Beaulieu, E., Goddéris, Y., Donnadieu, Y., Labat, D., Roelandt, C., 2012. High sensitivity of the continental-weathering carbon dioxide sink to future climate change. Nat. Clim. Chang. 2, 346–349.

Beauvais, A., Colin, F., 1993. Formation and transformation processes of iron duricrust systems in tropical humid environment. Chem. Geol. 106, 77–101.

Bridges, E.M., 1990. World Geomorphology. Cambridge University Press, Cambridge.

Burbank, D.W., Blythe, A.E., Putkonen, J., Pratt-Sitaula, B., Gabet, E., Oskin, M., Barros, A., Ojha, T.P., 2003. Decoupling of erosion and precipitation in the Himalayas. Nature 426, 652–655.

Burke, K.C., Durotoye, B., 1971. Geomorphology and superficial deposits related to late Quaternary climatic variation in south western Nigeria. Z. Geomorphol. NF 15 (4), 430–441.

Burke, K.C., Dewey, J.F., 1972. Orogeny in Africa. In: Dessauvagie, T.F.J., Whiteman, A.J. (Eds.), Africa Geology. University of Ibadan Press, Ibadan, Nigeria, pp. 583–608.

Campbell, I.A., 1991. Classification of rock weathering at writing-on-stone Provincial Park, Alberta, Canada. Earth Surf. Proc. Land. 16, 701–711.

Cooper, T.P., Dowding, P., Lewis F, O., Mulvin, L., O'Brien, P., Olley, J., O'Daly, G., 1991. Contribution of calcium from limestone and mortar to the decay of granite walling. In: Baer, N.S., Sabbioni, C., Sors, A.I. (Eds.), Science, Technology, European Cultural Heritage. Butterworth-Heinemann, Oxford, pp. 456–461.

Critelli, S., Di Nocera, S., Le Pera, E., 1991. Approcciometodologico per la valutazionepetrograficadelgrado di alterazionedegli gneiss del massicciosilano Calabria settentrionale. Geol. Appl. Idrogeol. 26, 41–70.

Dada, S.S., 2006. Proterozoic evolution of Nigeria. In: Oshi, O. (Ed.), The Basement Complex of Nigeria and Its Mineral Resources (A Tribute to Prof. M. A. O. Rahaman). AkinJinad & Co, Ibadan, Nigeria, pp. 29–44.

DeSwardt, A.M.J, Ogbukagu, I.K., Hubbard, F.H., 1965. 1:250,000 Geological Map of Sheet No. 60 Iwo, Geological Survey, Nigeria.

Diester-Hass, L., 1976. Late Quaternary climatic variation in Northwest Africa deduced from East Atlantic Sediment cores. Quat. Res. 6, 299–314.

Dorn, R.I., 1995. Digital processing of back-scatter electron imagery: a microscopic approach to quantifying chemical weathering. Geol. Soc. Am. Bull. 107, 725–741.

Dosseto, A., Vigier, N., Joannes-Boyau, R., Moffat, I., Singh, T., Srivastava, P., 2015. Rapid response of silicate weathering rates to climate change in the Himalaya. Geochem. Perspect. Lett. 1, 10–19.

Dreesen, R., Nielsen, P., Lagrou, D., 2007. The staining of blue stone limestones petrographically unravelled. Mater. Charact. 58, 1070–1081.

Faniran, A., 1969. Duricrust, relief and slope, populations in the northern Sydney district. Nig. Geog. J. 12 (1, 2), 53–62.

Faniran, A., 1970. The Sydney duricrusts: note on terminology and nomenclatures. Earth Sci. J. 4 (2), 117–128.
Faniran, A., 1971. Implications of deep weathering on the location of natural resources. Nig. Geog. J. 14 (2), 59–69.
Faniran, A., 1972. Depth and pattern of weathering in the Nigerian Precambrian basement complex rocks areas: a preliminary report. In: Dessauvagie, T.F.J., Whiterman, A.J. (Eds.), African Geology. Geology Department, University of Ibadan, Ibadan, Nigeria.
Faniran, A., 1975. Limitations of rock analysis: an example of laterites from Sydney Australia. J. Nig. Min. Geol. Metall. Soc. X, 15–23.
Formenti, P., Caquineau, S., Chevaillier, S., Klaver, A., Desboeufs, K., Rajot, J.L., Belin, S., Briois, V., 2014. Dominance of goethite over hematite in iron oxides of mineral dust from Western Africa: quantitative partitioning by X-ray absorption spectroscopy. J. Geophys. Res. 119 (22), 12740–12754.
Fripiat, J.J., Herbillon, A.J., 1971. Formation and transformations of clay minerals in tropical soils. Soils and Tropical Weathering: Proceeding of the Bandung Symposium 16 to 23 November 1969. UNESCO Natural Resources Research XI, Paris. pp. 15–24.
Gaillardet, J., Dupré, B., Louvat, P., Allègre, C.J., 1999. Global silicate weathering and CO_2 consumption rates deduced from the chemistry of large rivers. Chem. Geol. 159, 3–30.
Gandu, A.H., Ojo, S.B., Ajakaiye, D.E., 1986. A gravity study of the precambrian rocks in the Malumfashi area of Kaduna State, Nigeria. Tectonophysics 126, 181–194.
Gasse, F., 1975. L'évolution des lacs de l'Afar Central (Ethiopieet TFAI) du Plio-Pléistocène à l'Actuel. Thesis, 3 vols., University of Paris, Paris, France, Chapter 16.
Gbadegesin, A.S., Onafeso, O.D., 2010. The Physical Basis of Spatial Organisation in West Africa. In Maitrise de L'espace et Development en Afrique. Igue, J.O., Fo-douop, K., Aloko-N'Guessan, J. (Eds.), Collection Maitrise de l'espace et development, vol. 1, Karthala, Paris.
Goldstein, S.L., O'Nions, R.K., Hamilton, P.J., 1984. A Sm-Nd isotopic study of atmospheric dusts and particulates from major river systems. Earth Planet. Sci. Lett. 70, 221.
Grove, A.T., Warren, A., 1968. Quaternary landforms and climate on the south side of the Sahara. Geogr. J. 134 (2), 193–208.
Hansen, J.E., Sato, M., Lacis, A., Ruedy, R., Tegen, I., Matthews, E., 1998. Climate forcings in the industrial era. Proc. Natl. Acad. Sci. USA 95, 12753–12758.
Hansen, J., Sato, M., Russell, G., Kharecha, P., 2013. Climate sensitivity, sea level and atmospheric carbon dioxide. Philos. Trans. R. Soc. A 371, 20120294.
Hartog, P., McKenzie, P., 2004. The effects of alkaline solutions on limestone. Discover. Stone 3, 34–49.
Honeyborne, D.B., 1998. Weathering and decay of masonry. In: Ashurst, J., Dimes, F.G. (Eds.), Conservation of Building and Decorative Stone, Part I. Butterworth-Heinemann, Oxford, pp. 153–178.
Jeje, L.K., 1980. A review of geomorphic evidence for climate change since the Late Pleistocene in the rain-forest areas of southern Nigeria. Palaeogeogr. Palaeoclimatol. Palaeoecol. 31, 63–86.
Koita, M., Jourde, H., Koffi, K.J.P., Da Silveira, K.S., Biaou, A., 2013. Characterization of weathering profile in granites and volcanosedimentary rocks in West Africa under humid tropical climate conditions: case of the Dimbokro Catchment (Ivory Coast). J. Earth Syst. Sci. 122 (3), 841–854.
Lasisi, O.B., Aremu, D.A., 2016. New lights on the archaeology of Sungbo's Eredo south-western Nigeria. Dig It 3, 54–63.
Lea, F.M., 1970. The Chemistry of Cement and Concrete. E. Arnold Publishers, London.
Lewis, L.A., Berry, L., 2012. African Environments and Resources. Routledge, London.
Li, G., West, A.J., 2014. Evolution of Cenozoic seawater lithium isotopes: coupling of global denudation regime and shifting seawater sinks. Earth Planet. Sci. Lett. 401, 284–293.
Mabbutt, J.A., 1952. A study of granite relief from south-west Africa. Geol. Mag. 89 (2), 87–96.
Malhi, Y., Adu-Bredu, S., Asare, R.A., Lewis, S.L., Mayaux, P., 2013. African rainforests: past, present and future. Philos. Trans. R. Soc. B 368, 20120312.
Meybeck, M., 1987. Global chemical weathering of surficial rocks estimated from river dissolved loads. Am. J. Sci. 287, 401–428.
Millot, R., Gaillardet, J., Dupré, B., Allègre, C.J., 2002. The global control of silicate weathering rates and the coupling with physical erosion: new insights from rivers of the Canadian Shield. Earth Planet. Sci. Lett. 196, 83–98.
Mongelli, G., Cullers, R.L., Dinelli, E., Rottura, A., 1998. Elemental mobility during weathering of exposed lower crust: the kinzigitic paragneiss from the Serre, Calabria, southern Italy. Terra Nova 10, 190–195.
Nahon, D.B., 1986. Evolution of iron crusts in tropical landscapes. In: Colman, S.M., Dethier, D.P. (Eds.), Rates of Chemical Weathering of Rocks and Minerals. Academic Press, London, pp. 169–187.

Netterberg, F., Bennet, R.A., 2004. Blistering and cracking of airport runway surfacing due to salt crystallization. Proceedings of the 8th Conference on Asphalt Pavements for Southern Africa (CAPSA'04), Sun City, South Africa, September 12–16, 2004; paper 088. Available from: http://www.capsa-events.co.za/capsa04/Documents/088.pdf.

Obaje, N.G., 2009. Geology and Mineral Resources of Nigeria. Lecture Notes in Earth Sciences. Springer, Berlin.

Odler, I., 1998. Hydration, setting and hardening of Portland cement. In: Hewlett, P.C. (Ed.), Lea's Chemistry of Cement and Concrete. Butterworth-Heinemann, Oxford, pp. 241–297.

Olayinka, A.I., 1992. Geophysical siting of boreholes in crystalline basement areas of Africa. J. Afr. Earth Sci. 14, 197–207.

Onafeso, O.D., 2012. Analysis of changes in rainfall pattern and runoff predictions for the Lower River Niger Nigeria. Unpublished PhD thesis, Department of Geography, University of Ibadan, Nigeria.

Pope, G.A., 2013. Weathering in the tropics, and related extratropical processes. In: Shroder, J., Pope, G.A. (Eds.), Treatise on Geomorphology. Weathering and Soils Geomorphology, vol. 4, Academic Press, San Diego, CA, pp. 179–196.

Raymond, P.A., Cole, J.J., 2003. Increase in the export of alkalinity from North America's largest river. Science 301, 88–94.

Rodríguez-Navarro, C., Sebastián, E., Rodríguez-Gallego, M., 1997. An urban model for dolomite precipitation: authigenic dolomite on weathered building stones. Sediment. Geol. 109, 1–11.

Schneider, R.R., Price, B., Müller, P.J., Kroon, D., Alexander, I., 1997. Monsoon related variations in Zaire (Congo) sediment load and influence of fluvial silicate supply on marine productivity in the east equatorial Atlantic during the last 200,000 years. Paleoceanography 12 (3), 463.

Servant, M., 1973. Séquences continentales et variations climatiques: évolution du bassin du Tchad au Cénozoïque supérieur. Doctoral thesis, ORSTOM, University of Paris VI, 343 pp.

Shanahan, T.M., McKay, N.P., Hughen, K.A., Overpeck, J.T., Otto-Bliesner, B., Heil, C.W., King, J., Scholz, C.A., Peck, J., 2015. The time-transgressive termination of the African Humid Period. Nat. Geosci. 2, 1–5.

Shaw, C.T., 1981. The prehistory of West Africa. In: Ki-Zerbo, J. (Ed.), General History of Africa I: Methodology and African Prehistory. United Nations Educational, Scientific and Cultural Organization, Paris, France, pp. 611–633.

Skonieczny, C., Paillou, P., Bory, A., Bayon, G., Biscara, L., Crosta, X., Eynaud, F., Malaize, B., Revel, M., Aleman, N., Barusseau, J.P., Vernet, R., Lopez, S., Grousset, F., 2015. African humid periods triggered the reactivation of a large river system in Western Sahara. Nat. Commun. 6, 1–6.

Smith, B.J., Turkington, A.V., Warke, P.A., Basheer, P.A.M., McAlister, J.J., Meneely, J., Curran, J.M., 2002. Modelling the rapid retreat of building sandstones: a case study from a polluted maritime environment. In: Siegesmund, S., Vollbrecht, A., Weiss, T. (Eds.), Natural Stone, Weathering Phenomena, Conservation Strategies, Case Studies. Geological Society, London, pp. 347–362, Special Publications 205.

Townsend, H.E., 2002. Outdoor Atmospheric Corrosion, STP 1421. ASTM International, West Conshohocken, PA.

Turkington, A.V., Paradise, T.R., 2005. Sandstone weathering: a century of research and innovation. Geomorphology 67, 229–253.

Van der Sloot, H.A., 2000. Comparison of the characteristic leaching behavior of cements using standard (EN 196-1) cement mortar and an assessment of their long-term environmental behavior in construction products during service life and recycling. Cem. Concr. Res. 30, 1079–1096.

Von Blanckenburg, F., Bouchez, J., Wittmann, H., 2015. Stable runoff and weathering fluxes into the oceans over Quaternary climate cycles. Nat. Geosci. 8, 538–543.

White, A.F., Blum, A.E., 1995. Effects of climate on chemical weathering in watersheds. Geochim. Cosmochim. Acta 59 (9), 1729–1747.

Winkler, E.M., 1994. Stone in Architecture. Properties Durability. Springer, Berlin.

Young, R.W., Young, A.R.M., 1992. Sandstone Landforms. Springer, Berlin, p. 163.

CHAPTER

14

A Geologic Assessment of Historic Saint Elizabeth of Hungary Church Using the Cultural Stone Stability Index, Denver, Colorado

Casey D. Allen, Stacy Ester**, Kaelin M. Groom†, Roderick Schubert**, Carolyn Hagele**, Dana Olof**, Melissa James***

***The University of the West Indies, Cave Hill Campus, Barbados; **University of Colorado Denver, Denver, CO, United States; †Arizona State University, Tempe, AZ, United States**

OUTLINE

14.1 Introduction and Background 278
14.1.1 Auraria Campus 278
14.1.2 Saint Elizabeth's 280

14.2 Methods: Basics of the Cultural Stone Stability Index 281

14.3 Saint Elizabeth's CSSI Analysis 283
14.3.1 North-facing Panels 283
14.3.2 East-facing Panels 287
14.3.3 South-facing Panels 293
14.3.4 West-facing Panels 296
14.3.5 Overall Assessment 297

14.4 Implications and Conclusion 298

References 301

Urban Geomorphology. http://dx.doi.org/10.1016/B978-0-12-811951-8.00014-X

14.1 INTRODUCTION AND BACKGROUND

This chapter outlines a recent technique for assessing the geologic stability of cultural stone (e.g., buildings, monuments, bridges) based on observable forms and processes found in rock/stone decay science (weathering[1]). While techniques exist across disciplines to assess building stone, most are costly, time-consuming, and require special expertise and training to execute (Bruthans et al., 2014; Fitzner et al., 1992, 1997; Giesen et al., 2013; Griffin et al., 1991; Groom, 2014; Janbade et al., 2016; Janvier-Badosa et al., 2016; Jo et al., 2012; McKinley et al., 2006; Paradise, 1999; Smith et al., 2013; Thornbush, 2012; Warke et al., 2003). Further, even though geomorphology (and by natural extension, stone decay) undoubtedly plays an important role (Pope et al., 2002), aside from Fitzner et al. (2002); Fitzner (2004); and Fitzner's and Heinrich's (2001) work that hyperfocuses on stone decay minutiae, requires (expensive) specialized equipment and testing procedures, and remains difficult to translate into common vernacular more useful for everyday analyses, these assessments pay scant attention to *specific* stone decay (weathering) forms and processes—or worse, misinterpret and/or misrepresent specific decay processes (Kurtz et al., 2001; Lee and Chun, 2013; Siedel and Siegesmund, 2014). Building on the successes of Cerveny's (2005) and Dorn et al.'s (2008) pioneering work on the Rock Art Stability Index (RASI), its subsequent successes (Allen, 2008; Allen et al., 2011; Allen and Groom, 2013a, b; Allen and Lukinbeal, 2011; Cerveny et al., 2016; Groom, 2016), and following in the steps of Groom's (2017) adaptation of RASI for assessing hewn monuments in Petra, Jordan, the case study presented here utilizes the Cultural Stone Stability Index (CSSI) to assess the geologic stability of a significant historic building on the Auraria Campus (Denver, Colorado, USA): Saint Elizabeth of Hungary Roman Catholic Church.

Beginning with a brief historical synopsis of the Auraria Campus, including the structure in question, a basic overview of the CSSI technique is offered before expounding on findings that assess the geologic stability of Saint Elizabeth's, offering particular insight into major and minor stone decay forms and processes contributing to its deterioration. This section is offered in "technical report" style, giving the reader a feel for how the CSSI may be used in professional settings. The chapter concludes by noting some implications surrounding CSSI analyses, including perceived shortcomings, transferability across rock types and architectural styles, and other potential uses.

14.1.1 Auraria Campus

In the winter 1959 issue of the *Georgia Review*, Meaders (1959) wrote:

> In a manner of speaking, the godmother of Denver and Colorado, now celebrating their centennial year, was a Georgia lady and my wet nurse. My earliest memories of Auraria, Georgia, (for which the first white settlement in Colorado was named) are associated with a great consuming thirst, quenchable only by frequent draughts of water drunk from a gourd dipped into the only old oaken, moss-covered bucket in my experience and drawn by benefit of a creaking, groaning windlass from the far and icy depths of a North Georgia well (p. 406).

[1]Like Hall et al. (2012); Dorn et al. (2013); and others, we subscribe to the use of rock or stone decay in lieu of "weathering." Stone deterioration or rock deterioration are also acceptable. These terms more accurately reflect the breakdown of rock; and dispel the common weather-is-responsible-for-rock-breakdown misperception.

More than 40 years after her departure from Georgia to New Mexico, Meaders provides an apt starting point to introduce Colorado's Auraria. Settlement started in the waning quarter of 1858, after placer gold deposits were discovered in Cherry Creek, and a mining camp named Auraria was founded on its shores. Initially, three sites were originally plotted based on the gold's location: Auraria, Denver City, and Highlands (Fig. 14.1). After much protest and some financially driven campaigns, Auraria residents voted to merge with Denver City in April 1860.

At present, the Auraria space is a higher education center that provides an environment nurturing the human intellect—a space to slake a thirst for knowing through academic support, sustenance, and mental invigoration. Auraria's name belies its roots in the Latin signifier for gold, *aurum*. It seems even in the present-day, the notion of finding your dreams—that most personal gold—and then bringing the evasive glints to a knowable shimmering reality remains prevalent.

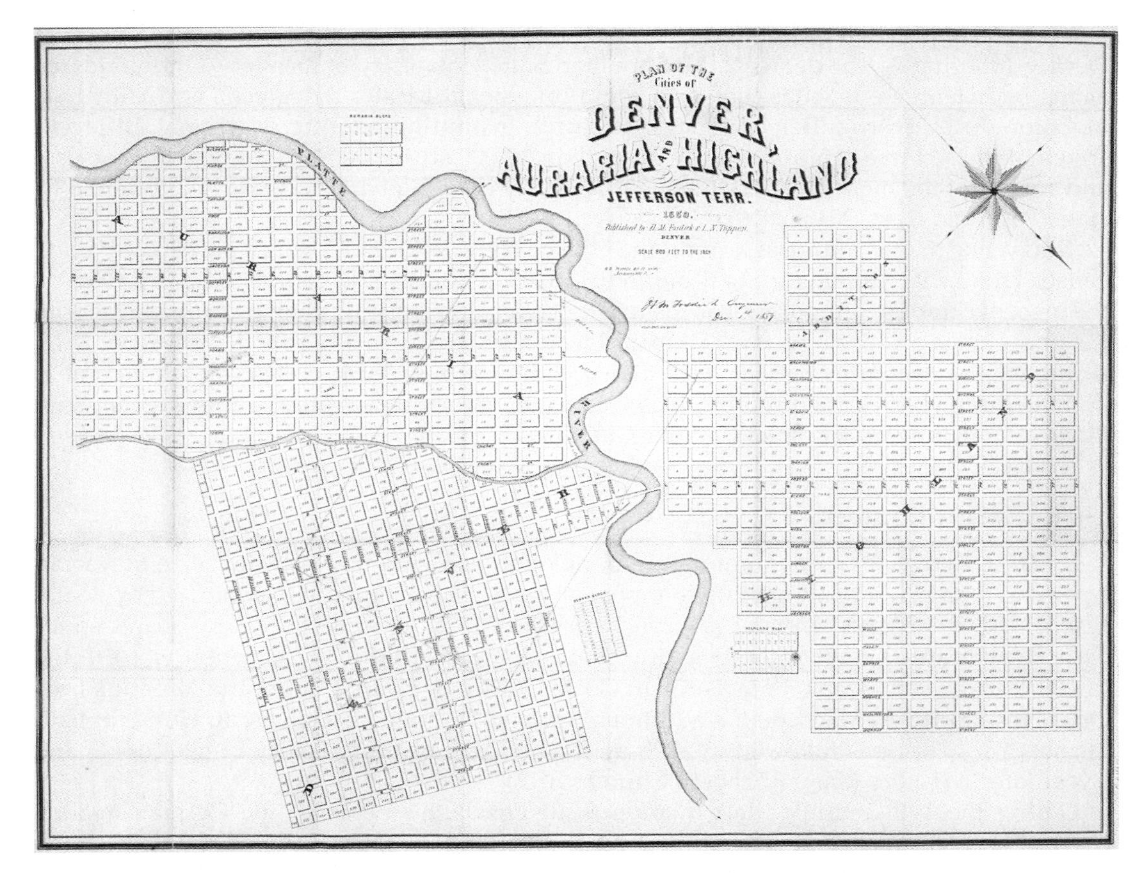

FIGURE 14.1 Map of Denver, Auraria, and the "Highlands" in 1859 by H.M. Fosdick. Highlands is to the right of the (large) Platte River, with Auraria (top-left), and Denver (angled, bottom-left), divided by Cherry Creek, where gold was first discovered in Colorado. *Source: Denver Public Library.*

By the mid-1920s, Auraria had become both a manufacturing area and distribution hub in addition to a growing Hispanic population, becoming a thriving walking-type of urban area, where shops, housing, manufacturing spaces, and warehouses all shared space and place (Page and Ross, 2015). Once America's Great Depression hit in the early 20th century, however, fortunes changed dramatically. For more than a quarter century after the Great Depression, Auraria saw an ensuing shift to other manufacturing and large warehousing areas specifically, leading to Auraria's decline and public perception of a blighted and shabby urban area (Page and Ross, 2015, 2016).

Beginning with the creation of Metropolitan State College of Denver in 1963, and the subsequent expansion of the University of Colorado's Denver Extension Center into an independent campus, the Denver Urban Renewal Authority and State of Colorado's Commission on Higher Education joined forces to find a suitable location for a center of higher education to serve the burgeoning downtown. In 1969, $12.6 million were provided by the US Department of Housing and Urban Development for purchase of the 169-acre Auraria urban renewal project. The result of this effort, the Auraria Higher Education Center, now hosts three distinct institutions of higher education, each serving a specific purpose: the Community College of Denver (offering 2-year degrees), Metropolitan State University of Denver (offering degrees at the baccalaureate level as well as a few professional Master's degrees), and University of Colorado Denver (a Tier I Carnegie Research Institution offering degrees at all levels). Together, these three campuses serve more than 50 000 students during any given semester and consist of the most diverse student body of any other higher education institution in the state (Page and Ross, 2015, 2016).

Today, like many campuses around the US, the Auraria Campus seems perpetually under construction, with seemingly a new building or remodeling project completed every year. Yet, the present campus configuration is the result of many design and urban planning ideas, with the final choice preserving a few extant and historical structures (Page and Ross, 2015, 2016). A few legacy structures remain, including the desanctified Saint Cajetan's Church and an unnamed rammed earthen sculpture as well as the structure assessed in this chapter: Saint Elizabeth of Hungary's Roman Catholic Church.

14.1.2 Saint Elizabeth's

By 1870, Denver's German immigrant population had grown to sufficient size to warrant a petition for a priest to address its community needs. The region's leadership, led by Bishop Machebeuf, established Denver's second parish in 1878 to serve the Auraria and southwest Denver area neighborhoods and a modest church of brick construction was erected. In 1887, "two Franciscans, Francis Koch, O.F.M. (order of Friars Minor) and Venatius Eder, OFM" responded to Machebeuf's additional request "to found a Franciscan House at Saint Elizabeth's." This was followed in 1890 by the building of a "two-story brick school" and, 1 year later, a rectory (Saint Elizabeth Church, 2017).

During the 19th century's last quarter, Saint Elizabeth's became "the German national church" and spiritual home for all of Denver (Noel and Wharton, 2016). The initial church soon became overcrowded, however, to the point that the old building was razed in 1898 to allow for the construction of the present Saint Elizabeth's structure (Noel and Wharton, 2016). A German Franciscan, Brother Adrian, had been assigned to the Parish and assisted the present structure's architect with the church's design. This new structure follows the Romanesque Revival style, and used locally sourced "rough-cut Castle Rock rhyolite..." with "...a dominant single corner spire soaring 162 ft...." (Noel and Wharton, 2016, p. 127). In 1936, a curved arcade, fountain, and friary were added. Included in the design by Jules Jacques Benoit Benedict was a private chapel and library, which retained "samples of the German stained glass and ornate woodwork gone from the modernized church sanctuary," and constructed through the beneficence of the May Bonfils Trust (Noel and Wharton, 2016, p. 127).

At present Saint Elizabeth's maintains its long tradition of direct community service "for the poor and hungry," which started with collections from grammar school children in 1890, who collected food and money for their community and its needy. In late 1907, Father Leo Heinrichs, OFM, extended this to providing daily sustenance each morning. Starting in the late 1970s, the pastor at that time reinvigorated this "tradition of feeding the hungry by organizing a bologna sandwich breadline behind the [C]hurch" (Saint Elizabeth Church, 2017). This continues at present 7-days-a-week from 11:00 a.m. until noon.

The structure, as it stands today, was consecrated as a sacred space on June 8, 1902, by Bishop Matz. Saint Elizabeth's Franciscans and generous German Catholics helped this community become the first in Colorado to be debt-free (Saint Elizabeth Church, 2017). In 1973, Saint Elizabeth's was added to the National Register of Historic Places. Today, due to a shortage of priests, Saint Elizabeth's serves as a mission church for the Denver Roman Catholic Cathedral, with Masses held daily and Sunday mornings.

14.2 METHODS: BASICS OF THE CULTURAL STONE STABILITY INDEX

Rock decay remains universal, as a finite number of processes yield a finite number of forms, regardless of the rock's creation, composition, structure, or use. A straightforward scientific index that efficiently assesses rock decay then, such as the RASI (Dorn et al., 2008), serves as a valuable analytical tool. Accounting for more than three-dozen forms of rock decay spread across six overarching categories, with only a few changes in terminology to reflect architecture vernacular, RASI can be adapted to a cultural/worked stone context, allowing for assessment of any stone type: bridges, sculptures, headstones, building stone, and others (Groom, 2017). When performing RASI, the trained researcher makes a quick drawing and takes a photograph of the facade (panel) to be assessed and then ranks each of the rock decay parameter on a scale of 0–3. The ratings for each element are then tallied and doubled, resulting in a final score for the facade in question with the risk for deterioration increasing with score (Table 14.1).

While the overall score remains useful for gaining a snapshot of general decay danger, each element can also be individually assessed by the site manager for its degree of decay contribution. For example, if a particular pillar or column has experienced severe deterioration, but the adjacent pillars remain in "good condition," the site steward can make an informed recommendation regarding repairs for one pillar, instead of spending unnecessary time and money on stone that does not need any maintenance. Like its cousin RASI, the CSSI was devised as an easily accessible tool for nonrock decay specialists to quickly assess stone architectural and monument features that may be geologically at risk, using nothing more

TABLE 14.1 The CSSI Scoring System with Accompanying Qualitative Interpretation

CSSI score range	Score interpretation
<20	Excellent condition
20–29	Good condition
30–39	Problems that could cause erosion
40–49	Urgent possibility of erosion
50–59	Great danger of erosion
60 and above	Severe danger of erosion

In reports, scoring ranges are often color-coded: from green (most stable, lowest CSSI score, and lowest decay risk) to yellow, orange, light red, and dark red (least stable, highest CSSI score, and highest decay risk), giving a qualitative value to the quantitatively-derived score. Panels assessed for this study follow this color-coding scheme to demonstrate its applicability for site managers.

than a writing utensil, paper, camera, and time. The CSSI is used in much the same manner as RASI and has been since before its inception (Groom, 2011).

As Groom (2017) noted, terminological changes in the CSSI from RASI do not include any additional rock decay processes as a general rule, but instead serve to widen the index's range to include cultural/worked stone. Specifically, Groom (2017, pp. 129-130) suggested that RASI's "Rock Coatings" category be modified "... due to the different roles of rock coatings between rock art and other cultural stone" because, as also noted by Dorn (1998), often rock coatings on rock art panels are beneficial, enhancing the rock art's stability. Groom (2017, pp. 129-130) explained that:

> Most petroglyphs, for example, are created by pecking or scraping through rock coatings to reveal the raw stone beneath the surface (Whitley, 2005). The contrast between the coated exterior and newly-exposed interior makes the art possible. Therefore, in RASI, two of the four rock coating elements have negative scores—indicating them as stabilizing agents. Alternatively, stone building [facades] and most other cultural stone are created with freshly quarried material, so any rock coating accumulation takes place after the stones are already in situ and beginning to decay. Also, since historic buildings and cultural stone often exist within cities and populated areas, as compared to the relative isolation of rock art sites, they may experience higher exposure to air pollution and urban traffic exhaust, leading to the development of harmful toxic rock coatings.

Because of the perceived need to include pollution and other factors that can negatively influence rock stability in urban settings (Inkpen et al., 2012a, b; Thornbush, 2012; Warke et al., 2003), the CSSI also, for example, has researchers rank the host stone's "carbonate coating" and "oxidation" with positive (i.e., detrimental) scores. Such additional inclusions do not affect the reliability or validity of the index, but serve to increase its applicability (Groom, 2017).

The point behind assessment tools like RASI and CSSI rests in their field-based, rapid, noninvasive, and cost-effective nature. They also promote flexibility in both scale (i.e., size and number of "panels" to assess) and personnel (i.e., nonspecialists can be trained quickly). Specifically, in this instance, a cadre of five researchers were trained in the CSSI following previously established protocols for RASI (Allen, 2008; Allen and Lukinbeal, 2011; Dorn et al., 2008), and assessed Saint Elizabeth of Hungary Roman Catholic Church located on the Auraria Campus in downtown Denver (USA). Following Groom's (2017) lead for hewn monuments, each edifice was divided into "panels" for ease of assessment. While each researcher was assigned a particular set of panels to assess, analysis of each panel as well as the overall structure was completed as a group. While CSSI has been used for years informally, this study marks just the second time it has been put into practical use, yielding the first geologic stability assessment of a historic building *in an urban setting*.

14.3 SAINT ELIZABETH'S CSSI ANALYSIS

Given the church's size, smaller sections based on the edifice's architecture were selected for analyses, following Groom's (2017) work. Saint Elizabeth's, like many churches, was built in relation to the Cardinal directions. For assessment ease (after Groom, 2017), panels were numbered according to their aspect (i.e., north, south, east, or west), and selected based on their architectural features. This resulted in 42 separate panels: 14 for the north aspect, three for the south aspect, 16 for the east aspect, and 9 for the west aspect.

Overall, and at first glance, the church presumably looks just like it did more than 100 years ago when construction was completed (Fig. 14.2). Closer inspection by a more trained eye, however, reveals several rock decay forms contributing to its overall deterioration. Although the entire edifice earned a CSSI rating of 26 (a "Good" status, average of 42 panels), some panels revealed very specific decay features, which should warrant a conservator's or site steward's attention. Although the building stone may be rough-hewn in appearance and contains small natural tafoni consistent with the Castle Rock rhyolite's explosive genesis, it has been expertly masoned and dressed. The church personnel do appear, however, to be doing a reasonably diligent job of monitoring the main decay and erosional concerns. Still, as newer cleaning techniques become available and parish funds permit, they would be a worthwhile investment for the building's maintenance. The following subsections offer a detailed analysis of each facade, highlighting potential weaknesses that could eventually lead to extreme degradation if left unaddressed.

14.3.1 North-facing Panels

With an average score of 26 (14 panels total), the north face of Saint Elizabeth's remains in "Good condition" (Fig. 14.3). The most dominant decay features are a combination of tafoni, flaking, and crumbly disintegration, which pervade the surfaces of both the stairs and the facades. Other notable decay features on the church's facades include oxidation, pollution, and a few small scaling (spalling) events. Still, most concerns with the north-facing facade remain relegated to the bottom couple of meters, with a few specific outliers. The single set of stairs on this aspect (N1), for example, show significantly more signs of decay than the rest of the church facade, including serious decay between mortar joints, resulting in 1–3 cm gaps between many of the stairs (Fig. 14.4).

Panel N1 (the stairs) also has obvious past and impending splintering, which contributed to large scaling events on the tops of several steps. These problems are likely exacerbated because of the difference in building materials: the main walls of the church are Castle Rock rhyolite and the stairs are sandstone.

FIGURE 14.2 **Saint Elizabeth of Hungary Roman Catholic Church as viewed from the southeast corner (looking northwest) showing the church's south-facing side and front entryway (east-facing, with the white statue, center).** While the overall building is structurally sound, decay processes are occurring on all four sides.

From afar, N9 (Fig. 14.5) clearly displays a darker color relative to the adjacent panels, as it did not receive recent cleaning like the other panels. The west-face of N10's column also hosts a pair of long, darkly colored vertical stripes, which appear to be an imprint left by a previously attached conduit of some sort (Fig. 14.6). Perhaps the conduit was attached during a cleaning event and later removed to reveal a stripe of darkened stone. Most concerning on N10, however, is that each base of its two columns show signs of potential future loss with precarious fissures at the corners adjacent to other panels, with N9 displaying the same feature. Other significant fissures stem from the vent openings at the bases of N9 and N11

FIGURE 14.3 The north-facing facade of Saint Elizabeth. Most of the aspects retain "Good condition" (*yellow color*) as measured by the CSSI, but several panels exhibit specific decay processes that have already occurred or will occur in the near future, especially N1 and N9.

FIGURE 14.4 Panel N1, covered with a heavy (anthropogenic) coating. Notice also the missing mortarwork above the vent, creating a gaping fissure as well as other multiple cracks in the mortar.

FIGURE 14.5 **Not cleaned like its neighboring panels, N9 hosts a much darker color (rock coating).** Most rock coatings enhance stability (Dorn, 1998) and removing them improperly can be detrimental to the host stone, opening the door for fiercer decay processes.

(Fig. 14.7) and flaking events having already occurred on the underside of N11's window arch (Fig. 14.8).

While seemingly stable overall, the north-facing facades also exhibit continuing decay forces that require constant attention and maintenance vigilance. Fissures in the sidewalk-level stones remain most apparent and offer great potential for exacerbated decay. Indeed, especially during the colder months, when deicing chemicals are used, the rock's internal individual structural integrity can be altered, increasing decay potential. Also of concern is the weakening of internal stone structure through the entry of moisture and Colorado's rapid freeze/thaw cycles. Further, while the north facade does not appear to be load-bearing, its

FIGURE 14.6 **Dark-color vertical stripes on the side of N10, perhaps from old piping.** Notice the difference in coloration between the portion of N10 that hosts the stripes and that to its immediate left. The left-side portion of N10 has been recently cleaned.

lowest level stones do bear the weight of above stones and this could be a contributing factor to the multiple column fissures.

14.3.2 East-facing Panels

While mostly stable with an overall score of 25 (Fig. 14.9), three "panels" on the east-facing facade scored particularly high: E5, E7, and E8, with average scores of 46, 49, and 52,

FIGURE 14.7 **An example of fissures around air vents. In this example from N11, fissures around the vent are visible.** Note also the scaling event—a small chunk of rock missing in the middle of a fissure independent of lithification on the stone to the left of the vent's grate.

FIGURE 14.8 **Flaking events already occurred and impending around N11's arched window.** Similar forms were also observed on the upper windows of N12 and N13.

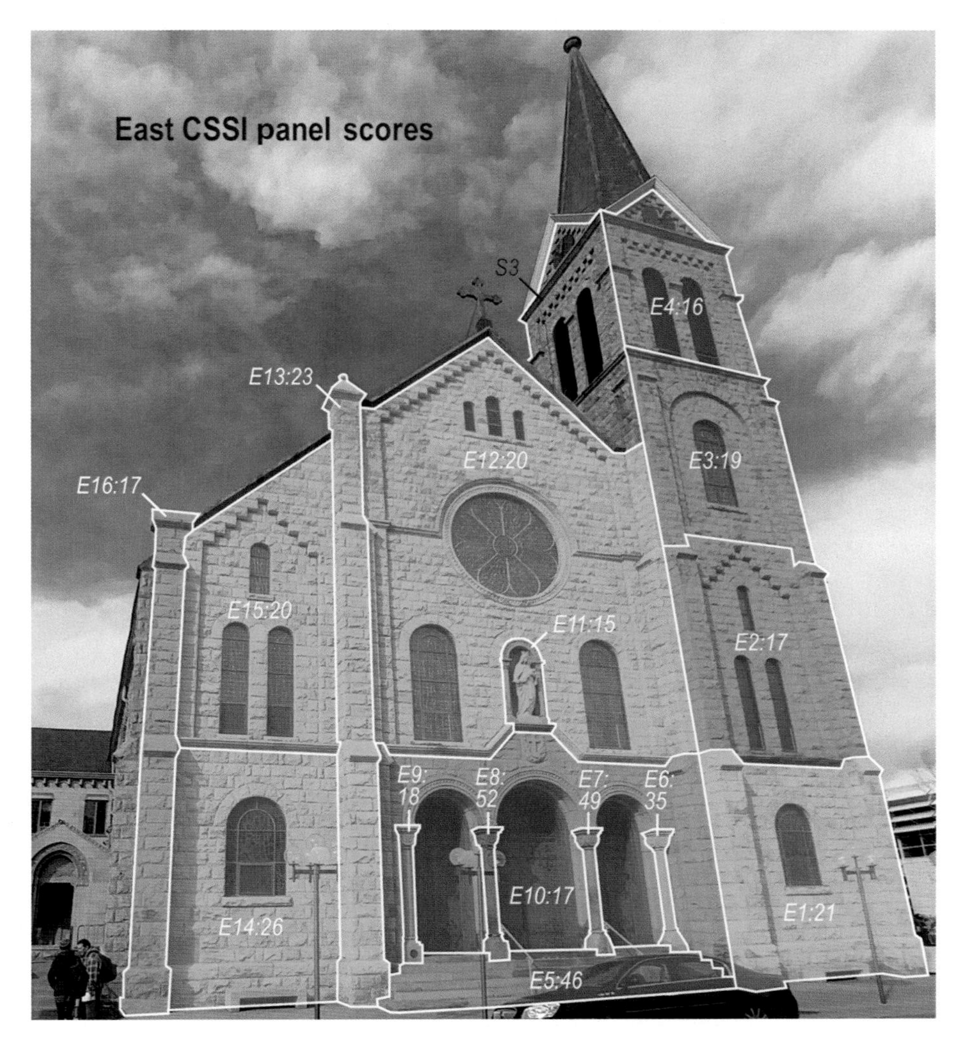

FIGURE 14.9 Saint Elizabeth's east-facing facade. Earning an overall score of 25 ("Good condition" status), the east facade has some looming problems that, if not addressed promptly, may result in large stone loss and the need for very costly repairs. In particular, E5, E7, and E8 remain at high risk, especially E7 and E8—the highest scoring panels on the entire structure and in "Urgent danger of erosion," specifically, at their bases (compare with Figs 14.13–14.16).

respectively, earning a ranking of "Urgent possibility of erosion" (E5 and E7) and "Great danger of erosion" (E8).

These three "panels"—stairs (E5, Fig. 14.10) and two main columns on either side of the main doors (E7 and E8) that comprise Saint Elizabeth's main parishioner entryway—host the bulk of decay concerns. This is perhaps no surprise, as Saint Elizabeth's holds regular Mass.

Still, decay on the stairs and columns show significantly more signs of deterioration than the rest of the church's east-facing facade. Differing rock type may also play a role in the

FIGURE 14.10 **The stairs of Saint Elizabeth's main entryway on the east-facing facade.** As Saint Elizabeth is still a functioning church, it is no surprise that these stairs remain in "Urgent danger of erosion." Specifically, notice the impending splintering of individual stairs and loose/cracked mortar, similar to N1 (compare Figs 14.4 and 14.5).

entryway's decay because the stairs (E5) are comprised of the same weakened sandstone as N1 and the columns (E7 and E8) were hewn from Pikes Peak granite, but never "polished" (Fig. 14.11A), and the columns' bases were originally Castle Rock rhyolite, but deteriorated portions have been replaced with concrete fashioned to look like the original stone (Fig. 14.11B). This mismatching of stone type, although perhaps aesthetically and architecturally pleasing, leaves inherent weaknesses exposed, especially at the bases of the columns where different rock types are in contact with each other, promoting strong differential decay.

For example, where different rock types come into contact with each other, such as column bases and capitals, erosion of the stones has necessitated the use of concrete repairs to create and in some instances replace the original carved stone surfaces (Fig. 14.11). These repairs have bonded with mixed results to the underlying Castle Rock rhyolite. While newer additions, they still exhibit significant surface decay and decomposition, most likely due to the use of pavement-clearing salts during winter months. The newer encasements may protect at first, but they also allow the decay agents to collect on the Castle Rock rhyolite surfaces, since the bond between underlying stone and repaired surface remains less than absolute. Additionally, Denver's extreme seasonal climate shifts, especially in terms of precipitation fluctuation and rapid temperature changes, create cyclical conditions that are challenging for any building material, but perhaps more so for rough-hewn and nonpolished rock joined together with a coarse-grained mortar and concrete.

Small future scaling (spalling) on the base of E9 (a column) was also observed; and the main entrance hosts many fissures at each base (Fig. 14.12). This is a high-traffic area with continual use of the stairs, and stair-railing installation sites also seem to be contributing to

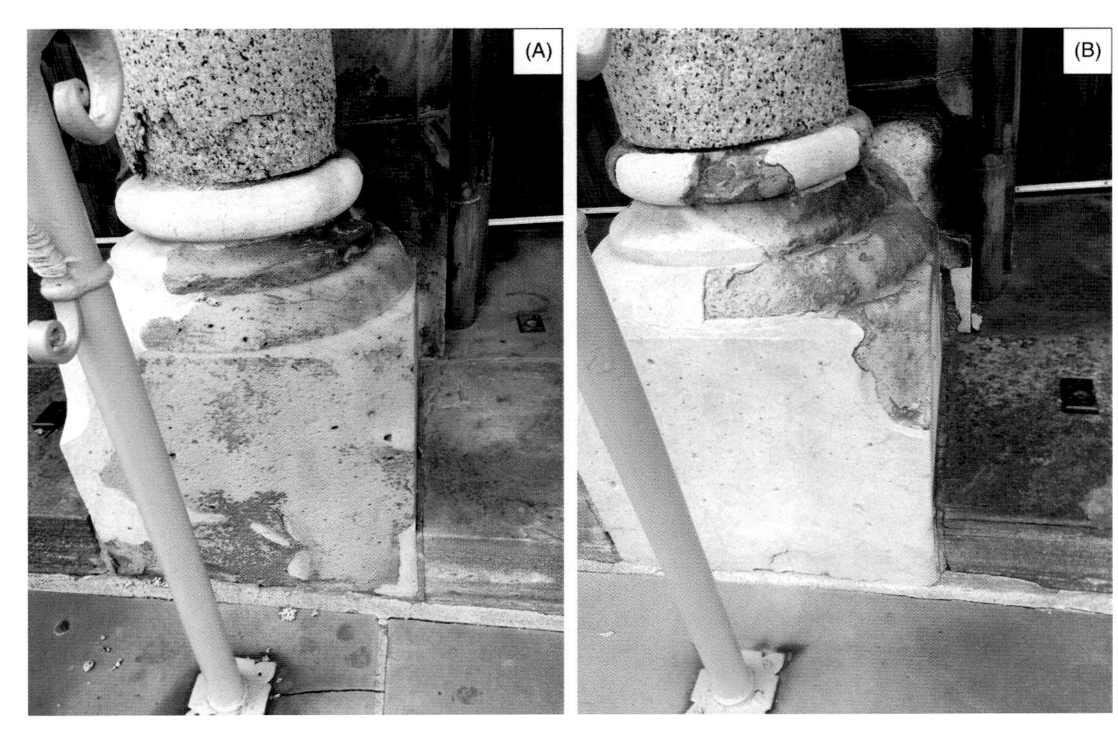

FIGURE 14.11 (A) Flaking of E7 (the column) accompanied by crumbly and granular disintegration. Note the small pieces of granite strewn across the steps at the column's base. The columns, while made of usually strong Pikes Peak granite, have only been smoothed, not polished or veneered, leaving them exposed to the elements. Indeed, the rock's matrix has been weakened so much that even someone casually leaning against this column exacerbates the crumbly and granular disintegration process. Notice also the fissure at the stair railing's base (bottom-center of image), perhaps brought on by the rail's oxidation, extending into the mortar of the steps (E5). (B) Examples of scaling, flaking, crumbly disintegration, and granular disintegration around E8. The lighter-colored stone is concrete, poured and sculpted to fit the original column base specifications. The replaced concrete area includes faux-tafoni as part of the stone dressing in an attempt to generate the same "feel" of Castle Rock rhyolite.

oxidation and fissures, which may eventually lead to larger events than crumbly and granular disintegration, ultimately compromising the railings as well as the staircase itself (Fig. 14.11). Additionally, the stones closest to the ground (within ~2 m of the ground), show more signs of decay than upper areas, with the stairs showing serious decay in their mortar work as well as splintering.

Higher up, as well as away from the entryway, most panels remain in "Good" or "Excellent" condition. Some fissures independent of stone lithification, such as the ledge over the main entrance remain present (Fig. 14.13), but with that exception and window cavities (Fig. 14.8), the upper and side panels of the east aspect remain considerably more stable than those near to the ground and around the entryway.

Of particular note on the church's east-facing aspect, is the statue of Saint Elizabeth herself. For the CSSI assessment, this beautifully carved statue (and its accompanying alcove) was given its own panel number (E11) due to the cultural significance. Although this statue (and her alcove)

FIGURE 14.12 **An example of fissures independent of stone lithification around column bases.** This particular fissure (with a pencil for scale) occurs on E8. Just to the right of the pencil is a very light discoloration (*the white areas*) where salt efflorescence and subflorescense is occurring. Often occurring through capillary action, efflorescence, and subflorescence exert pressure on pore spaces, exacerbating interior vulnerabilities of the rock.

earned a low CSSI score, there are still concerns about the amount of basal flaking, discoloration (possibly due to oxidation) around her wreath, and a missing pendant from her crown (Fig. 14.14). Still, for a century-old painted concrete statue, the condition remains remarkable.

In all, while the immediate east-facing facade appears stable, any impending losses around the main entrance must be taken seriously. For example, the most dominant decay features—a combination of tafoni, flaking, crumbly and granular disintegration, and (impending) fissures—pervade the surfaces of both the stairs and the facades. Other minor, but notable, decay features include missing and disintegrated mortar, oxidation, pollution (especially

FIGURE 14.13 **Fissure independent of lithification extending through a ledge on E10.** Left untreated, this fissure will only widen, leading to (small) flaking and subsequent scaling. This ledge also exhibits flaking (right-center and far-right-center).

near the facade's base), and some small scaling (both impending and already occurred) as well as flaking (impending and already occurred) and anthropogenic coatings (most likely pollution) at higher levels.

14.3.3 South-facing Panels

Of the four aspects, the south-facing scored the lowest CSSI rating (23, Fig. 14.15). This may be due to the aspect itself, where continual sun, even during the winter months, serves as a buffer in keeping away snow and ice better than the other, nonsun-facing aspects.

FIGURE 14.14 **The statue of Saint Elizabeth adorning the alcove above the church's main entrance.** This statue, now over 100 years old, is made of what appears to be concrete (no reference could be found of any records on its creation, even after questioning church personnel, other than its relative age), and is regularly painted to keep its brilliant white color. A close examination via telephoto lens reveals paint chipping off the sculpture, especially around its base. Notice also the missing crown prong.

As with other facades, however, minor areas of decay remain, such as tafoni, flaking, and granular and crumbly disintegration. Darkened spots of discoloration, likely the result of pollution, also span the south-facing facade, and in some areas the (likely protective, Dorn, 1998) coating has been removed. Small plants also line the base of S1, but do not pose any immediate decay threat.

FIGURE 14.15 **Saint Elizabeth's south-facing facade, displaying "Excellent" and "Good" CSSI scores.** While very stable overall, the church's south aspect did have a few minor concerns, such as fissures, flaking on the underside of window arches, and very small plant growth at the facade's base.

Similar to other facades on adjoining aspects, mortar separation—including whole chunks of missing mortar—remains a key deterioration factor on the south-facing facade. Indeed, the separating and missing mortar in doorway and window cavities on this side of the church appear to be worse than other facades, especially as they occur on several archways. The most conspicuous of these on panel S1, is a vertical fissure extending (upward) from the easternmost window arch's keystone along the mortar line of four separate stone layers (Fig. 14.16). A similar event has occurred adjacent to S1's window (Fig. 14.17).

Panel S2 is well-protected by a semicircular exterior hallway structure, so the base was inaccessible, but there was little evidence of the same mortar separating/disappearing as on panel S1. Panel S3 was difficult to assess from the ground and required use of high-magnification binoculars to assess. Still, similar to S2, no mortar separation/disappearing

FIGURE 14.16 **Beginning at the top of S2's highest window and extending across four bricks, missing mortar creates a gap in the masonry.** The open space increases the probability of fissure formation and its resultant flaking and scaling events. This area remains perhaps the most at-risk section of S1.

was observed, although a rock coating was present on some of the decoration, likely from pollution. Additionally, unlike the other three sides of the church, some stones during the earlier morning hours exude condensate (i.e., appear moist to the eye and damp to the touch) that, when coinciding with the inherent tafoni in Castle Rock rhyolite, could prove to be a future decay culprit, especially during winter months when temperatures can range from below freezing to well above freezing in a given 24-h period. While a few minor problems exist, overall, the south-facing aspect appears to be in very stable condition, aligning with the CSSI's "Good condition" status.

14.3.4 West-facing Panels

While earning a "Good condition" score of 28, the west facade still exhibits minor decay concerns (Fig. 14.18). The most obvious and contributing factor to the west facade's deterioration rests in its location, next to a high-traffic and delivery road even though it remains a bit more secluded than the other facades at first glance. This leads to the main overarching

FIGURE 14.17 **Creating similar conditions as seen on the underside of other window arches (compare Figs 14.10 and 14.17), the missing mortar along S1's window edge will ultimately lead to flaking and scaling events if left untreated.** Specifically, note the large crack between the stone of the arch and the windowpane, extending from the old mortar work (bottom-center) all the way to the beginning of the curve of the archway.

decay-contributing factor: constant vehicular interaction. As west panels remain adjacent to a frequently used road, pollution becomes a major issue, and a dark coating up to 1-m high from the base is present all along the west facade (Fig. 14.19). The presence of asphalt, and its subsequent need to be repaired regularly in the sometimes-extreme Colorado climate, may also affect the future stability of the stone as it abuts against the panels' bases.

As with other panels' aspects, mortar cracking also remains an issue, but here, owing to the more secluded location, garbage (e.g., cigarette butts) is often stuffed into the Castle Rock rhyolite's inherent tafoni by passersby. Another minor decay feature that affects the west facade are fissures independent of stone lithification, resulting in future flaking and scaling events like those found on other facades around the church.

14.3.5 Overall Assessment

From a rock-decay perspective, although small problems potentially affecting the building stones' future do exist, each facade appears to be overall stable. Judging from the building size and pervasive use of mortar for sealing the space between stones and repairing broken/missing pieces of decorative and dressed stone, the parish community most likely engages in significant

FIGURE 14.18 **Saint Elizabeth's west facade.** Although the west aspect earned the highest average CSSI score (28), panels with aspects offset slightly to the north or south did not display as high a ranking as those facing directly west, revealing that aspect may play a role in stone decay processes (Groom, 2014).

repairs when funds are available. That said, missing mortar work scattered across each facade needs attention. Continued monitoring of any decay will aid in deterring deterioration events at Saint Elizabeth's; and it is recommended that repairs occur in a timely manner.

14.4 IMPLICATIONS AND CONCLUSION

The RASI has been proven as a noninvasive, cost-effective, and easy-to-understand field-based research and assessment tool over the past decade (Allen et al., 2011; Allen and Groom, 2013a, b; Allen and Lukinbeal, 2011; Cerveny, 2005; Cerveny et al., 2016; Dorn et al., 2008; Groom, 2016, 2017); and the CSSI follows suit. Even so, RASI's strengths come with accompanying shortcomings (see Groom, 2017, especially Chapter 3.2.1, for a comprehensive overview). Similarly, although the CSSI remains the same as RASI with the exception of few terminological differences, it nonetheless exhibits specific benefits and challenges, the largest being its underlying conditional assumptions. Groom (2017, p. 130) elaborated:

FIGURE 14.19 **Panels W1 (two windows), W2 (singular window, above W1), part of W3 (to the left of W2), and W4 (left of drainpipe).** This image shows the thick, dark, and continually present anthropogenic coating of Saint Elizabeth's west facade. The coating extends the aspect's entire length and about 1 m above ground level.

Unlike RASI, CSSI compares current conditional statuses against assumed non-decayed baselines, since most cultural stone resources were created from "new" material. With RASI, researchers need to recognize that there will be a certain degree of "inherited decay"—rock decay that took place before the rock art was created—in their final scores since most rock art exists on preexposed surfaces. In contrast, the "virgin surfaces" of built monuments and building facades foster the assumption that all decay present has occurred after the completion of the resource. This assumption provides researchers with a controlled timeline of decay—allowing them to estimate factors such as rates of decay and date of decay initiation, which are much more difficult to calculate in natural settings. That said, the possibility remains that assumed baselines can skew results if the original surface was different than presumed (e.g., if stone dressings imitating textural deterioration were applied intentionally).

Additionally, panel definition remains distinct between RASI and CSSI. For example, when using RASI, panels are assigned by the researcher based on previously identified records or following the protocol established by the entity in charge (e.g., archaeological staff). Further, rock art tends to be found on fairly flat surfaces, such as cliff faces and boulder facets, which allows for panel division based on Cardinal direction (i.e., aspect—the direction the panel faces) or where motifs are most abundant (Groom, 2016). In the CSSI's case, because it is used to assess any built stone structure, including not just buildings but bridges, monuments, statues, and even gravestones, field preparation and site mapping to determine "panels" can be complicated but also flexible, as Groom (2017, p. 131) has suggested:

> When dealing with large building [facades], statues, or other more detailed cultural stone, site mapping and preparation can be a little more complicated, but also more flexible. CSSI researchers have the ability to define panels/features in whatever way best suits allotted field time, available resources, or desired precision. For example, a square building could be divided into four panels by aspect (i.e., "north side," "east side," "south side," "west side") or the same building could be divided by feature (e.g., "north side window arch 1," "north side window arch 2"). That same building could be assessed by aspect first, and then any specific characteristics of particular interest or importance can be analyzed individually. A large building [facade] could be just as easily divided into a handful of quadrants or dozens of individual elements, depending on the design and intention of the research. While studies with more panels will provide a more detailed analysis, they are much more time intensive and risk becoming counterproductive.

Groom (2017, p. 131) continued: "The intention of techniques like RASI and CSSI are to provide cost-effective rapid field assessments. If a building were divided into too many elements, a CSSI investigation would be prohibitively slow and potentially defeat the purpose of the work." In the case of Saint Elizabeth's, panels were defined based on architecture, but limited also by feasibility. For example, assessing the higher panels required finding different vantage points—on the tops of other buildings, on various floors of adjacent buildings—and nearly always required binoculars. Additionally, panel division for Saint Elizabeth's could have been brick-by-brick or horizontally, rather than vertically. Therein lies a benefit of the CSSI: the scale and intensity of assessment can be determined by the site manager based on need. That is, if a brick-by-brick assessment is needed—on the entire structure or just a single part—then the CSSI can be adjusted to accommodate. Indeed, one key to the CSSI's success rests in this situational adaptability.

Using the case study of a historic building in downtown Denver, CO, USA, this chapter focused on outlining the CSSI as a new technique for assessing the inherent geologic weaknesses in stone as well as highlighting its application for site managers. In a single afternoon, a research team of six people were able to assess Saint Elizabeth of Hungary Roman Catholic Church for its geologic stability (based on current and impending stone decay forms and processes) and then expound upon the findings in a succinct report, offering the site manager insight into areas most in need of monitoring and repair. Although other rock/stone decay assessments exist, most remain expensive, time-consuming, invasive (which can be detrimental to the host stone), and require a specialist to perform the assessment and interpret the results. The CSSI—like its RASI cousin (Dorn et al., 2008)—exhibits the opposite of these traits. As a technique for rapid, field-based assessment of rock deterioration, both the RASI and the CSSI have proven themselves valuable. Add to that the straightforward terminology/vernacular and quick training times for nonspecialists and a powerful tool emerges. With future assessments planned in the US, Europe, and the Middle East, the CSSI sets itself apart as a useful, cost-effective, and adaptable technique that, with appropriate training, can be used by almost anyone to evaluate current and impending rock weaknesses for any stone structure, regardless of size, building material, or location.

References

Allen, C.D., 2008. Using Rock Art as an Alternative Science Pedagogy. Arizona State University, Tempe, AZ.

Allen, C.D., Groom, K.M., 2013a. Evaluation of Grenada's "Carib Stones" via the Rock Art Stability Index. Appl. Geogr. 42, 165–175.

Allen, C.D., Groom, K.M., 2013b. A geologic assessment of Grenada's carib stones. Int. Newslett. Rock Art 65, 19–24.

Allen, C.D., Lukinbeal, C., 2011. Practicing physical geography: an actor-network view of physical geography exemplified by the Rock Art Stability Index. Prog. Phys. Geogr. 35 (2), 227–248.

Allen, C.D., Cutrell, A.K., Cerveny, N.V., Theurer, J., 2011. Advances in rock art research. La Pintura 37 (1), 4–6, 13.

Bruthans, J., Soukup, J., Vaculikova, J., Filippi, M., Schweigstillova, J., Mayo, A.L., Masin, D., Kletetschka, G., Rihosek, J., 2014. Sandstone landforms shaped by negative feedback between stress and erosion. Nat. Geosci. 7 (8), 597–601.

Cerveny, N., 2005. A weathering-based perspective on rock art conservation. Dissertation, Geography, Arizona State University, Tempe, AZ.

Cerveny, N.V., Dorn, R.I., Allen, C.D., Whitley, D.S., 2016. Advances in rapid condition assessments of rock art sites: Rock Art Stability Index (RASI). J. Archeol. Sci. Rep. 10 (2016), 871–877.

Dorn, R.I., 1998. Rock Coatings (Developments in Earth Surface Processes, number 6), first ed. Elsevier, Amsterdam, The Netherlands.

Dorn, R.I., Whitley, D.S., Cerveny, N.V., Gordon, S.J., Allen, C.D., Gutbrod, E., 2008. The Rock Art Stability Index: a new strategy for maximizing the sustainability of rock art as a heritage resource. Herit. Manag. 1 (1), 37–70.

Dorn, R.I., Gordon, S.J., Allen, C.D., Cerveny, N., Dixon, J.C., Groom, K.M., Hall, K., Harrison, E., Mol, L., Paradise, T.R., Sumner, P., 2013. The role of fieldwork in rock decay research: case studies from the fringe. Geomorphology 200, 59–74.

Fitzner, B., 2004. Documentation and evaluation of stone damage on monuments. In: Proceedings of the 10th International Congress on Deterioration and Conservation of Stone. June 27–July 2, 2004, Stockholm, Sweden. .

Fitzner, B., Heinrichs, K., 2001. Damage diagnosis at stone monuments: weathering forms, damage categories and damage indices. Acta-Univ. Carol. Geol. 45 (1), 12–13.

Fitzner, B, Heinrichs, K, Kownatzki, R, 1992. Classification and mapping of weathering forms. In: Proceedings of the 7th International Congress on Deterioration and Conservation of Stone. June 15–18, 1992, Lisbon, Portugal. .

Fitzner, B., Heinriches, K., Kownatzki, R., 1997. Weathering forms at natural stone monuments: classification, mapping and evaluation. Int. J. Restor. Build. Monum. 3 (2), 105–124.

Fitzner, B., Heinrichs, K., la Bouchardiere, D., 2002. Damage index for stone monuments.

Giesen, M.J., Ung, A., Warke, P.A., Christgen, B., Mazel, A.D., Graham, D.W., 2013. Condition assessment and preservation of open-air rock art panels during environmental change. J. Cult. Herit. 15 (1), 49–56.

Griffin, P., Indictor, N., Koestler, R., 1991. The biodeterioration of stone: a review of deterioration mechanisms, conservation case histories, and treatment. Int. Biodeterior. 28 (1–4), 187–207.

Groom, K.M., 2011. Historical assessment of rock decay on Dry Wash Bridge, Petrified Forest National Park. National Park Service, Rocky Mountain Cooperative Ecosystem Study Unit Project Summary Report.

Groom, K.M., 2014. Analysis of environmental influences on dressed stone decay: a case study of tafoni development on a hewn Djinn block in Petra, Jordan. Thesis, University of Arkansas.

Groom, K.M., 2016. Fading imagery: a mixed method analysis of rock art deterioration in the Arkansan Ozarks. Int. Newslett. Rock Art 74 (1), 14–20.

Groom, K.M., 2017. Rock art management and landscape change: mixed field assessment techniques for cultural stone decay. Dissertation, University of Arkansas.

Hall, K., Thorn, C., Sumner, P., 2012. On the persistence of 'weathering'. Geomorphology 149, 1–10.

Inkpen, R., Viles, H., Moses, C., Baily, B., 2012a. Modelling the impact of changing atmospheric pollution levels on limestone erosion rates in central London, 1980–2010. Atmos. Environ. 61, 476–481.

Inkpen, R., Viles, H., Moses, C., Baily, B., Collier, P., Trudgill, S., Cooke, R., 2012b. Thirty years of erosion and declining atmospheric pollution at St Paul's Cathedral. Atmos. Environ. 62, 521–529.

Janbade, P., Thakur, N., Tandon, B., 2016. Quantifying the damage and decay for conservation projects: identification, classification and analysis of the decay and deterioration in stone. Science and Art: A Future for Stone. p. 343.

Janvier-Badosa, S., Brunetaud, X., Beck, K., Al-Mukhtar, M., 2016. Kinetics of stone degradation of the Castle of Chambord in France. Int. J. Archit. Herit. 10 (1), 96–105.

Jo, Y.H., Lee, C.H., Chun, Y.G., 2012. Material characteristics and deterioration evaluation for the 13th century Korean stone pagoda of Magoksa temple. Environ. Earth Sci. 66 (3), 915–922.

Kurtz, J., Harry, D., Netoff, D.I., 2001. Stabilization of friable sandstone surfaces in a desiccating, wind-abraded environment of south-central Utah by rock surface microorganisms. J. Arid Environ. 48 (1), 89–100.

Lee, C.H., Chun, Y.G., 2013. Effect of climate conditions on weathering of stone cultural heritages in Korea. Int. J. Digit. Content Technol. Appl. 7 (12), 451.

McKinley, J.M., Warke, P.A., Lloyd, C.D., Ruffell, A.H., Smith, B.J., 2006. Geostatistical analysis in weathering studies: case study for Stanton Moor building sandstone. Earth Surf. Process. Landf. 31 (8), 950–969.

Meaders, M.I., 1959. Auraria was my friend. Georgia Rev. 13 (2), 402–408.

Noel, T.J., Wharton, N.J., 2016. Denver Landmarks and Historic Districts. University Press of Colorado, Boulder, CO.

Page, B., Ross, E., 2015. Envisioning the urban past: GIS reconstruction of a lost Denver district. Front. Digit. Humanit. 2 (3), 1–18.

Page, B., Ross, E., 2016. Legacies of a contested campus: urban renewal, community resistance and the origins of gentrification in Denver. Urban Geogr. 38 (9), 1293–1328.

Paradise, T.R., 1999. Analysis of sandstone weathering of the Roman theater in Petra, Jordan. Ann. Dept. Antiq. Jordan 43, 353–368.

Pope, G.A., Meierding, T.C., Paradise, T.R., 2002. Geomorphology's role in the study of weathering of cultural stone. Geomorphology 47 (2), 211–225.

Saint Elizabeth of Hungary Roman Catholic Church, 2017. Saint Elizabeth's website. Available from: http://stelizabethdenver.org/welcome/.

Siedel, H., Siegesmund, S., 2014. Characterization of stone deterioration on buildings. Stone in Architecture: Properties, Durability. Springer, 349-414.

Smith, B., Curran, J., Warke, P., Adamson, C., Stelfox, D., Savage, J., 2013. Stone-built heritage inventory and "performance in use" condition assessment of stonework. Quart. J. Eng. Geol. Hydrogeol. 46 (4), 391–404.

Thornbush, M.J., 2012. A site-specific index based on weathering forms visible in Central Oxford, UK. Geosciences 2 (4), 277–297.

Warke, P.A., Curran, J.M., Turkington, A.V., Smith, B.J., 2003. Condition assessment for building stone conservation: a staging system approach. Build. Environ. 38, 1113–1123.

Whitley, D.S., 2005. Introduction to Rock Art Research. Left Coast Press, Walnut Creek, CA.

CHAPTER

15

Photographic Technique Used in a Photometric Approach to Assess the Weathering of Pavement Slabs in Toronto (Ontario, Canada)

Mary J. Thornbush

University of Oxford, Oxford, United Kingdom

OUTLINE

15.1 Introduction 303

15.2 A New Method 305

15.2.1 A Novel Application 307

15.2.2 Measuring Surface Roughness 309

15.3 Results with Discussion 310

15.3.1 The Photometric Approach 311

15.4 Conclusions 313

References 313

15.1 INTRODUCTION

Quantitative photographic (photometric) approaches do not need to be complicated and expensive. Digital photographic cameras have become commonplace and can be acquired cheaply, even at relatively high digital resolutions of 1200 dots per inch or better. These cameras are also being manufactured in smaller sizes, making them more easily portable and field-applicable. They can be transported safely for vast distances and mounted for precise measurement in the field. Indeed, studies that have deployed photographic surveys to track landscape change at various scales have utilized digital cameras to photograph and rephotograph landforms and landscapes throughout the world according to diverse techniques now available.

Urban Geomorphology. http://dx.doi.org/10.1016/B978-0-12-811951-8.00015-1

Contemporary approaches to the monitoring of pavements have employed laser scanning and photogrammetry to derive three-dimensional (3-D) information about surface decay. Ouyang and Xu (2013), for instance, promoted the use of a 3-D camera and laser light as an automated inspection system. Their 3-D system was able to capture a depth resolution of 0.5 mm and transverse resolution of 1.56 mm per pixel at a height of 1.4 m above ground level. Used in an innovative manner, high-precision laser scanners (triangulation types) are capable of providing 3-D information of the areal surface layer based on texture characterization (Bitelli et al., 2012). Triangulated meshes (which are more affordable) have also been deployed to measure curvature (and surface roughness) of rock faces from a distance of 3 m (Lai et al., 2014). Terrestrial laser scanners are capable of measuring surface roughness at all scales; with georeferencing at the mm-range, detection can occur as low as 6 mm (at 95% confidence level) applied in situ over ranges of 50 m to assess landscapes (Lague et al., 2013). So long as georeferencing is possible (in coordinate space), a model or geographic information system (GIS) database can be established based on data output derived even from single photographs (Collier et al., 2001; Inkpen et al., 2008). Applications have included rock art (and a variety of types, including engravings, pictographs, petroglyphs considered) based on multiscale models of cave geometry derived from 3-D digital surveys using a terrestrial laser scanner, high-resolution digital cameras, and total station (González-Aguilera et al., 2009). Data acquired at the planetary scale, from a Lunar Orbiter laser altimeter, provided a means by which to assess surface roughness based on elevation derived from gridded data records adopted as image pixels to develop regional roughness maps (Cao et al., 2015).

Authors using digital photogrammetry to measure the surface roughness of rocks via digital stereoimages derived using a high-resolution digital camera (Kodak DCS 420 nonmetric) versus a Rolleiflex 6006 metric film camera and the least squares method, discovered a root-mean-square (*RMS*) roughness of 0.001–0.067 mm and 0.010–0.056 mm, respectively, compared to a laser profilemeter (Lee and Ahn, 2004). Line laser imaging in 3-D has been shown by the Georgia Department of Transportation to measure ridge-to-valley depth as a micromilling resurfacing method that has an error of less than 0.4 mm, so it has the potential to operate better than photogrammetric approaches to quality control of pavement surfaces by identifying cracks and rough spots (Tsai et al., 2014). Furthermore, rock glaciers in Austria have been explored using terrestrial (ground-based or close-range) photogrammetry to detect and quantify surface change.

Other instruments have been selected as tools to measure surface roughness. Fischer and Luettge (2007), for example, chose to quantify surface topography using vertical scanning interferometry because of its high vertical resolution and large field-of-view. The application of this instrument on slate surfaces enabled these authors to ascertain the influence of organic matter degradation in controlling rock surface reactivity and topography.

Surface roughness is an important parameter to examine because of its connection to rock weathering. Using high-resolution X-ray computed tomography (CT scanning), for instance, Jia et al. (2014) showed that 3-D visualized images were capable of conveying macroscopic fracturing in rocks resulting from well-developed roughness. It is possible to apply many of these remote-sensing techniques to rock interiors as well as surfaces, as was evident with borehole and surface electrical resistivity tomography to generate images of buried structures, even in the presence of concrete pavement slabs in an urban setting (Tsokas et al., 2011). X-ray imaging apparatus was applied to cylindrical samples of pervious concrete in order to determine porosity (affecting pavement permeability and thought to be the most important pavement property) in 2- and 3-D (Ahn et al., 2014), with higher results for the latter (3-D X-ray). Work performed on granite in Antarctica has also pointed to the importance of porosity (as well as grain size) by influencing thermal stress building up throughout the day and seasonally (Hall et al., 2008). Furthermore, poor compaction (allowing for the presence of air voids) permits moisture to enter paving materials (asphalt as well as concrete) and cause rapid pavement deterioration (Kassem et al., 2008). X-ray CT was used with ground-penetrating radar (GPR) and both techniques were found to be efficient in the evaluation of asphalt pavement compaction that could affect weathering. Radar imaging has been applied to carbonate rocks to assess backscatter signatures that are used as indicators of surface roughness known to be diagnostic of weathering (Budkewitsch et al., 1996). Benavente et al. (2003) concluded, based on spectrophotometric data, that surface roughness affects color change, so it is an important variable to study in lightness research.

This chapter focuses on horizontal surfaces derived from limestone in the form of concrete pavements, which cover many North American cities, including this study's locale: the city of Toronto. The purpose here is to provide an affordable, easy-to-use technique for the 3-D quantification of weathering in order to assess urban landscape change. The focus is cross-temporal, as a single street (namely, Gladstone Avenue) is used as a case study in which to apply the integrated digital photography and image processing (IDIP) method developed by Thornbush (2010) and applied outdoors as the O-IDIP method. The technique is apparently versatile when calibrated and can be deployed to acquire information about lightness (*L*) and chromatic change (*a* and *b*) based on color channels (*a* being the green-red spectrum and *b*, blue-yellow) using the CIELAB color system. Other studies have also used CIELAB color space (Cerimele and Cossu, 2009, to develop an image segmentation process), with the lightness parameter of different surface finishes found not to be affected by roughness (Sanmartín et al., 2011). The current study focuses on lightness quantification, as the color of pavement is nearly white, and a two-point (black–white) calibration procedure based on a color chart (outlined in the next section) is all that is required for its outdoor application.

Specifically, this study aims to present a new application of the O-IDIP method based on a pavement study performed in downtown Toronto. Its objectives are to: (1) outline the development of the O-IDIP method as a photometric approach; (2) convey its current application linked with already published studies in order to provide context; and (3) examine the findings comparatively as a quantitative method to measure surface roughness based on linear correlations of histogram-output data. Some limitations of this photometric approach are addressed; but first, the method will be conveyed in the context of existing published studies.

15.2 A NEW METHOD

The IDIP method was originally devised in the laboratory using exposure discs (Thornbush and Viles, 2004a). These limestone sensors had been exposed up to 5 years at urban sites and one background (cleaner) site located in central Oxford, United Kingdom (Thornbush and Viles, 2006). To be precise, the method requires that a digital camera be mounted on a tripod for consistent height and base stability. Having the camera mounted facilitates movement from one site to another in outdoor surveys, so this remained an important aspect of the O-IDIP method. However, in one instance (Thornbush, 2010), a viewpoint was necessary from an observer's perspective so that the use of a tripod was excluded. The sampling height for photographs is normally within 1.3 m tripod height above the ground and approximately 1-m distance from the photographic object, which usually has been building walls.

The method was previously deployed as part of a repeat photographic (rephotographic) survey, with the original photographic set attained in 1997 (Thornbush and Viles, 2008). These earlier photographic surveys were not digital and printed photographs needed to be scanned for digitization. However, this step was skipped with the incorporation of a digital camera. Most of this work was executed outdoors, but still in the IDIP phase prior to 2008 before it was finally calibrated using a simple grayscale and X-Rite SP68 Sphere Spectrophotometer (Thornbush, 2008). A color chart was subsequently added in photographs for chromatic calibration based on *a* and *b* color channels. Calibration procedures were developed, including one for green-red (*a*) calibration using an X-Rite ColorChecker to assess algal growth (Thornbush, 2013a). A Gretagmacbeth ColorChecker Color Rendition Chart was employed later in the development of a five-point soiling index (Thornbush, 2014a). At this time, the method was finally developed to encompass a three-point (black-white and green-red) procedure that calibrated for lightness and chroma (Thornbush, 2014b).

While original studies were based on lightness measurement to assess soiling, later applications of the method quantified color change and surface decay. Whereas the former (color change) represents degradation, the latter is linked with deterioration. It is important to quantify lightness and color change in order to establish a rate of surface degradation, which could eventually either lead to enhanced surface roughness (which in this case is associated with surface deterioration and decay). Therefore, lightness-based estimations of surface roughness (because uneven surfaces have more disparate lightness and, hence, an augmented standard deviation or SD *L* value) are useful to assess damage and the progression of surface decay, which in this case is conveyed by increasingly uneven (or roughened) surfaces. One method derived from IDIP was the decay mapping in Adobe Photoshop (DMAP) approach devised by Thornbush and Viles (2007a). This approach examined lightness change on a boundary wall based at a certain level of brightness ($L = 77\%$). The lightness component was found to be sensitive to shadows and outdoor lighting, so an overcast condition was recommended in later research (Thornbush, 2014c). Last, the method was most recently used to measure surface roughness as an application to rock decay and not just the soiling of surfaces and color modification (Thornbush, 2014d).

Both IDIP and O-IDIP use Adobe Photoshop to derive histograms of digital photographs after they have been suitably calibrated. However, it has been noted that it is possible to use other computer software with histogram functionality, such as Corel DRAW and more (Thornbush, 2010), such as ImageJ software used for image processing by Jeong et al. in Chapter 10 of this volume. The numeric output from these histograms includes mean, median, and SD values of lightness and chromatic variables (out of 255 for Lab Color Mode, then converted to a proportion). Because shadows are deliberately avoided, and only flat surfaces are photographed in these close-up surveys, any remaining variations in SD values in particular are attributable to surface roughness. Early studies focusing on lightness attributed variations in mean and median values to soiling patterns (Thornbush and Viles, 2004b). Most recently, the color chart has been used as a way to select samples of the surface area photographed, with areas located immediately behind the color chart in photographic pairs (providing comparison across time in one photographic survey) cropped and selected for study (Thornbush, 2014e). In this case, pixels are tracked and can typically represent less than 10% of the total photographed area (Thornbush, 2016). Nevertheless, the method still permits for areal quantification that traditional point-sample instruments (such as the spectrophotometer) may not allow. It is part of a photogeomorphological approach that was adopted by the author in various studies in the city of Oxford, United Kingdom, as already delineated in this section, that have been published since 2004 (Thornbush, 2013b) and more recently in a quantitative approach (Thornbush, 2008, 2010).

15.2.1 A Novel Application

The application of the IDIP method outdoors through calibration, resulting in the O-IDIP method, has been used mostly on flat vertical surfaces, such as buildings and walls. This has been an oblique close-up deployment of the method to assess limestone soiling and decay. Oblique ground-based digital photography has dominated the literature on weathering studies, including close-range multispectral imagery used in digital 3-D mapping based on red-green-blue color space to create false color images (Olariu et al., 2003). This study represents a new application of the technique to horizontal concrete surfaces (applied perpendicular to the ground surface from a tripod mount, Fig. 15.1). In the literature, pavements commonly refer to asphalt paved surfaces, mainly roads (El Gendy et al., 2011). Very few published studies address concrete pavements and walkways or sideways in particular and this study fills a literature gap that is important to modern development in cities, especially those located in North America where concrete is abundantly used in the built environment.

There are several known existing applications of monitoring horizontal surfaces. Biological crusts, for instance, can develop on rock surfaces as well as soils as biological soil crusts or (BSCs) (also noted by Jeong et al. in Chapter 10 of this volume; as well as Allen, 2005; Allen et al., 2009; Belnap and Lange, 2001; Bowker et al., 2014, 2016; Rodríguez-Caballero et al., 2012; Viles, 2008; Viles et al., 2008) that can affect runoff and erosion on land surfaces. This indicates that there is a greater applicability of this study beyond rock weathering research; and soil erosion research may be an area of research relevance. Others have already examined the potential to measure soil erosion using low-cost digital cameras to make robust measurements in the field through the application of a model, even by those unskilled in photogrammetry and image processing (Filin et al., 2013). Digital photography and image-processing techniques have allowed for the accurate measurement of vegetation cover in the Arctic (Chen et al., 2010), with further benefits of its objectivity, efficiency, and versatility. Additional applications of the photometric approach, conveying its importance and relevance, include planetary research. Photometric techniques, for instance, have been deployed along with topographic and spectral analysis to study the lunar surface (Wöhler et al., 2014). A photometric study of radiance was performed for Mars, and the data showed good agreement with a thermal emission spectrometer on-board the NASA Mars Global Surveyor orbiter (Esposito et al., 2007). Most recently, surface roughness has also been measured using grayscale images of specimen surfaces in finish turning (Shahabi and Ratnam, 2016).

Relevant vertical, multiscale applications using high-resolution optical imagery can also extend to modeling natural hazards, including floods, rockfalls, and avalanches (Hollaus et al., 2011) due to reduced flow velocity as surfaces are roughened. Joints can be characterized in 3-D along fault lines, for instance, at the landscape scale (Mah et al., 2013). Others have

FIGURE 15.1 **Typical sampling of the sidewalk during the photographic field survey.**

deployed a microroughness meter along 20-cm horizontal transects (repeated each 20 m of fault height) along a scarp in the Apennines, Italy (Giaccio et al., 2002). Digital stereophotogrammetry can also be executed using a digital camera with a macrolens and a mount system along with specification of location, which can be attained using a global position system, and some modern 3-D digital cameras already have this function integrated. In this way, it is possible to apply digital photogrammetry to geological objects, even fossils as well as natural cliffs and surface exposures (Fujii et al., 2006).

When both terrestrial laser scanning (laser imaging detection and ranging or LIDAR) and 3-D mosaics of high-resolution rectified digital photographs were used at a fault zone located in the Italian Alps (Bistacchi et al., 2011), the roughness of faults evident at the m-mm scale showed a clear relationship with precursor joints, but some anisotropy with the Hurst exponent (0.6–0.8). At nanoscale resolution, it is possible to characterize limestone degradation using microscopy (scanning force microscopy) and optical measurements to derive *RMS* and power spectral density of surface roughness (Orihuela et al., 2014). Dai and Ng (2014) have used a transmission X-ray microscope with 3-D cohesive zone modeling to characterize nanoscale (30 nm resolution) internal frost microdamage to cement.

15.2.1.1 Three-dimensional (3-D) Quantification

The O-IDIP method represents a photometric approach to quantifying surface change. Until recently, the technique has been used for areal quantification (along a 2-D plane). Its 3-D application is possible, however, through consideration of SD *L* values that capture variation in lightness and, therefore, surface lightness change. Values associated with SD *L* demonstrate another aspect of weathering. They have been known, for instance, to be affected by shadows on the surface that are representative of surface imperfections (decay). For this latter reason, Thornbush (2014d, e) used this measure as a way to quantify changes in surface texture or surface roughness.

Shadows were effectively eliminated with the application of the technique in overcast outdoor lighting conditions (Thornbush, 2008). Thornbush (2010) found, for example, that shadows (rather than soiling patterns) were responsible for inflating SD *L* values because these values reflect photographic contrast (rather than brightness) affected by shadows. However, contrast could be exploited as a means of indirectly acquiring a 3-D view of flat surfaces, since it could be augmented by surface texture and roughness (shadows cast over a bumpy surface). This can be affected by granular disintegration, abrasion, biological growths, and so on that roughen the surface either through subtractive or additive weathering processes. Continued rock weathering enhances surface roughness, for instance, making surfaces increasingly prone to microbial colonization (Miller et al., 2012), so that rock weathering (stone decay) and surface roughness are interrelated.

15.2.2 Measuring Surface Roughness

The field-based application of IDIP for quantitative (histogram-based) measurement of flat surfaces requires that surfaces be dry (as wetting will enhance their reflectivity) and photographed without direct sunlight in order to minimize the impact of shadows on data output. However, previous research comparing overcast and clear sky conditions found that the latter (clear sky) resulted in less spread of measured *RMS* roughness values and produced more consistent results (Thornbush, 2014d). Although % mean- and median-*L* values (derived from calibrated histogram-based image processing outputs, using the 10-step procedure outlined by Thornbush, 2014e) are known to be highly correlated (e.g., $r = 0.9488$ according to Thornbush, 2014e), they vary from SD *L* values, which have been found not to be most strongly correlated with *RMS* roughness.

The current study chose to perform the digital photographic survey in an overcast outdoor condition so as not to produce error in lightness measurements due to cast shadows. Mean, median, and SD *L* values derived from histograms in Lab Mode in Adobe Photoshop (Version 7) are reported and considered here to decipher any linear trends. The impact of the calibration procedure first introduced by Thornbush (2008) and further developed with the color chart deployed in this study is also examined in terms of its effects on the dataset (uncalibrated versus calibrated). This calibration procedure samples "true color" by inclusion of a color chart in one of the photographic pairs and then sampling (using the sampler tool) black-and-white color swabs. Adjustments are made so that black approximates $L = 0\%$ and white 100%. Data output from selected (cropped) portions of photographic pairs were overlain in order to select an identical areal sample. Profile-based roughness measurements have already been compared to calibrated O-IDIP values and indicate that mean and median L values have negative linear correlations with *RMS* roughness (measured using a digital depth gauge to acquire four profiles—on at least 10 surfaces—at 5 mm intervals along 19 cm long profiles, after McCarroll, 1997), with median values being the most highly correlated with (*RMS*) surface roughness ($r = -0.78$ according to Thornbush, 2014d). The findings of this study are provided on a comparative basis to linear correlations used for analysis.

15.3 RESULTS WITH DISCUSSION

Photogrammetry was originally recognized for geological mapping based on overlapped aerial photographs and provided the basis for producing regional-scale topographic maps (Slavinski and Morris, 2005). Close-range digital photogrammetry operates on the same principles, where photographs taken at known locations with a fixed camera position (orientation) can render 3-D images (Slavinski and Morris, 2005).

Terrestrial (ground-based or close-range) photogrammetry has been used in order to detect and quantify surface change on Austrian rock glaciers (for high-mountain mapping), for instance, and this monitoring is possible even with cheap digital (consumer) cameras and computer software (Kaufmann, 2012). So, the expense can be mostly attributed to the powerful software that is necessary to run photogrammetric analysis, with computer vision required for improved image extraction and analysis based on least-squares matching. Furthermore, it is possible to take this approach in a rephotogrammetric approach to monitor glacial retreat (as in 1986, 1999, 2003, and 2008 by Kaufmann, 2012). The use of consumer-grade digital cameras makes photogrammetric techniques appealing, especially if automated (Ahmed et al., 2011). According to these authors, digital imaging systems can be used with or without laser profilers, so that alternative low-cost technology can be applied, for instance, to monitor road surfaces.

Conventional ground-control collection methods may be inadequate for accuracies required beyond the subcentimeter range (Matthews et al., 2004), even though adequate for aerial photographic scales (e.g., 1:1500 or greater). However, recent technological advances in affordable digital cameras and computer software have allowed for the quantification of image distortion, meaning that simple digital cameras can be deployed as a type of virtual surveying instrument (Matthews et al., 2004). The authors have more recently (Matthews et al., 2011) applied close-range photogrammetry to digitally record (and virtually capture) 13 layers of excavation to expose the Laetoli footprints in Tanzania. They obtained stereoscopic images using a digital single lens reflex camera mounted on a monopod to derive 3-D aerial orthophotographs of the site. The authors praised the technique for its portability and efficiency (with a final dataset produced in minutes).

Close-range digital photogrammetry has been employed to produce digital elevation models (DEMs); however, the quality of DEM measurements limits the detection of erosion and deposition up to 5.3 mm (Gessesse et al., 2010), which is not much different from the application of the current study (at a base sampling point distance of 5 mm). It is also difficult

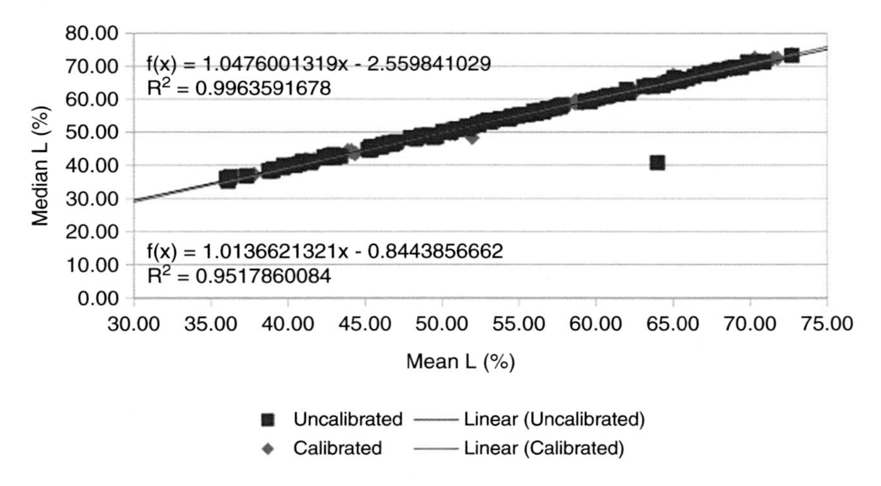

FIGURE 15.2 **Mean versus median *L* for uncalibrated versus calibrated lightness values.**

to automate the process due to various challenges outlined by the authors, making it also comparable to the current technique, minus the expense and expertise required to operate photogrammetric hardware and software. Ahmad et al. (2013) have also applied DEMs (e.g., Shuttle Radar Topographic Mission) at spatial resolutions of 90 m for surface roughness mapping in order to identify areas influenced by neotectonics. Using multiscale analysis, it is possible to compute surface roughness based on DEMs (Hani et al., 2012).

In the current study, % mean and median *L* values were more highly correlated than in previous research ($r = 0.9982$, with only one outlier) and an almost perfect positive linear correlation existed between the measured (calibrated) variables (Fig. 15.2). The calibration noticeably improved this correlation (from $r = 0.9756$) between measured lightness based on the O-IDIP method and the data stamp on the concrete slabs (in years). However, it should be noted that the calibration procedure improved the linear correlation of both % mean and median *L* values, but not % SD *L* values (Fig. 15.3). This could be due to more adjustments (a total of 5836 in the current study) required to contrast (5418) and not brightness (418) as part of the calibration process. Similarly, Thornbush (2014e) found that the 10-step calibration procedure required more adjustments to contrast (994 versus only 41 for brightness in Table 1, p. 546) under overcast conditions, but not under a clear sky, indicating that photographing when overcast requires more contrast adjustments.

15.3.1 The Photometric Approach

In the current study, median *L* values were found to be most indicative (calibrated or not) of lightness change (specifically, lightness loss or soiling). Median filtering has been utilized by other authors to assess surface roughness (Hu et al., 2009), alongside grayscale equalization and histogram conversion amplification. Other published works have shown that surface roughness, at least at the nanoscale, is directly related to reduced lightness (loss of brightness) and color variations on limestone (Orihuela et al., 2014), which could be considered as analogous to SD *L* values.

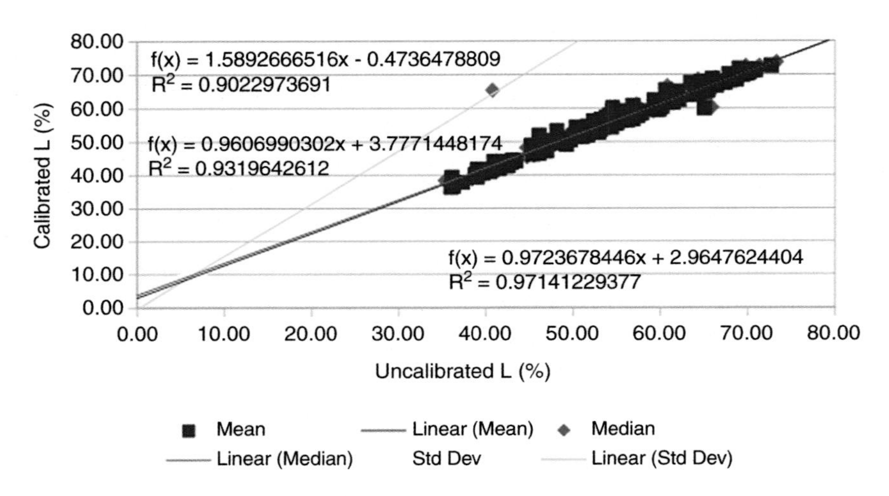

FIGURE 15.3 **Uncalibrated versus calibrated lightness values.**

Another study deployed a small "unmanned" aerial vehicle over a glacier to build a high-resolution orthomosaic based on a standard compact camera carried on-board (Rippin et al., 2015). Since ice surface roughness was found to be higher in areas with a greater channel density, these areas should have lower reflectance values. When applied to the current study, this would indicate that where surface roughness is high, brightness (lightness) should be low. An older photometric study considered the use of multiple scattering (in different parts of a rough surface) to suppress shadowing effects (Shkuratov et al., 2005), suggesting that surface roughness can be used to compensate for any albedo effects.

Research assessing soil roughness at fine spatial scales based on a point laser profiling instrument used at a resolution of 2 mm (Croft et al., 2009), discovered a strong relationship with directional reflectance (of $R^2 = 0.94$). Even though this was performed on soil surfaces, and based on laser profiling for surface roughness with the use of a spectroradiometer for reflectance measurement, the correlation likewise shows a strong linear correlation, although this study (and previous research, as by Thornbush, 2014d) has revealed a negative linear correlation, with lightness reduced as surface roughness (*RMS*) increases. This compares with other studies (Katami et al., 1996), that similarly conveyed an inverse relationship between surface reflectance (relative lightness) and surface roughness (measured through the arithmetical average roughness height or *Ra*). In the latter study, a short exposure to an acidic environment caused loss of lightness to be early (within 9 months of exposure), but little differences of *Ra* in later (3–9 compared to 1–3 months) exposures. This mimics findings by Thornbush and Viles (2007b), where weight loss of limestone sensors was evident in just 13–24 days of immersion in acidic solutions in a climatic simulation experiment. Therefore, it would seem that the surface roughness parameter is associated with decay (e.g., dissolution, granular disintegration, abrasion) rather than soiling (loss of lightness).

Surface roughness determination of building stones (Spanish granites) discovered steeply increasing trends at edges and corners of rock specimens (López-Acre et al., 2010). This finding helps to place the current sampling near stamp marks located toward the edge of pavement slabs in this study. Future research could compare the results when samples are obtained closer to slab centers (away from edges and corners). In addition, between 6.49 and 8.30% of the total image size was used on which to base the analysis in the current study, as the area behind the color chart was only employed (after Thornbush, 2014d,e). Still, a greater surface area could have been utilized to augment sample size. The rationale for this approach, however, was to closely replicate the selection of pixels in photographic pairs, which could have distorted the calibration process if not consistently measured. If this method is applied without calibration, a larger sample surface area could be tested.

As a caveat, although it is possible to apply digital photography and image processing as a photometric approach for data acquisition, verification via ground-truthing and/or robotic follow-up (Helper et al., 2010) remains indispensable. Even though this application of O-IDIP to measure surface roughness was based solely on image processing of digital photographs, other studies already field-verified the method (Thornbush, 2014d) in its vertical application (to walls), which should not be too different from the current application to horizontal surfaces.

15.4 CONCLUSIONS

When using the O-IDIP method to measure surface roughness on flat surfaces, it is evident based on the findings of this study that they should be executed in clear sky conditions in order to reduce the number of contrast adjustments required in the calibration process. Moreover, % median *L* values appear to be the most tightly correlated (than % mean *L* and especially % SD *L* values) with lightness loss since time of exposure. The calibration procedure improves % mean *L* more than median or SD values of lightness quantification, however with the latter actually being worsened. This suggests that the calibration procedure can be dropped if % median *L* values in particular are used. This would allow for sampling of a larger surface area derived from digital photographs rather than point-sampling performed by many instruments in use today that require filling in gaps between sampling points that could expound error in image sampling and representation.

References

Ahmad, S.R., Mahmood, S.A., Qureshi, J., 2013. DEMbased computation of topographic surface roughness to reveal incision in Gilgit-Baltistan Pakistan. Pak. J. Sci. 65 (1), 167–172.

Ahmed, M., Haas, C.T., Haas, R., 2011. Toward low-cost 3D automatic pavement distress surveying: the close range photogrammetry approach. Can. J. Civ. Eng. 38, 1301–1313.

Ahn, J., Jung, J., Kim, S., Han, S.-I., 2014. X-ray image analysis of porosity of pervious concretes. Int. J. GEOMATE 6 (1), 796–799.

Allen, C.D., 2005. Micrometeorology of a smooth and rugose biological soil crust near Coon Bluff Arizona. J. Arizona-Nevada Acad. Sci. 38 (1), 21–28.

Allen, C.D., Dorn, J.D., Dorn, R.I., 2009. Fire in the desert: initial gullying associated with the Cave Creek Complex fire, Sonoran Desert Arizona. Yearb. Ass. Pac. Coast Geogr. 71 (1), 182–195.

Belnap, J., Lange, O.L., 2001. Biological Soil Crusts: Structure, Function, and Management. Springer, Berlin.

Benavente, D., Martínez-Verdú, F., Bernabeu, A., Viqueira, V., Fort, R., García del Cura, M.A., Illueca, C., Ordóñez, S., 2003. Influence of surface roughness on color changes in building stones. Color Appl. Res. 28 (5), 343–351.

Bistacchi, A., Griffith, A., Smith, S.A.F., di Toro, G., Jones, R., Nielsen, S., 2011. Fault roughness of seismogenic depths from LIDAR and photogrammetric analysis. Pure Appl. Geophys. 168, 2345–2363.

Bitelli, G., Simone, A., Girardi, F., Lantieri, C., 2012. Laser scanning on road pavements: a new approach for characterizing surface texture. Sensors 12, 9110–9128.

Bowker, M.A., Maestre, F.T., Eldridge, D., Belnap, J., Castillo-Monroy, A., Escolar, C., Soliveres, S., 2014. Biological soil crusts (biocrusts) as a model system in community, landscape and ecosystem ecology. Biodivers. Conserv. 23 (7), 1619–1637.

Bowker, M.A., Belnap, J., Büdel, B., Sannier, C., Pietrasiak, N., Eldridge, D.J., Rivera-Aguilar, V., 2016. Controls on distribution patterns of biological soil crusts at micro-to global scales. Biological Soil Crusts: An Organizing Principle in Drylands. Springer International, New York, NY, 173-197.

Budkewitsch, P., DIorio, M.A., Harrison, J.C., 1996. C-band radar signatures of lithology in Arctic Environments; preliminary results from Bathurst Island, Northwest Territories. Current Research. Geological Survey of Canada, Ottawa, ON, 67–72 pp..

Cao, W., Cai, Z., Tang, Z., 2015. Lunar surface roughness based on multiscale morphological method. Planet. Space Sci. 108, 13–23.

Cerimele, M.M., Cossu, R., 2009. A numerical modelling for the extraction of decay regions from color images of monuments. Math. Comput. Simulat. 79, 2334–2344.

Chen, Z., Chen, W., LeBlanc, S.G., Henry, G.H.R., 2010. Digital photograph analysis for measuring percent plant cover in the Arctic. Arctic 63 (3), 315–326.

Collier, P., Inkpen, R.J., Fontana, D., 2001. Mapping decay: integrating scales of weathering within a GIS. Earth Surf. Proc. Land. 26 (8), 885–900.

Croft, H., Anderson, K., Kuhn, N.J., 2009. Characterizing soil surface roughness using a combined structural and spectral approach. Eur. J. Soil Sci. 60, 431–442.

Dai, Q., Ng, K., 2014. Transmission X-ray microscope nanoscale characterization and 3D micromechanical modeling of internal frost damage in cement paste. J. Nanomech. Micromech. 4 (1), 1.

El Gendy, A., Shalaby, A., Saleh, M., Flintsch, G.W., 2011. Stereo-vision applications to reconstruct the 3D texture of pavement surface. Int. J. Pavement Eng. 12 (3), 263–273.

Esposito, F., Giuaranna, M., Maturilli, A., Palomba, E., Colangeli, L., Formisano, V., 2007. Albedo and photometric study of Mars with the Planetary Fourier Spectrometer on-board the Mars Express mission. Icarus 186, 527–546.

Filin, S., Goldshledger, N., Abergel, S., Arav, R., 2013. Robust erosion measurement in agricultural fields by colour image processing and image measurement. Eur. J. Soil Sci. 64, 80–91.

Fischer, C., Luettge, A., 2007. Converged surface roughness parameters: a new tool to quantify rock surface morphology and reactivity alteration. Am. J. Sci. 307 (7), 955–973.

Fujii, Y., Hori, S., Takahashi, M., Takemura, T., Lin, W., 2006. Difference of the surface roughness due to the orientation of the rock fabrics in Inada granite, measured by digital photogrammetry. J. Jpn. Soc. Eng. Geol. 47 (5), 252–258.

Gessesse, G.D., Fuchs, H., Mansberger, R., Klik, A., Rieke-Zapp, D.H., 2010. Assessment of erosion, deposition and rill development on irregular soil surfaces using close range digital photogrammetry. Photogramm. Rec. 25 (131), 299–318.

Giaccio, B., Galadini, F., Sposato, A., Messina, P., Moro, M., Zreda, M., Cittadini, A., Salvi, S., Todero, A., 2002. Image processing and roughness analysis of exposed bedrock fault planes as a tool for paleoseismological analysis: results from the Campo Felice fault (central Apennines Italy). Geomorphology 49, 281–301.

González-Aguilera, D., Muñoz-Nieto, A., Gómez-Lahoz, J., Herrero-Pascual, J., Gutierrez-Alonso, G., 2009. 3D digital surveying and modelling of cave geometry: application to Paleolithic rock art. Sensors 9, 1108–1127.

Hall, K., Guglielmin, M., Strini, A., 2008. Weathering of granite in Antarctica: II thermal stress at the grain scale. Earth Surf. Proc. Land. 33, 475–493.

Hani, A.F.M., Sathyamoorthy, D., Asirvadam, V.S., 2012. Computing surface roughness of individual cells of digital elevation models via multiscale analysis. Comput. Geosci. 43, 137–146.

Helper, M.A., Lee, P., Bualat, M., Adams, B., Deans, M., Fong, T., Heggy, E., Hodges, K.V., Hurtado, Jr., J.M., Young, K., 2010. Robotic follow-up to human geological and geophysical field work: experiments at Haughton Crator, Devon Island, Canada, November 2010 annual meeting, Boulder, CO. Abstracts with Programs – Geological Society of America 42 (5), 66–67 pp.

Hollaus, M., Aubrecht, C., Höfle, B., Steinnocher, K., Wagner, W., 2011. Roughness mapping on various vertical scales based on full-waveform airborne laser scanning data. Remote Sens. 3, 503–523.

Hu, Z., Zhu, L., Teng, J., Ma, X., Shi, X., 2009. Evaluation of three-dimensional surface roughness parameters based on digital image processing. Int. J. Adv. Manuf. Technol. 40, 342–348.

Inkpen, R., Duane, B., Burdett, J., Yates, T., 2008. Assessing stone degradation using an integrated database and geographical information system (GIS). Environ. Geol. 56, 789–801.

Jeong, A.S.Y., Walker, I.J., Dorn, R.I., 2018. Urban geomorphology of an arid city: case study of Phoenix, Arizona. In: Thornbush, M.J., Allen, C.D. (Eds.), Urban Geomorphology: Landforms and Processes in Cities. Elsevier, San Diego, CA (Chapter 10).

Jia, L., Chen, M., Jin, Y., 2014. 3D imaging of fractures in carbonate rocks using X-ray computed tomography technology. Carbon. Evapor. 29, 147–153.

Kassem, E., Walubita, L., Scullion, T., Masad, E., Wimsatt, A., 2008. Evaluation of full-depth asphalt pavement construction using X-ray computed tomography and ground penetrating radar. J. Perform. Constr. Fac. 22 (6), 408–416.

Katami, T., Takahara, Y., Nishikawa, H., Katou, K., 1996. Convenient methods for appreciating effects of acid rain on stone building materials. J. Jpn. Soc. Atmos. Environ. 31 (3), 125–131.

Kaufmann, V., 2012. The evolution of rock glacier monitoring using terrestrial photogrammetry: the example of Äusseres Hochebenkar rock glacier (Austria). Austrian J. Earth Sci. 105 (2), 63–77.

Lague, D., Brodu, N., Leroux, J., 2013. Accurate 3D comparison of complex tomography with terrestrial laser scanner: application to the Rangitikei canyon (N-Z). ISPRS J. Photogramm. 82, 10–26.

Lai, P., Samson, C., Bose, P., 2014. Surface roughness of rock faces through the curvature of triangulated meshes. Comput. Geosci. 70, 229–237.

Lee, H-S., Ahn, K-W., 2004. A prototype of digital photogrammetric algorithm for estimating roughness of rock surface. Geosci. J. 8 (3), 333–341.

López-Acre, P., Varas-Muriel, M.J., Fernández-Revuelta, B., Álvarez de Buergo, M., Fort, R., Pérez-Soba, C., 2010. Artificial weathering of Spanish granites subjected to salt crystallization tests: surface roughness quantification. Catena 83, 170–185.

Mah, J., Samson, C., McKinnon, S.D., Thibodeau, D., 2013. 3D laser imaging for surface roughness analysis. Int. J. Rock Mech. Min. 58, 111–117.

Matthews, N.A., Noble, T.A., Breithaupt, B.H., 2004. From dinosaur tracks to dam faces: a new method for collecting three-dimensional data, November 2004 annual meeting, Denver, CO. Abstracts with Programs – Geological Society of America 36 (5), 384 p.

Matthews, N.A., Noble, T.A., Musiba, C., Breithaupt, B.H., 2011. Stepping from the past to the future: close-range photogrammetry at the Laetoli footprint site in Tanzania, October 2011 annual meeting, Minneapolis, MN. Abstracts with Programs – Geological Society of America 43 (5), 140 p.

McCarroll, D., 1997. A template for calculating rock surface roughness. Earth Surf. Proc. Land. 22, 1229–1230.

Miller, A.Z., Rogerio-Candelera, M.A., Dionísio, A., Macedo, M.F., Saiz-Jiménez, C., 2012. Evaluación de la influencia de la rugosidad superficial sobre la colonización epilitica de calizas mediante téchicas sin contacto. Assessing the influence of surface roughness on the epilithic colonisation of limestones by non-contact techniques. Mater. Construcc. 62 (307), 411–424.

Olariu, M.I., Xu, X., Mohamed, A., Aiken, C.L.V., Ammann, L., 2003. The use of oblique close-range multi-spectral imagery in digital 3-D mapping of the Jackfork turbidites at Big Rock Quarry, Arkansas, November 2003 annual meeting, Boulder, CO, USA. Abstracts with Programs – Geological Society of America 35 (6), 260 p.

Orihuela, M.F., Abad, J., González Martínez, J.F., Fernández, F.J., Colchero, J., 2014. Nanoscale characterisation of limestone degradation using Scanning Force Microscopy and its correlation to optical appearance. Eng. Geol. 179, 158–166.

Ouyang, W., Xu, B., 2013. Pavement cracking measurements using 3D laser-scan images. Meas. Sci. Technol. 24, 1–9.

Rippin, D.M., Pomfret, A., King, N., 2015. High resolution mapping of supra-glacial drainage pathways reveals links between micro-channel drainage density, surface roughness and surface reflectance. Earth Surf. Proc. Land. 40, 1279–1290.

Rodríguez-Caballero, E., Cantón, Y., Chamizo, S., Afana, A., Solé-Benet, A., 2012. Effects of biological soil crusts on surface roughness and implications for runoff and erosion. Geomorphology 145–146, 81–89.

Sanmartín, P., Silva, B., Prieto, B., 2011. Effect of surface finish on roughness, color, and gloss of ornamental granites. J. Mater. Civil Eng. 23 (8), 1239–1248.

Shahabi, H.H., Ratnam, M.M., 2016. Simulation and measurement of surface roughness via grey scale image of tool in finish turning. Precis. Eng. 43, 146–153.

Shkuratov, Y.G., Stankevich, D.G., Petrov, D.V., Pinet, P.C., Cord, A.M., Daydou, Y.H., Chevrel, S.D., 2005. Interpreting photometry of regolith-like surfaces with different topographies: shadowing and multiple scattering. Icarus 173, 3–15.

Slavinski, H.A., Morris, W.A., 2005. Evaluating the application of close range photogrammetry for geological mapping, October 2005 annual meeting, Salt Lake City, UT. Abstracts with Programs – Geological Society of America 37 (7), 108 p.

Thornbush, M., 2008. Grayscale calibration of outdoor photographic surveys of historical stone walls in Oxford, England. Color Res. Appl. 33 (1), 61–67.

Thornbush, M.J., 2010. Measurements of soiling and colour change using outdoor rephotography and image processing in Adobe Photoshop along the southern façade of the Ashmolean Museum, Oxford. Smith, B.J., Gomez-Heras, M., Viles, H.A., Cassar, J. (Eds.), Limestone in the Built Environment: Present-Day Challenges for the Preservation of the Past, 331, Geological Society, London, Special Publications, pp. 231–236.

Thornbush, M., 2013a. Digital photography used to quantify the greening of north-facing walls along Broad Street in central Oxford, UK/ L'utilisation de la photographie numérique pour quantifier le verdissement de la façade septentrionale longeant Broad Street dans le centre d'Oxford, Royaume-Uni. Géomorphologie 2, 111–118.

Thornbush, M.J., 2013b. Photogeomorphological studies of Oxford stone—a review. Land. Anal. 22, 111–116.

Thornbush, M.J., 2014a. A soiling index based on quantitative photography at Balliol College in central Oxford UK. J. Earth Ocean Atmos. Sci. 1 (1), 1–15.

Thornbush, M.J., 2014b. Orientational effects on soiling measurements at the Sheldonian Theatre in central Oxford, UK. J. Build. Res. 1 (1), 1–27.

Thornbush, M.J., 2014c. The contribution of climbing plants to surface acidity and biopitting evident at the University of Oxford Botanic Garden UK. Int. J. Adv. Earth Environ. Sci. 2 (2), 12–21.

Thornbush, M.J., 2014d. A new (digital) technique for areal measurements of stonewall surface roughness. Am. J. Geosci. 4 (2), 24–31.

Thornbush, M.J., 2014e. Measuring surface roughness through the use of digital photography and image processing. Int. J. Geosci. 5 (5), 540–554.

Thornbush, M.J., 2016. A photogeomorphological study examining 50 years of pavement soiling in Toronto, Canada. Z. Geomorphol. 60 (3), 163–174.

Thornbush, M., Viles, H., 2004a. Integrated digital photography and image processing for the quantification of colouration on soiled surfaces in Oxford England. J. Cult. Herit. 5 (3), 285–290.

Thornbush, M.J., Viles, H.A., 2004b. Surface soiling pattern detected by integrated digital photography and image processing of exposed limestone in Oxford, England. Saiz-Jimenez, C. (Ed.), Air Pollution and Cultural HeritageA.A. Balkema Publishers, London, pp. 221–224.

Thornbush, M., Viles, H., 2006. Changing patterns of soiling and microbial growth on building stone in Oxford England after implementation of a major traffic scheme. Sci. Total Environ. 367 (1), 203–211.

Thornbush, M.J., Viles, H.A., 2007a. Photo-based decay mapping of replaced stone blocks on the Boundary Wall of Worcester College, Oxford. Prˇikryl, R., Smith, B.J. (Eds.), Building Stone Decay: From Diagnosis to Conservation, 271, Geological Society, London, Special Publications, pp. 69–75.

Thornbush, M.J., Viles, H.A., 2007b. Simulation of the dissolution of weathered versus unweathered limestone in carbonic acid solutions of varying strength. Earth Surf. Proc. Land. 32 (6), 841–852.

Thornbush, M.J., Viles, H.A., 2008. Photographic monitoring of soiling and decay of roadside walls in Oxford, England. Environ. Geol. 56 (3–4), 777–787.

Tsai, Y., Wu, Y., Lewis, Z., 2014. Full-lane coverage micromilling pavement-surface quality control using emerging 3D line laser imaging technology. J. Transp. Eng. 140 (2), 1–8, 04013006.

Tsokas, G.N., Tsourlos, P.I., Vargemezis, G.N., Pazaras, N. Th., 2011. Using surface and cross-hole resistivity tomography in an urban environment: an example of imaging the foundations of the ancient wall in Thessaloniki North Greece. Phys. Chem. Earth 36, 1310–1317.

Viles, H.A., 2008. Understanding dryland landscape dynamics: do biological crusts hold the key? Geogr. Compass 2 (3), 899–919.

Viles, H.A., Naylor, L.A., Carter, N.E.A., Chaput, D., 2008. Biogeomorphological disturbance regimes: progress in linking ecological and geomorphological systems. Earth Surf. Proc. Land. 33 (9), 1419–1435.

Wöhler, C., Grumpe, A., Berezhnoy, A., Bhatt, M.U., Mall, U., 2014. Integrated topographic, photometric and spectral analysis of the lunar surface: application to impact melt flows and ponds. Icarus 235, 86–122.

CHAPTER

16

Conclusion

Mary J. Thornbush, Casey D. Allen***

*University of Oxford, Oxford, United Kingdom;

**The University of the West Indies, Cave Hill Campus, Barbados

OUTLINE

16.1 Introduction	317	16.3 Conclusions	319
16.2 Future Studies	318	References	320

16.1 INTRODUCTION

With the expansion of human settlements into urban environments, natural landscapes are increasingly being disturbed and modified to accommodate urbanization. This volume has included 14 chapters that cover a wide variety of case studies and examples of urban expansion affecting geomorphology in locations that humans are occupying. This control of the landscape in cities has led to the "commodification" of land through privatization, as evident in the Andalién River system in Chile, South America (Espinosa et al., 2018). There has also been ongoing engineering work in cities that has led to an increasing number of engineered landscapes in the built environment where urban development has occurred, as for instance: in Palma, the capital of the Mediterranean island of Mallorca (Petrus et al., 2018); in the Warta River, Poznań, Poland (Zwoliński et al., 2018); and in the Okanagan basin, BC, Canada (MacDuff and Bauer, 2018). Where airports are being built, this requires not only land clearance, but also leveling, resulting in broad landscape changes that can affect waterways, agricultural plots, and the overall natural environment (Pijet-Migoń and Migoń, 2018). In addition to the human alteration of environments where cities are expanding, resource exploitation and the impacts associated with resource extraction exist. Górska-Zabielska and Zabielski (2018), for example, presented the case of abiotic resources in Pruszków, Poland that has become of interest to the urban geotourist and, from this reference point, which could be deployed for environmental protection. This conveys the message that where there are problems, there are also potential solutions. This is also evidenced by the city of Johannesburg in

Urban Geomorphology. http://dx.doi.org/10.1016/B978-0-12-811951-8.00016-3

South Africa (Knight, 2018), where mine waste dumps have been reoccupied and repurposed, urban greening is being encouraged, and the city's history is being recalled based on relic sites.

Nevertheless, the extraction of natural resources for human exploitation has continued to plague cities, especially where there is transnational exploitation of natural resources and indigenous populations have become marginalized and displaced, such as in the Andalién River system, Chile, South America, where cases of sustainable urban development have been promoted to reduce flood risk (Espinosa et al., 2018). The challenges associated with expanding human society, including increased use of natural resources and land encroachment, have led to dramatic modifications of natural landscapes, as for instance noted by Gamache et al. (2018), who examined (four) towns in the San Juan Mountains, USA, where there has been an abundance of modification associated with the extraction of natural resources, including water use, as well as slope stabilization, river modification, and built environments arranged to accommodate recreational activities.

16.2 FUTURE STUDIES

Impending work needs to continue to address a variety of landscapes as in this volume, which has included mountainous, fluvial, coastal, island, desert, and karstic systems. This needs to come from various continents around the world, particularly where population growth is concentrated and urbanization is happening at unprecedented rates in developing countries. An example of this is Nigeria in sub Saharan Africa (Onafeso and Olusola, 2018). It is also pertinent that authors investigate not only problems, but also solutions (Knight, 2018), so that an understanding of how to solve impeding problems can be grasped. There is an abundance, for instance, of fluvial examples where human occupation and disturbance has affected natural ecosystems (Petrus et al., 2018) and indigenous populations (Espinosa et al., 2018). Studies on rivers in Chile in South America, some African countries (Nigeria, South Africa), Europe (Poland), and Canada (BC) have conveyed the modern issues concerning river floodplains and their habitats (Espinosa et al., 2018; Gamache et al., 2018).

In addition to work conducted on incipiently natural landscapes, some of the chapters here have conveyed contemporary impacts on existing built environments, such as that of stone decay research studies. Allen et al. (2018) and Thornbush (2018), for example, respectively examined a historic building in Colorado, USA and calcareous materials found on pavements in Toronto, Canada. Others have examined the built environment from the perspective of impacted landscapes, such as Jeong et al. (2018), who studied desert landforms and processes in the urban metropolis of Phoenix, Arizona, USA, which included pediments, alluvial fans, aeolian sand sheets, and desert pavements affected by rock varnish accretion, dirt cracking, desert pavement formation, rock fall, debris flows, high magnitude flooding, and pedimentation. Of concern within the realm of environmental urban geomorphology is the need to examine waste-based pollution studies, including mine waste dumps (Knight, 2018; Machado and Rodrigues, 2018). Finally, it is important to continually address the way that problems in disaster-prone areas can be exacerbated, and become increasingly vulnerable, by rapid population growth and development in recent decades, resulting in a multitude of natural hazards, such as giant landslides, large-area floods, earthquakes, sinkhole collapse, and land subsidence (Fort and Adhikhari, 2018).

16.3 CONCLUSIONS

Amassing populations and their infringement on natural landscapes results in many alterations and impacts, particularly in the developing world where population growth is centered and urbanization is rampant. Countries in South America, Africa, and Asia are particularly experiencing increasing vulnerability to change and are, consequently, becoming more disaster-prone. With this growth is the continued demand for resources and the subsequent creation of waste often achieved when consumptive practices are driven by resource extraction activities. The case of airport building is demonstrative of new forms of land clearance and leveling that are required for urban occupation. In the past, this was associated with land clearance for farming, but now clearance is needed for transportation and housing. So, the shift from agricultural land use to covering surfaces with impervious materials for road building and occupation in cities is less attuned to natural systems and potentially more disruptive and destructive to both the natural environment and ultimately anthropogenic activities. Now, natural land surfaces are being covered with concrete, blocking natural drainage and impeding pedogenesis in cities. Such surfaces are compressed, as under high-rise buildings and motorways, to such an extent that they can no longer be reclaimed for agricultural use if needed in the future. The way that we are currently developing urbanscapes is such that there is no going back. For this reason, future generations will need to find novel solutions to the problems that are being created today.

The popularity of living in cities and the amassing of people there has led to more attention in geomorphology and other research disciplines that focus on urban growth and development. The city has emerged as its own entity, much as in the way that landforms morph and evolve. Cities are increasingly focal points in research that examine human impacts on landscapes, human-environment interactions, and environmental sustainability. It is important that geomorphologists take this opportunity to engage with the literature that has been, for example, established on smart cities and the implications that this may have for the organization, design, and growth of urban landscapes.

Last, geomorphologists are accustomed to engaging with broadening topical areas, as for instance environmental geomorphology, but also climatic geomorphology, which has become more prevalent with recognition of the global issue surrounding climate change during the Anthropocene. People, however, are also contributing to the development of urban climates the world over, which could become particularly impactful at the global scale. Climatic considerations are especially relevant to geomorphology and the way that past and future landscapes evolve. It is not just earthquakes that have the potential to seriously modify urban landscapes. Other climatic events, such as megastorms, hurricanes, and dust storms (especially in dry areas), could damage cities and affect an increasing number of people living in these areas. Vertical growth, while ideal for housing rapidly growing urban populations, can be hazardous where such phenomena as for example megastorms are concerned, with extreme wind speeds and precipitation. Tracking natural hazards that more readily impact expanding urban areas is becoming more important as urban migration persists amid a contemporary landscape dynamic that necessitates continued monitoring and adaptation.

References

Allen, C.D., Ester, S., Groom, K.M., Schubert, R., Hagele, C., Olof, D., James, M., 2018. A geologic assessment of historic Saint Elizabeth of Hungary Church using the Cultural Stone Stability Index, Denver, Colorado. In: Thornbush, M.J., Allen, C.D. (Eds.), Urban Geomorphology: Landforms and Processes in Cities. Elsevier, San Diego, (Chapter 14).

Espinosa, P., Horacio, J., Ollero, A., de Meulder, B., Jaque, E., Muñoz, M.D., 2018. When urban design meets fluvial geomorphology: a case study in Chile. In: Thornbush, M.J., Allen, C.D. (Eds.), Urban Geomorphology: Landforms and Processes in Cities. Elsevier, San Diego, (Chapter 9).

Fort, M., Adhikhari, B.R., 2018. Pokhara (central Nepal): a dramatic yet geomorphologically active environment vs. a dynamic, rapidly developing city. In: Thornbush, M.J., Allen, C.D. (Eds.), Urban Geomorphology: Landforms and Processes in Cities. Elsevier, San Diego, (Chapter 12).

Gamache, K., Giardino, J.R., Zhao, P., Owens, R.H., 2018. Bivouacs of the Anthropocene: urbanization, landforms and hazards in mountainous regions. In: Thornbush, M.J., Allen, C.D. (Eds.), Urban Geomorphology: Landforms and Processes in Cities. Elsevier, San Diego, (Chapter 11).

Górska-Zabielska, M., Zabielski, R., 2018. Geotourism development in an urban area. Based on the local geological heritage (Pruszków, central Mazovia, Poland). In: Thornbush, M.J., Allen, C.D. (Eds.), Urban Geomorphology: Landforms and Processes in Cities. Elsevier, San Diego, (Chapter 3).

Jeong, A.S.Y., Walker, I.J., Dorn, R.I., 2018. Urban geomorphology of an arid city: case study of Phoenix, Arizona. In: Thornbush, M.J., Allen, C.D. (Eds.), Urban Geomorphology: Landforms and Processes in Cities. Elsevier, San Diego, (Chapter 10).

Knight, J., 2018. Transforming the physical geography of a city: an example of Johannesburg, South Africa. In: Thornbush, M.J., Allen, C.D. (Eds.), Urban Geomorphology: Landforms and Processes in Cities. Elsevier, San Diego, (Chapter 8).

MacDuff, A., Bauer, B.O., 2018. Urban stream geomorphology and salmon repatriation in Lower Vernon Creek, British Columbia (Canada). In: Thornbush, M.J., Allen, C.D. (Eds.), Urban Geomorphology: Landforms and Processes in Cities. Elsevier, San Diego, (Chapter 5).

Machado, C.A., Rodrigues, S.C., 2018. Environmental contamination by technogenic deposits in the urban area of Araguaína, Brazil. In: Thornbush, M.J., Allen, C.D. (Eds.), Urban Geomorphology: Landforms and Processes in Cities. Elsevier, San Diego, (Chapter 7).

Onafeso, O., Olusola, A., 2018. Urban stone decay and sustainable built environment in the Niger River basin. In: Thornbush, M.J., Allen, C.D. (Eds.), Urban Geomorphology: Landforms and Processes in Cities. Elsevier, San Diego, (Chapter 13).

Petrus, J.M., Ruiz, M., Estrany, J., 2018. Interactions among geomorphology and urban evolution since the antiquity in a Mediterranean city. In: Thornbush, M.J., Allen, C.D. (Eds.), Urban Geomorphology: Landforms and Processes in Cities. Elsevier, San Diego, (Chapter 2).

Pijet-Migoń, E., Migoń, P., 2018. Landform change due to airport building. In: Thornbush, M.J., Allen, C.D. (Eds.), Urban Geomorphology: Landforms and Processes in Cities. Elsevier, San Diego, (Chapter 6).

Thornbush, M.J., 2018. Photographic technique used in a photometric approach to assess the weathering of pavement slabs in Toronto (Ontario, Canada). In: Thornbush, M.J., Allen, C.D. (Eds.), Urban Geomorphology: Landforms and Processes in Cities. Elsevier, San Diego, (Chapter 15).

Zwoliński, Z., Hildebrandt-Radke, I., Mazurek, M., Makohonienko, M., 2018. Anthropogeomorphological metamorphosis of urban area on post-glacial landscape; case study Poznań City. In: Thornbush, M.J., Allen, C.D. (Eds.), Urban Geomorphology: Landforms and Processes in Cities. Elsevier, San Diego, (Chapter 4).

Author Index

A
Abad, J., 308, 311
Abba, S.I., 265
Abergel, S., 307
Abichou, A., 10
Abiye, T., 140
Abrahams, A.D., 181, 189, 191, 193
Acín, V., 164
Acuña, V., 152
Adams, B., 313
Adamson, C., 278
Adegun, O.B., 140
Adelsberger, K.A., 183
Adhikari, B.R., 236, 245, 246, 249
Adhikari, J., 238, 239, 253
Adhikhari, B.R., 3, 318
Adu-Bredu, S., 270
Afana, A., 307
Agosin, M., 154
Ahmad, S.R., 311
Ahmed, M., 310
Ahn, J., 304
Ahn, K-W., 304
Aiken, C.L.V., 307
Ajakaiye, D.E., 265
Alcántara Peña, P., 17, 25
Aleman, N., 269
Alexander, I., 266
Alho, P., 14
Aliste, E., 157, 158, 160
Allègre, C.J., 266, 268
Allen, C., 178
Allen, C.D., 4, 183, 201, 278, 281, 282, 298, 300, 307, 318
Allen, H., 210
Almendras, A., 157, 158, 160
Al-Mukhtar, M., 278
Alomar Esteve, G., 24
Alonso, E.E., 108
Álvarez de Buergo, M., 312
Alves, C., 262, 272
Alvey, A.A., 142
Ambani, A.-E., 140
Ammann, L., 307
Amouroux, D., 140
Andermann, C., 236, 246, 249
Anderson, K., 312
Anderson, R.S., 181
Andersson, S.P., 181
Ando, H., 251
Andrès, Y., 140
Angel, S., 152
Annegarn, H., 140
Annegarn, H.J., 140
Antczak-Górka, B., 71
Apolinarska, K., 71
Applegarth, M.T., 179, 193
Arai, I., 252
Arav, R., 307
Aremu, D.A., 269
Arenas, F., 156
Arita, K., 235, 236
Armitage, N., 140
Armstrong, B., 216
Arnold, V., 138, 140
Arribas, A., 24
Asare, R.A., 270
Asher, M., 153
Ashraf, M.A., 118, 119
Asirvadam, V.S., 310
Atapattu, K, 153
Attal, M., 236, 250
Atwood, W.W., 207, 212
Aubrecht, C., 307
Augustijn, P.W.M., 142, 143

B
Baddock, M.C., 200
Bahat, D., 190
Bahdasarau, M., 38, 44
Baily, B., 282
Baird, C., 119, 124
Bakatula, E.N., 140
Balkwill, K., 130
Ballarín, D., 164
Bandauko, E., 129, 140, 143
Baral, H., 232, 240, 241, 248, 251, 255
Barcelø Crespí, M., 27
Bardají, T., 13, 14
Bardina, M., 152
Barger, N.N., 183

Barnes, B., 139
Barnosky, A.D., 179, 200
Baros, Z., 103, 109
Barriendo, M., 14
Barros, A., 268
Bartkowski, T., 58, 65, 71
Barusseau, J.P., 269
Basant, K., 236, 250
Basheer, P.A.M., 272
Bastian, L., 268
Bauer, B.O., 3, 317
Baynes, F.J., 269
Bayon, G., 266, 268, 269
Beall, J., 139
Beauvais, A., 266, 267
Beck, K., 278
Becker, P., 139
Bejarano, E., 28
Belin, S., 267
Bell, F.G., 134, 140
Belnap, J., 179, 183, 201, 307
Ben, Ghozzi, F., 10
Benavente, D., 304
Benessaiah, K., 10
Benito, G., 14, 31
Bennet, R.A., 272
Bennett, E.M., 10
Bentlin, F.R.S., 118
Berezhnoy, A., 307
Bermell, S., 266, 268
Bernabeu, A., 304
Bernat Roca, M., 26
Bernhardt, A., 236, 245, 246, 249
Bernhardt, E., 208
Berry, L., 265
Bertolani, F.C., 116
Bes de Berc, S., 236, 250
Betancourt, J., 178
Bhandary, N., 248
Bhatt, M.U., 307
Bhattarai, M., 246
Biaou, A., 267, 268
Biedrowski, Z., 58, 61, 64
Bielawski, P., 40, 43
Biscara, L., 269
Bistacchi, A., 151
Bitelli, G., 304
Bitri, A., 236, 250
Blair, R., 208, 211, 212
Blanc, F., 10
Blei, A., 152
Bloch, R., 139
Blott, S., 85, 89
Blum, A.E., 266
Blythe, A.E., 268
Bo, M.W., 104
Bobbink, I., 153
Bobbins, K.L., 135, 140
Bollinger, L., 236, 246, 250, 251
Booysen, H.J., 139, 140, 142
Bory, A., 269
Bose, P., 304
Bostwick, T.W., 179, 186, 189, 199
Botasaneanu, L., 164
Bouchez, J., 268
Bourcier, P., 10
Bowen, J.T., 104, 108
Bowker, M.A., 179, 183, 201, 307
Bowles, Z., 193
Boyer, D.E., 190
Brady, N.C., 124
Brauer, A., 71
Bravard, J.P., 151, 165
Brazel, A.J., 199
Breithaupt, B.H., 310
Bremner, L., 139
Brenot, A., 266
Bridges, E.M., 265, 271
Brierley, G.J., 152
Briois, V., 267
Brisson, J., 140
Brodu, N., 304
Broecker, W.S., 185
Brown, A.G., 200, 201
Brown, E.J., 151
Brozovic, N., 152
Brunetaud, X., 278
Brunsden, D., 189
Bruthans, J., 278
Bualat, M., 313
Bubeck, S., 143
Buckle, H., 139, 140
Büdel, B., 183, 307
Budkewitsch, P., 304
Bullard, J.E., 187, 200, 201
Burbank, D.W., 268
Burbank, W.S., 183, 190
Burby, R.J., 29
Burdett, J., 304
Burke, J., 189
Burke, K.C., 265, 270
Burns, M.J., 140
Burton, K.W., 266
Butcher, S., 139
Buytaert, W., 31
Byrne, J., 142, 143

C

Cabero, A., 13, 14
Cai, Z., 304
Calle, M., 14
Calsamiglia, A., 30
Calvo-Cases, A., 30
Camarassa, A.M., 14
Campbell, I.A., 264
Cantón, Y., 307
Cao, W., 304
Caplan, L., 238
Caquineau, S., 267
Carayon, N., 10
Carignan, J., 266
Carmona, P., 14
Carter, C., 214
Carter, N.E.A., 306
Cartwright, A.P., 135
Castillo-Monroy, A., 307
Cau Ontiveros MÁ, 10, 24, 25
Cearreta, A., 10, 179, 200
Cerbin, S., 71
Cereceda, P., 155
Cerimele, M.M., 305
Cerveny, N., 298
Cerveny, N.V., 4, 278, 281, 282, 298, 300
Chachaj, J., 58, 62
Chamizo, S., 307
Chapman, T.P., 138
Chappell, J., 14
Chaput, D., 306
Charlton, R., 151
Chase, C.G., 179
Chaudhary, V.B., 179, 183, 201
Chávez, M.E., 24, 25
Chazarenc, F., 140
Chen, M., 304
Chen, W., 307
Chen, Z., 307
Chevaillier, S., 267
Chevrel, S.D., 312
Chiaro, G., 252
Chikowore, G., 129, 140, 143
Chimuka, L., 140
Chin, A., 31
Chiotti, L., 183
Chirisa, I., 129, 140, 143
Chiverrell, R.C., 200, 201
Chmal, R., 58, 60, 61, 63
Choa, V., 104
Chornesky, E., 208
Christaras, B., 39
Christgen, B., 278
Christopher, A.J., 138
Chu, J., 104
Chu, N.-C., 266
Chun, Y.G., 278
Church, M., 222
Cidell, J.L., 104, 108
Cincio, Z., 58, 62
Cittadini, A., 307
Ciuk, E., 60
Civco, D.L., 152
Clarno, A., 139
Cline, M.L., 190, 199
Coates, D.R., 1, 2, 221
Coelho, M.R., 116
Cohen, J., 31
Colangeli, L., 307
Colchen, M., 248
Colchero, J., 308, 311
Cole, J.J., 266
Coleman, T.J., 140
Colin, F., 266, 267
Collier, P., 282, 304
Collins, B.D., 31
Collins, S., 208
Coney, P.J., 179
Contreras, M., 157, 158, 160
Cook, J.P., 190, 199
Cooke, R., 1, 2, 282
Cooke, R.U., 103, 189, 221
Cooper, D.J., 195
Cooper, T.P., 272
Coratza, P., 71
Cord, A.M., 312
Corner, J., 152
Correa-Araneda, F., 164
Cossu, R., 305
Crankshaw, O., 139
Craul, P.J., 117
Critelli, S., 269
Croft, H., 312
Cross, W., 214
Crosta, X., 269
Crush, J., 142
Crutzen, P.J., 10
Cukrowska, E., 140
Cukrowska, E.M., 140
Cullers, R.L., 269
Cunningham, R.B., 181
Curran, J., 278
Curran, J.M., 272, 278, 282
Cutrell, A.K., 278, 282, 298
Czubla, P., 44

D

Da Gloria Motta Garcia, M., 39
Da Silveira, K.S., 267, 268
Dada, S.S., 265
Dahal, R.K., 248
Dai, Q., 308
Damians, Manté, A., 27
Darby, S.E., 165
Dare-Edwards, A.J., 187
D'Auria, V., 140
Dávid, L., 103, 109
Davies, P.M., 152
Daydou, Y.H., 312
De Bruyn, I.A., 134, 140
De La Corte Bacci, D., 39
De Meulder, B., 3, 140, 153, 317, 318
De Pauw, N., 142
De Wet, T., 139
Deans, M., 313
Dearman, W.R., 269
Dębski, A., 67
Del Lama, E.A., 39
Del Monte, M., 39
Delile, H., 10
Demsey, K.A., 195
Deng, Y., 222
Dennielou, B., 266, 268
Depke, T.J., 219
Depoi, F., 118
Desboeufs, K., 267
DeSwardt, A.M.J., 265
Dewey, J.F., 265
Dhital, M.R., 251
Di Nocera, S., 269
Di Toro, G., 151
Díaz, E., 164
Dibble, H.L., 183
Diester-Hass, L., 270
Dietze, E., 183
Dietze, M., 183
Díez, J., 151
Dinelli, E., 269
Dionísio, A., 309
DIorio, M.A., 304
Dippenaar, M.A., 139, 140, 143
Dister, E., 165
Dobson, A., 208
Doherty, G., 152
Dohrenwend, J.C., 191
Domenech, S., 164
Domingues, T.C. de G., 118, 124
Doornkamp, J.C., 103, 189
Dorn, J.D., 307
Dorn, R.I., 3, 4, 179, 180, 183, 186, 189, 191, 193, 199, 201, 264, 278, 281, 282, 286, 294, 298, 300, 307, 318
Dos, S., 118
Dosseto, A., 268
Douglas, I., 103, 104, 108
Douglass, J., 193
Dowding, P., 272
Dowling, R., 39
Dowling, R.K., 38, 39, 52
Drackner, M., 139, 140
Dreesen, R., 272
Du Toit, W.J.F., 130, 131
Duane, B., 304
Dufour, A., 268
Duke, C., 208
Dunn, L.G., 212
Dupré, B., 266, 268
Duquesnoy, M., 150
Durotoye, B., 270
Dwidedi, S., 249
Dyjor, S., 60

E

Eagar, J.D., 199
Eamer, J.B.R., 187
Edgeworth, M., 179, 200
El Gendy, A., 307
El-Assal, M., 10
Eldridge, D., 307
Eldridge, D.J., 307
Ellery, K., 130
Ellery, W.N., 130, 138, 140
Ellis, E.C., 179, 200
Ellis, M., 10, 179, 200
Elosegi, A., 151
Elvidge, C.D., 186
Empereur, J.-Y., 10
Errázuriz, A.M., 155
Escartín Bisbal, J., 10
Escolar, C., 307
Espinosa, P., 3, 317, 318
Esposito, F., 307
Esther, S., 4, 318
Estrany, J., 13, 30, 317, 318
Etoubleau, J., 266, 268
Evans, I.S., 222
Ewald, K.C., 165
Ewan, J., 189
Ewan, R.F., 189
Eynaud, F., 269

F

Fahl, U., 143
Faist, A.M., 183
Faniran, A., 266
Fatti, C.E., 140
Fedotova, A., 214

Ferguson, B.K., 140
Fernández, F.J., 308, 311
Fernández, G.Á.V., 116
Fernández-Revuelta, B., 312
Filin, S., 307
Filippi, M., 278
Finkel, R.C., 179, 201
Fischer, C., 304
Fitzner, B., 278
Flaux, C., 10
Fleming, K., 14
Fletcher, T.D., 140, 152
Flintsch, G.W., 307
Florsheim, J.L., 31
Flynn, M.K., 139
Fong, T., 313
Fontana, D., 304
Fontanals R., 14
Formenti, P., 267
Formisano, V., 307
Fornós, J., 13, 14
Fornós, J.J., 10, 24
Fort, M., 3, 235, 236, 238, 245, 246, 248, 249, 318
Fort, R., 304, 312
Fortesa, J., 30
Fragkias, M., 152
France-Lanord, C., 246
Frazier, P.S., 187
Fredi, P., 39
Freire, E., 39
French, S.P., 29
Frêne, C., 156
Fripiat, J.J., 263
Fryirs, K.A., 152
Fuchs, H., 310
Fuchs, M., 183
Fuggle, R.F., 130, 131
Fujii, Y., 308
Fuller, J.E., 195
Funk, J.R., 108

G

Gabet, E., 268
Gadhoum, A., 10
Gaillardet, J., 246, 266, 268
Gajurel, A., 236, 250
Galadini, F., 307
Gałuszka, A., 179, 200
Gałzka, D., 44
Gamache, G., 214
Gamache, K., 3, 210, 214, 318
Gamache, K.R., 210
Gandu, A.H., 265
García del Cura, M.A., 304
García Delgado, C., 17, 24, 25
García Riaza, E., 11, 24
García-Comendador, J., 30
García-Pichel, F., 179
Gasse, F., 269
Gautam, P., 251
Gautier, J.N., 165
Gbadegesin, A.S., 262, 264
Genske, D.D., 134, 140
Gérente, C., 140
German, C.R., 266
Gerten, D., 18
Gessesse, G.D., 310
Gessner, M.O., 151
Geyer, H.S., 129
Giaccio, B., 307
Giaime, M., 10, 24
Giardino, J.R., 3, 199, 208, 210, 213, 214, 219, 318
Giesen, M.J., 278
Gilbert, G.K., 193
Gili Suriñach, S., 18
Gillette, D.A., 183
Giménez, J., 13, 14, 29
Gimeno, M., 164
Giordan, D., 39
Girardi, F., 304
Girault, F., 246
Giri, S., 251
Giuaranna, M., 307
Gober, P., 198
Goiran, J.-P., 10
Gold, B., 208
Goldreich, Y., 130, 131
Goldshledger, N., 307
Goldstein, S.L., 266
Gomer, D., 165
Gómez-Lahoz, J., 304
González Casasnovas, C., 24
González Martínez, J.F., 308, 311
González-Aguilera, D., 304
González-Hernández, F.M., 13, 14
Gonzalo, L.E., 164
Goodchild, M.F., 222
Gordon, J.E., 39, 151
Gordon, M., 179, 186, 189, 199
Gordon, S.J., 4, 278, 281, 282, 298, 300
Górska, M., 44
Górska-Zabielska, M., 2, 46, 51, 317
Goslar, T., 71
Gosseye, J., 140
Götz, G., 143
Goudie, A., 221
Goudie, A.S., 10, 31, 103, 200, 208, 218
Goy, J.L., 13, 14
Graf, W.L., 188
Graham, D.W., 278

Grams, P.E., 218
Granado, D., 164, 165, 169
Granados-Aguilar, R., 213
Granger, D.E., 179, 201
Grangier, L., 39
Gray, M., 151
Griffin, P., 278
Griffith, A., 151
Griffiths, P.G., 190
Grimalt Gelabert, M., 27
Grimalt, M., 13
Grinsted, A., 14
Grocholski, W., 60
Groom, K.M., 4, 278, 281–283, 298–300, 318
Grossenbacher, K., 190
Grousset, F., 269
Grove, A.T., 270
Grumpe, A., 307
Guerrero, A.V., 18
Guerrero, V., 18
Guglielmin, M., 304
Guijarro, J.A., 15
Guneralp, B., 152
Gupta, A., 1
Gupta, H.V., 223
Gurung, D.R., 246, 248
Gurung, H.B., 232, 234, 236, 238, 239, 245
Gutbrod, E., 4, 278, 281, 282, 298, 300
Gutiérrez Lloret, S., 26
Gutierrez-Alonso, G., 304

H

Haas, C.T., 310
Haas, R., 310
Haegle, C., 4, 318
Haines, P.E., 14
Hall, K., 179, 304
Hamilton, C., 136
Hamilton, P.J., 266
Han, S.-I., 304
Hani, A.F.M., 310
Hanisch, J., 250
Hansen, J., 271
Hansen, J.E., 270
Harden, C.P., 31
Hardy, C.H., 131
Harms, T.A., 179
Harpham, T., 139
Harris, R.C., 189
Harrison, E., 193
Harrison, J.C., 304
Harrison, P., 129, 143
Harry, D., 278
Hart, D.M., 138
Hartnett, H.E., 199
Hartog, P., 272
Haykin, S., 223
Haywood, A., 10
Heggy, E., 313
Heil, C.W., 269
Hein, K.A.A., 134, 140
Heinriches, K., 278
Helper, M.A., 313
Hengl, T., 222
Henry, G.H.R., 307
Herbert, C.W., 139
Herbillon, A.J., 263
Herckes, P., 199
Héroin, E., 151
Herrero-Pascual, J., 304
Herrick, J.E., 183
Hesnard, A., 10
Hetz, K., 142, 143
Hidalgo, R., 154, 156, 161
Higaki, D., 248
Hildebrandt-Radke, I., 58, 62, 71, 73, 74, 317
Hillaire-Marcel, C.B.G., 13, 14
Hirsch, P.J., 181
Hodges, K.V., 313
Hoelzmann, P., 236, 245, 246, 249
Höfle, B., 307
Holland, T., 10
Hollaus, M., 307
Holt, W.E., 179
Honeyborne, D.B., 272
Honey-Rosés, J., 152
Horacio, J., 3, 151, 164, 317, 318
Hori, S., 308
Hormann, K., 236
Hose, T.A., 39
Hoshi G., 246, 248
Hostmann, M., 165
House, P.K., 195
Houser, C., 208, 214
Howard, A.D., 181, 189, 191
Howe, E., 214
Hradunova, A., 38, 44
Hrychanik, M., 38, 44
Hsu, K.l., 223
Hu, Z., 311
Hubbard, F.H., 265
Huber, A., 156
Hughen, K.A., 269
Hume, M., 83
Hungr, O., 189, 190
Hurtado, Jr. J.M., 313
Huyghe, P., 236, 250
Hyeres, A.D., 200

I

Ibisate, A., 164, 165, 169
Idso, S.B., 199
Illies, J., 164
Illueca, C., 304
Ilyés, Z., 103, 109
Indictor, N., 278
Ingram, R.S., 199
Inkpen, R., 282, 304
Inkpen, R.J., 304
Iroumé, A., 156
Iverson, R.M., 189

J

Jacobson, R., 208
Jäger, S., 250
Jakob, M., 189, 190
Jakubowski, T.H., 43
James, M., 4, 318
Janbade, P., 278
Janvier-Badosa, S., 278
Jaque, E., 3, 155, 156, 163, 317, 318
Jara, D., 158
Jasiewicz, J., 58, 71
Jayangondaperumal, R., 236, 250
Jeandel, C., 179, 200
Jeje, L.K., 270
Jenkins, H.C., 11
Jensen, J.R., 222
Jeong, A.S.Y., 3, 318
Jerbania, I., 10
Jevrejeva, S., 14
Jia, L., 304
Jin, Y., 304
Jo, Y.H., 278
Joannes-Boyau, R., 268
John, B., 250
Johnson, N.C., 179, 183, 201
Johnston, P., 14
Jones, D., 189
Jones, R., 151
Joshi, S., 248
Jouanne, F., 236, 250
Jourde, H., 267, 268
Journaux, A., 199
Józefowiczowa, K., 67
Jung, J., 304

K

Kaiser, Ch., 39
Kaleta, J., 40
Kaniecki, A., 57, 58, 65–67, 70, 72, 73
Kara, M., 71
Karabanov, A., 38, 44
Karasaki, K., 190
Karczewski, A., 58, 71
Kargel, J.S., 248
Kasarda, J.D., 101
Kassem, E., 304
Katagiri, T., 252
Katami, T., 312
Katou, K., 312
Kaufmann, V., 310
Kawaguchi, E., 241
Kazumi Dehira, L., 39
Kelkar, K., 199
Kellett, J., 187
Kelley, S., 193
Kelley, S.B., 191, 193
Kerntke, M., 250
Kesel, R.H., 179, 191
Kgomongoe, M., 139, 140
Khanal, N.R., 246, 248
Kharecha, P., 271
Khatiwada, B., 252
Kicińska-Świderska, A., 39
Kim, S., 304
King, J., 269
King, M., 179, 186, 189, 199
King, N., 312
King, T., 139
Kingsland, S., 208
Kirchner, J.W., 179, 201
Kiyota, T., 252
Klaver, A., 267
Kleber, A., 183
Kleidon, A., 18
Kletetschka, G., 278
Klik, A., 310
Klinger, Y., 236, 250, 251
Kneen, M.A., 140
Knight, J., 3, 135, 140, 317, 318
Kóčka-Krenz, H., 57, 65, 67, 68
Koestler, R., 278
Koffi, K.J.P., 267, 268
Koirala, A., 250
Koirala, B.P., 246
Koita, M., 267, 268
Kondolf, G., 84, 85, 93
Kondracki, J., 39, 56
Kopczyński, K., 52
Korb, C.C., 118
Korup, O., 236, 245, 246, 249
Korzeń, J., 39, 43
Kostrzewski, A., 55
Kowalczyk, A., 38, 40, 43
Kownatzki, R., 278

Kozarski, S., 58, 61, 64, 71
Kożuchowski, K., 38
Kranz, R., 208
Kroon, D., 266
Krygowski, B., 58, 60–62, 64, 71
Krzyczkowski, H., 40
Krzywicki, T., 38, 44
Kuhn, N.J., 312
Kunkel, A., 60
Kurtz, J., 278
Kurzawska, A., 72
Kuwano, R., 252
Kwangwama, N.A., 129, 140, 143

L

La Bouchardiere, D., 278
Laborel- Deguen, F., 10
Laborel, J., 10
Lacis, A., 270
Ladislas, S., 140
Laflamme, R.A., 222
Lafont, R., 10
Lagos, M., 155, 156
Lagrou, D., 272
Lague, D., 304
Lai, P., 304
Lambeck, K., 14
Lange, O.L., 183, 307
Lantieri, C., 304
Larson, P.H., 179, 191, 193
Lasisi, O.B., 269
Lavé, J., 236, 250
Lawson, N., 103, 104, 108
Le Campion, J., 10
Le Fort, P., 248
Le Pera, E., 269
Le Roux, A., 142, 143
Lea, F.M., 272
LeBlanc, S.G., 307
Lee, C.H., 278
Lee, H-S., 304
Lee, P., 313
Leonard, G., 248
Leroux, J., 304
Lesage, E., 142
Lewin, M., 39, 43
Lewis F, O., 272
Lewis, L.A., 265
Lewis, S.L., 270
Lewis, Z., 304
Li, G., 268
Lian, O., 187
Lijewski, T., 38
Lin, W., 308
Lindenmayer, D.B., 181
Lindsay, G., 101
Liong, S.-Y., 223
Lisiecki, L.E., 38
Liu, T., 185
Livingstone, I., 187
Lloyd, C.D., 278
Lóczy, D., 103
Lollino, G., 39
Lomax, J., 183
Long, K., 90, 92
Lopez, A., 107
López Castro, J.L., 10
Lopez, S., 269
López-Acre, P., 312
Louvat, P., 266, 268
Lozano-García, M.S., 178
Lucas-Borja, M.E., 30
Luedke, R.G., 183, 190
Luettge, A., 304
Lukinbeal, C., 282
Lusilao-Makiese, J.G., 140

M

Ma, X., 311
Mabbutt, J.A., 266
Mabin, A., 139
MacDonald, G.K., 10
MacDuff, A., 3, 92, 96, 317
Macedo, M.F., 309
Macfarlane, A., 239
Machado, C.A., 3, 116, 119–124, 318
Machado, C.R., 39
Machado, M.J., 14
Madden, C., 236, 250
Maestre, F.T., 307
Magirl, C.S., 190
Mah, J., 307
Maharjan S.B., 246, 248
Mahmood, S.A., 311
Majecka, A., 38, 44
Makohonienko, M., 58, 71, 73, 74, 317
Malaize, B., 269
Malan, N., 139, 142
Malavoi, J.R., 151, 165
Malhi, Y., 270
Mall, U., 307
Malthus, T.R., 31
Malungani, T., 139, 140
Mamchyk, S., 38, 44
Manchileo, D., 158
Mandy, N., 135, 136
Manera, C., 28
Mansberger, R., 310

Marcé, R., 152
Marcus, M., 200
Marcus, M.G., 199
Mardones, M., 156–158
Margottini, C., 39
Marks, L., 38, 44
Marriner, N., 10, 24
Martin, S., 39
Martínez Solares, J.M., 29
Martínez-Verdú, F., 304
Martins, L., 39
Marunteanu, C., 39
Masad, E., 304
Masin, D., 278
Mateos, R., 13, 14
Mateos Ruiz, R.M., 24
Mathee, A., 139
Mather, K.F., 207, 212
Matthews, E., 270
Matthews, N.A., 310
Maturilli, A., 307
Mayaux, P., 270
Mayo, A.L., 278
Mazel, A.D., 278
Mazhindu, E., 129, 140, 143
Mazoyer, M., 10
Mazurek, M., 73, 74, 317
Mazurek, S., 41
McAlister, J.J., 272
Mcauliffe, J.R., 178
McCarroll, D., 309
McCarthy, T.S., 134, 138, 140
McCulloch, J., 140
McElroy, B.J., 208
McHarg, I., 152
McKay, N.P., 269
McKenzie, P., 272
McKinley, J.M., 278
McKinnon, S.D., 307
McPherron, S.P., 183
McTainsh, G.H., 200
Meaders, M.I., 278, 279
Meierding, T.C., 278
Melton, M.A., 189
Meneely, J., 272
Meng, X., 14
Menk, J.R.F., 116
Merchel, S., 236, 246, 249
Mesri, G., 108
Messina, P., 307
Meuser, H., 118
Meybeck, M., 262
Meyer, H., 153
Mezcua, J., 29
Michael, D.R., 181
Michałowski, A., 71
Mierow, D., 239
Migoń, FP., 39
Migoń, P., 109, 317
Mikułowski, B., 38
Milecka, K., 58, 63
Miller, A.Z., 309
Miller, C.L., 222
Millot, R., 268
Mobley, M.L., 208
Moffat, I., 268
Mohamed, A., 307
Mohd, J.M., 118, 119
Mohr, C., 156
Mojski, J.E., 39, 44
Molnar, P., 181
Monchambert, J.-Y., 10
Mongelli, G., 269
Montanar, A.G., 18
Moore, C.B., 186
Moore, G.E., 208, 212
Moore, J.C., 14
Mora, D., 164
Moranta Jaume, L., 24
Morhange, C., 10, 24
Moro, M., 307
Morris, W.A., 310
Moses, C., 282
Mosselman, E., 165
Mostafavi, M., 152
Mucina, L., 130
Mueller, E.R., 218
Mugnier, J.L., 236, 250
Müller, P.J., 266
Mulvin, L., 272
Munné, A., 152
Muñoz, M.D., 3, 154, 317, 318
Muñoz-Nieto, A., 304
Murray, M.J., 139, 140
Murray, P.S., 200
Murthy M.S.R., 246, 248
Murton, J., 200, 201
Musiba, C., 310
Musselman, Z.A., 218
Muszyński, A., 71
Mutz, M., 151

N

Nagy, M.L., 179
Nahon, D.B., 266
Naicker, K., 140
Naidoo, S., 139
Nastar, M., 139, 143

Nations, D., 179
Naylor, L.A., 306
Nel, A.L., 131
Nengovhela, A.C., 140
Nesbitt, R.W., 266
Netoff, D.I., 278
Netterberg, F., 272
Network, N., 223
Neudorf, C., 187
Neupane, Y., 249
Nevin, A., 139
Newell, J., 142, 143
Newsome, D., 39
Newson, M., 152
Ng, K., 308
Nhlengetwa, K., 134, 140
Nielsen, P., 272
Nielsen, S., 151
Nijhuis, S., 153
Nishikawa, H., 312
Nitychoruk, J., 38, 44
Noble, T.A., 310
Noel, T.J., 280
Nowacki, Ł., 38, 44
Nowaczyk, N., 71
Nowicki, Z., 40, 43
Nyćkowiak, M., 58, 63

O

Obaje, N.G., 265
Oberlander, T.M., 186, 191
Oberlin, C., 10
O'Brien, P., 272
O'Daly, G., 272
Odler, I., 272
Ogbukagu, I.K., 265
Ogola, J.S., 140
Oi, H., 248
Ojelede, M.E., 140
Ojha, T.P., 268
Ojo, S.B., 265
Okamura, M., 248
Olariu, M.I., 307
Olayinka, A.I., 265
Ollero, A., 3, 151, 152, 164, 165, 169, 317, 318
Olley, J., 272
Ollier, C.D., 180
Olson, D., 4, 318
Olszewski, D.I., 183
Olusola, A., 4, 318
Onafeso, O., 4, 318
Onafeso, O.D., 262, 264, 270
O'Nions, R.K., 266
Onken, J.A., 195
Orbdlik, P., 165
Ordóñez, S., 304
Orihuela, M.F., 308, 311
Ortega-Guerrero, B., 178
Orts, A., 214
Oskin, M., 268
Ostapuk, P., 179, 186, 189, 199
Otto-Bliesner, B., 269
Ouyang, W., 304
Overpeck, J.T., 269
Owens, J.W., 187
Owens, R.H., 3, 318

P

Pagán, E.O., 179, 186, 189, 199
Page, B., 280
Page, K.J., 187
Paillou, P., 269
Palacio-Prieto, J.L., 38, 39
Palani, S., 223
Palmer, M., 208
Palmer, M.A., 152
Palmer, R.E., 193
Palomba, E., 307
Pandey, R., 236, 250
Pandey, R.N., 238
Pant, M.R., 236, 250
Pant, S.R., 251
Papis, J., 71
Paradise, T.R., 263, 278
Parent, J., 152
Parnell, S., 139, 143
Parnell, S.M., 138
Parsons, A.J., 181, 189, 191, 193
Pasławski, Z., 57, 60
Paudel, L., 248
Paudyal, K., 232, 240, 241, 248, 251, 255
Pawłowski, S., 58, 60, 65
Pazaras, N. Th., 304
Pearthree, P.A., 180, 189, 190, 195, 199
Pécher, A., 248
Peck, J., 269
Pelletier, J.D., 191
Penck, W., 193
Pérez, A., 179
Pérez Cueva, A., 14
Pérez, L., 154
Pérez-Soba, C., 312
Perrier, F., 246
Peter, A., 165
Petermann, P., 165
Peterson, G.D., 10
Petrov, D.V., 312
Petrus, J.M., 28, 317, 318
Peulvast, J.P., 236
Péwé, R.H., 199

Péwé, T.L., 199
Pfab, M., 130
Pfeifer, L., 10
Phillips, H., 138
Pica, A., 39, 71
Piégay, H., 151, 165
Pielach, M., 38, 44
Pietrasiak, N., 307
Pijet-Migoń, E., 109, 317
Pike, R., 222
Pinet, P.C., 312
Pinkham, R., 153
Pinyol, N.M., 108
Pirie, G.H., 138
Plater, A.J., 200, 201
Pleskot, K., 71
Pleuger, E., 10
Pochocka-Szwarc, K., 38, 44
Poirier, C., 179, 200
Pokhrel, R.M., 252
Pomar, L., 11
Pomfret, A., 312
Pons Esteva, A., 29
Ponzevera, E., 266, 268
Pope, G.A., 264, 278
Potere, D., 152
Poudel, K., 248
Poudel, K.R., 239
Pounds, N.J.G., 11
Pożaryski, W., 60
Pozebon, D., 118
Pradhan, P.M., 250
Pradhananga, U.B., 250
Pratt-Sitaula, B., 268
Price, A.E., 214
Price, B., 266
Price, D.M., 187
Prieto, B., 305
Pritchard, J.M., 199
Prominski, M., 153
Prone, A., 10
Ptolemy, R., 82, 83
Putkonen, J., 268
Puzrin, A.M., 108
Pye, K., 85, 89

Q

Quintana, F., 161
Qureshi, J., 311

R

Rajaure, S., 246
Rajendran, C.P., 250
Rajendran, K., 250
Rajot, J.L., 267
Ramez, P., 151
Ramírez-Hernández, J., 218
Ransome, F., 214
Ratajczak, T., 41
Ratnam, M.M., 307
Raudsepp-Hearne, C., 10
Raymo, M.E., 38
Raymond, P.A., 266
Real de Asua, R., 165, 169
Reddy, R.A., 130
Redeker, C., 152
Rees, W., 208
Regmi, D., 248
Regmi, N.R., 210
Regolini, G., 39
Reilly, M.K., 152
Reuter, H.I., 222
Revel, M., 268, 269
Reynard, E., 39, 52, 71
Reynolds, S.J., 179, 180
Rhoads, B.L., 195
Richards, S.M., 180
Richardson, C.D., 152
Riebe, C.S., 179, 201
Rieke-Zapp, D.H., 310
Riera-Frau, M.M., 25, 26
Rihosek, J., 278
Rijal, S., 232, 240, 241, 248, 251, 255
Riley, A., 153
Rimal, B., 232, 239–241, 248, 251, 255
Rimal, L.N., 250
Rippin, D.M., 312
Rivera-Aguilar, V., 307
Rizza, M., 236, 250
Robert, J.H., 214
Robinson, J., 129, 139, 143
Rodier, X., 136
Rodrigo, E.S., 14
Rodrigues, M.L., 39
Rodrigues, S.C., 3, 318
Rodríguez-Burgueño, J.E., 218
Rodríguez-Caballero, E., 307
Rodríguez-Gallego, M., 272
Rodriguez-Lloveras, X., 31
Rodríguez-Navarro, C., 272
Rodríguez-Perea, A., 13
Rogachevskaya, L.M., 117
Rogerio-Candelera, M.A., 309
Rogerson, C.M., 139
Rogerson, J.M., 139
Rohde, S., 165
Rojas, O., 151, 156, 157, 159, 161, 162
Roman-Quetgles, J., 25, 26
Romero, H, Vidal, C, 161
Romero Recio, M., 10

Rose, J., 14, 200, 201
Ross, E., 280
Rosselló Bordoy, G., 24–26
Rosselló Geli, J., 27
Rosselló, R., 27
Rossi, M., 116
Rottura, A., 269
Roudart, L., 10
Rousseau, D.P.L., 142
Rowley, T., 210, 213
Rubin, M., 129, 143
Rubinowski, Z., 39
Ruedy, R., 270
Ruffell, A.H., 278
Rugel, G., 236, 246, 249
Ruiz, J.M., 14
Ruiz, M., 317, 318
Rule, S.P., 139
Russell, G., 271
Rutherford, M.C., 130
Rutherfurd, I.D., 152
Rutkowski, M., 42
Rychel, J., 38, 44
Rylova, T., 38, 44
Rzepa, G., 41
Rzodkiewicz, M., 71

S

Sabater, S., 152
Sadr, K., 136
Saiz-Jiménez, C., 309
Sakrikar, S.M., 250
Salazar, C., 164
Saleh, M., 307
Salvi, S., 307
Samson, C., 304, 307
Sánchez Fabre, M., 164
Sanhueza, C., 158
Sanjurjo-Sánchez, J., 262, 272
Sanmartín, P., 305
Sannier, C., 307
Sapkota, S., 236, 250
Sapkota, S.N., 236, 246, 250, 251
Sapkota, T., 241
Sartoretto, S., 10
Sathyamoorthy, D., 310
Sato, M., 270, 271
Savage, J., 278
Schäffler, A., 139, 142, 143
Schälchli, U., 213
Scheidel, W., 24
Schmeeckle, M.W., 193
Schmidt, J.C., 218
Schmitt-Mercury, S., 10
Schneider, D.W., 152
Schneider, E., 165
Schneider, R.R., 266
Scholz, C.A., 269
Schubert, R., 4, 318
Schultz, C., 139, 140
Schulz, W., 45
Schurmans, U.A., 183
Schwanghart, W., 236, 245, 246, 249
Schweigstillova, J., 278
Scullion, T., 304
Sebastián, E., 272
Seddon, D., 238, 239
Segura, F., 14
Selby, M.J., 181
Sellwood, B.W., 11
Seong, Y.B., 179, 183, 191, 193, 201
Servant, M., 269
Seto, K.C., 152
Shaalan, C., 10
Shafroth, P.B., 218
Shahabi, H.H., 307
Shakya, S.R., 241
Shalaby, A., 307
Shanahan, T.M., 269
Shane, G., 152
Shannon, K., 140, 153
Shaw, C.T., 269, 270
Shaw, J.R., 195
Shepard, B., 82, 83
Sheridan, M.F., 179
Shi, X., 311
Shkuratov, Y.G., 312
Shorten, J.R., 135, 143
Shugar, D., 187
Sieben, J., 213, 214
Siedel, H., 278
Siegesmund, S., 278
Sihlongonyane, M.F., 129, 143
Sikorski, A., 67
Silva, B., 305
Silva, P.G., 13, 14
Simon, D., 142
Simone, A., 304
Singh, T., 268
Sivapalan, M., 18
Skoczylas, J., 52
Skonieczny, C., 269
Skwara, M., 43
Slaney, P., 90, 91, 92, 93
Slatt, R.M., 199
Slavinski, H.A., 310
Słomka, T., 39
Słowiński, M., 71

Smith, B., 278
Smith, B.J., 272, 278, 282
Smith, J.R., 183
Smith, S.A.F., 151
Solé-Benet, A., 307
Soliveres, S., 307
Sorooshian, S., 223
Sosa-Nájera, S., 178
Soukup, J., 278
Soulet, G., 266
Spencer, J.E., 179, 180
Sposato, A., 307
Srinivasan, V., 18
Srivastava, P., 268
Stacey, T.R., 134, 140
Stach, A., 55
Stankevich, D.G., 312
Stankowski, W., 71
Steinnocher, K., 307
Stelfox, D., 278
Stimberg F. D., 153
Stoermer, E., 10
Stokman, A., 153
Stolle, A., 236, 245, 246, 249
Stop, Pruszków, 52
Stork, N., 232, 240, 241, 248, 251, 255
Story, A., 142
Strini, A., 304
Strong, C.L., 200
Stuckless, J.S., 179
Stump, E., 179
Stutz, A.J., 31
Suguio, K., 117
Summerhayes, C., 179, 200
Sumner, A., 179
Surian, N., 165
Surridge, A.D., 130
Sutfin, N.A., 195
Swart, A., 139
Swetnam, T., 178
Swilling, M., 139, 142
Szabó, J., 103
Szałata, A., 58
Szczuciński, W., 71

T

Takahara, Y., 312
Takahashi, M., 308
Takemura, T., 308
Tandon, B., 278
Tang, Z., 304
Tapponnier, P., 236, 250, 251
Tarolli, P., 200, 201
Tegen, I., 270
Teng, J., 311
Tengö, M., 10
Termes, M., 152
Ter-Stepanian, G., 116, 117
Tessier, E., 140
Thakur, N., 278
Thapa, B., 248
Theurer, J., 278, 282, 298
Thibodeau, D., 307
Thom, B.G., 14
Thomas, D.S., 200, 201
Thorn, C.E., 179
Thornbush, M., 56, 200, 305, 306, 309
Thornbush, M.J., 179, 253, 278, 282, 305, 306, 309, 311, 312, 317, 319
Thorndycraft, V.R., 200, 201
Thornes, J.B., 10
Tiwari, D.R., 236, 250
Tjallingii, R., 71
Tkalich, P., 223
Tobolski, K., 71
Todero, A., 307
Todes, A., 129, 139, 140, 143
Tofelde, S., 236, 246, 249
Toffah, T.N., 135, 140
Tomaschek, J., 143
Tomaszewski, E., 58, 62
Tomczak, E., 41
Tomlinson, R.B., 14
Tooth, S., 200, 201
Topolski, J., 56, 57, 59
Topping, D.J., 218
Torab, M., 10
Toucanne, S., 266, 268
Tous Meliá, J., 17
Townsend, H.E., 272
Troć, M., 58, 60, 63, 72
Troy, T.J., 18
Trudgill, S., 282
Tsai, Y., 304
Tsokas, G.N., 304
Tsourlos, P.I., 304
Tucker, R.F., 134
Turkington, A.V., 263, 272, 278, 282
Turton, A., 139, 140
Tutu, H., 140
Twidale, C.R., 181, 191
Tyson, P.D., 130, 131

U

Udas, G.M., 238
Ung, A., 278
Upadhayaya, P.K., 241
Upreti, B., 236, 250

Upreti, B.R., 241
Urrutia, H., 158
Usuki, N., 248

V
Vacchi, M., 10, 24
Vaculikova, J., 278
Valero, F., 152
Van der Sloot, H.A., 272
Van Der Woerd, J., 236, 250
Van Devender, T.R., 178
Van Winkle, R.S., 214
Van Zee, J.W., 183
Vanrolleghem, P.A., 142
Varas-Muriel, M.J., 312
Vargemezis, G.N., 304
Varis, O., 139, 140, 143
Vega, Á., 152
Venter, J., 138, 140
Vergari, F., 39
Vernet, R., 269
Vidal, C., 156–158
Vietz, G.J., 152
Vigier, N., 266, 268
Viles, H., 272, 282, 305
Viles, H.A., 103, 181, 183, 208, 218, 306, 307, 312
Viljoen, M.F., 139, 140, 142
Viljoen, M.J., 134
Viljoen, R.P., 134
Vincent, K.R., 195
Viqueira, V., 304
Vitek, J.D., 210, 213, 214, 219
Vivent, D., 10
Vodyanitskii, Y.N., 118
Voermanek, H., 153
Vogel, C., 140
Von Blanckenburg, F., 268
Von Fürer-Haimendorf, C, 238

W
Wackernagel, M., 208
Wagner, W., 307
Wainwright, J., 10, 200, 201
Waldheim, C., 152
Walendowski, H., 49
Walker, I.J., 3, 187, 318
Wallace, T.C., 179
Walsh, C.J., 140, 152
Walubita, L., 304
Warke, P., 278
Warke, P.A., 272, 278, 282
Warren, A., 270
Warshawsky, D.N., 142
Watanabe, T., 248
Waters, C.N., 179, 200
Webb, R.H., 190, 199
Webster, J., 83
Weiersbye, I., 140
Weil, R.R., 124
Wells, S.G., 183
Werritty, A., 151
West, A.J., 268
Wharton, N.J., 280
White, A.F., 266
Whitley, D.S., 4, 278, 281, 282, 298, 300
Wilcocks, J.R.N., 131
Wilkinson, B.H., 208
Williams, M., 10
Wimberley, F.R., 140
Wimsatt, A., 304
Winkler, E.M., 272
Winkler, T., 129, 139, 143
Wittmann, H., 268
Włodarski, W., 71
Wohl, E., 31
Wohl, E.E., 195
Wöhler, C., 307
Wöjcik, Z., 39
Wolch, J.R., 142, 143
Wolman, M., 93
Woronko, B., 38, 44
Woszczyk, M., 71
Woyda, S., 41
Wu, Y., 304
Wyckoff, W., 206, 207
Wyrzykowski, J., 38

X
Xu, B., 304
Xu, X., 307

Y
Yagi, H., 248
Yamanaka, H., 235, 236
Yates, T., 304
Yibas, B., 140
Yokoyama, Y., 14
Yoshida, M., 235, 236
Yoshino, K., 248
Yoshinori, I., 39
Youberg, A., 190, 199
Young, A.R.M., 262
Young, D., 94, 96
Young, K., 313
Young, R.G., 151
Young, R.W., 262

Yu, B.Y., 179, 183, 201
Yule, D., 236, 250
Yusoff, I., 118, 119

Z

Zabielski, R., 2, 317
Zack, T., 143
Zagórski, Z., 67
Zalasiewicz, J., 10, 179, 200
Zaldokas, D., 90, 91, 92, 93
Zazo, C., 13, 14
Zbucki, Ł., 38, 44
Zeller, S., 153
Zhao, P., 3, 199, 318
Zhu, L., 311
Zreda, M., 307
Żurawski, M., 61
Zwartz, D., 14
Zwoliński, Z., 317
Zwoliński, Z.B., 55, 73, 74
Żynda, S., 58, 64
Żyromski, M., 52

Subject Index

A
Acid rain, 271
Adaptive management program, 169
Adobe Photoshop (DMAP) approach, 306
Aerotropolis, 101
Afforestation, 30
Agriculture, 10, 39, 115
  chemical products (fertilizer) to enhance, 115
  irrigated, 188
  in the Mesopotamia region, 10
Ahaggar, 270
Airports, 101
  area occupied by high-capacity modern airports, 102
  building, 101
    landform change due to, 103
  geomorphic change, 110
    artificial and partly artificial islands, 108–109
    coastline alteration, 103, 105–106
    land leveling, 106–107
    land reclamation, 104
    small-scale alteration of relief, 109–110
    types of, 104
  ground infrastructure, 101
  Hong Kong Airport, 103, 109
  Incheon International Airport in Seoul, 106
  at Kastrup in Copenhagen, 110
  by size, 107
  space, 109
Air transportation, 101
Åland Islands, 45, 48, 49
Alluvial complex, 2
Alluvial fans, 10
Alpine Centre shopping mall, 86
Alpine environment, 206
Alpine mountain towns, 206
Aluminum (Al), 119
Amazon region, 116
American Nettie, 212
America's Great Depression, 280
Amphitheater Landslide, 215
Anadromous salmon populations, 83
Ancient Mazovian Metallurgy, 37, 42
Andalién River floodplain, 161
Andalién River system, 156, 317, 318
Animas Rivers, 207, 219
ANN. *See* Artificial neural networks (ANNs)
Annapurna Range, 3, 232
Anthropocene, 1, 10
  epoch, 1
Anthropogenesis, 10
Anthropogenic, 1
  activity, 11, 116
  cityscape, 1
  climate change, 1
  influence, 10
  interferences, 117
  materials, 117
Apartheid, 129, 138
APP. *See* Areas of permanent preservation (APP)
Aquifer, 40
  acidic water contaminate, 140
  canopy to the bottom of, 210
  in glaciated regions, 210
  Quaternary, 43
  water retention capacity of, 167
Araguaína
  chemical composition of toxic material, 123
  technogenic deposits
    main materials in, 124
    soil contamination in urban area, 119
    transversal profile in Cimba neighborhood, 122
    types, geographical distribution, and, 120, 121
  urban area, 119
Archaeological
  data show an initial location of Palma, 24
  evidences, 2
  excavation, 72
  findings, 17
  sites, 10, 18, 42, 136
    Neolithic palaeoenvironmental deterioration, 270
  surveys, 273
Areas of permanent preservation (APP), 119
Armala valley, 253
  sinkhole formation, 253
Arsenic (As), 117, 118
Artificial and partly artificial islands, 108
Artificial horizons, 118
Artificial islands, 3, 101, 109
  to host Osaka Kansai International Airport, 109

Artificial neural networks (ANN), 223, 224, 226
  parameters and categories, 224
  prediction of urban suitability, 228
Association of Polish Mechanics, 37
Atlantic coast, 269
Atlantic Ocean-flowing Vaal/Orange River system, 131
Auraria Campus, 278
  map of, 279
Aurum, 279
Avalanches, 216, 307
  debris, 220
  shed on US Hwy 550 south of Ouray, 220

B

Balearic Archipelago, 11
Balearic Islands, 18
Ballena, 195
Baltic Sea, 49
Bays, 10
Bear, 214
Beaumont Hotel, 215
Bed and bank characterization, 85
  grain-size template, 85
  pebble counts, 85
  spawning habitat, 85
  using GRADISTAT software, 85
    grain size statistics for reach, 89
Beigan Island (Taiwan), 107
Berlin, 56
Biodepletion, 264
Biodiversity, 96, 130, 143, 152, 169
Biofiltration, 140
Biological soil crusts (BSCs), 179, 183
Biomes, 270
Biotite, 269
Bivouacs, 206
Bogdanka, 56
Bog iron ore, 41
Boron (B), 117
Braamfontein, 136
British Columbia, 81
Brussels Airport, 102
Bureau of Land Management, 208

C

Cadmium (Cd), 117
Calcite, 123, 272
Camelback Mountain, 199
Camp Bird Mine, 212, 218
Canada, 81, 317
  engineering projects, 81
  Okanagan River, 81
Carbon dioxide, 272
  chemical weathering of silicate rocks and, 268
  emission, 271
Carbonic acid, 271
Carbon monoxide, 140
Carex riparia, 140
Cargo terminals, 101
Cascade, 214
Cascade Creek channel, 219
Cementitious materials like limestone, 272
Central European Plain, 55, 56
Central Mazovia, 37
Chad basin, 269
Channel sediments, 63
Chek Lap Kok, 109
Chemical alterations, 118, 266, 273
Chemical decay of rocks, 264, 273
Cherry Creek, 279
Chicago, 152
Chile, 150
  adaptive management program, 169
    focus on, 169
  Andalién-Nonguén confluence, 157
  Andalién River system, 158, 161
  area of the Nonguén Stream floodplain, 161
  bio- and geodiversity, 151
  Chilean rivers, 154
  comparative map, 150
  Concepción, expansion of, 157, 158
  design exercise, 161–169
    conceptualization problem by Espinosa, 162
    current state of fluvial geomorphology within study area, 162
    interpretative map of urban tissue, 164
  disaster as urban developer, 154–156
    new Regulatory Plan (PRC-60), 1960, 154
  drainage network, 151
  efforts on hydrogeomorphological processes, 152
  flooding, 161
  fluvial geomorphology, 151
  fluvial territory, defined by, 151, 165–166
    alternatives designed for study area, 167
  geomorphology, 151
    and its urbanization, 150
  governmental development policy "Operación Sitio", 161
  hydrogeomorphological (IHG) index, 164
    Nonguén Stream, four reaches, 164
    parameters, 164
  interpretative map of
    Andalién watershed and Nonguén subwatershed
      and their relationship with urban areas and nature reserve, 158
    Andalién watershed/Nonguén subwatershed, 158
    risk exposure, 159
      study area, 159
  landscape

based urban solutions for redefining border area, 151
oriented urban design, 151
urbanism, concept of, 152
lateral and vertical mobility of river, 151
lower Andalién River, 154
main economic activities, 154
mapping explorations
Chile characterization and Andalién watershed location, 155
methods of flood defense, and water management to, 153
morphodynamic activation techniques, 153
Emsher River in Dortmund, 153
Isar River in Munich, 153
Shunter River in Braunschweig, 153
multipurpose Green System, 169
schematic, 170
natural landscape
and cities, 150
as integral component of design process, 152
neoliberalism, 161
Nonguén floodplain, 157
Nonguén Stream, 154, 157, 158
photo-interpretation analysis, 151
geomorphological dynamics of river, 151
projects focused on geomorphology restoration, 153
Baxter, 153
Berkeley (California), 153
Blackberry Creek, 153
Cerrito Creeks, 153
daylighting, 153
Quail Creek, 153
Strawberry Creek, 153
rainfall and anthropogenic intervention, 150
redefinition of boundaries (the edge), 166
systematic mapping of design exercise, 168
river
biodiversity, 152
exploitation, 152
flooding dynamics, 152
restoration-rehabilitation (RR), 150
site-specific context, 156–161
correlative map associating urban growth, and river complexity, 160
sociospatial capital adjustments, 154
strain on riparian ecology, 152
urban design, 151
concept of restoration in formulation, 153
urban growth correlative map
after 1955, wetlands occupation, 157
urban history, 150
urbanization, 151
Wallmapu, 150
water management system, 167
multifunctionality, key actions, 167
Natural Water Retention Measures, 167
water urbanism, 153
Chromium (Cr), 117
Chübu Centrair International Airport in Nagoya, 108
Chwaliszewo, 67
CIELAB color system, 305
Cimba Creek, 122
City of Araguaína, 119
Claylike texture, 122
Clay pit ponds, 43
Clay sandstones, decay of, 262–264
Coanegra, 18
Coastline alteration, 105
Colorado, 206, 207, 216, 279
Colorado Mineral Belt, 208
Colorado's Auraria, 279
Columbia River system, salmon stocks in, 83
Commodification of land, 317
Common Carp, 96
Construction and demolition wastes, 121
Coolidge Dam, 188
Copper (Cu), 117, 211
Core stones, 181
Cross-disciplinary approach, 1
Cryogenic processes, 217
CSSI. *See* Cultural Stone Stability Index (CSSI)
Cultural Stone Stability Index (CSSI), 4, 278, 281, 282
basics of, 281–282
scoring system with accompanying qualitative interpretation, 282
versus Rock Art Stability Index (RASI), 281, 278, 282, 298, 300
Cultures, 18, 139, 262
eastern to the western Mediterranean, 18
heritage, 37
Curtail treatment, 118
Cut-off hillside at Beigan Island, Taiwan, 108
Cybina and Bogdanka Streams, 65
Cybina River, 56, 70

D

3-D camera, 304
Debris, 68, 122, 183, 190, 192
avalanche, 220
construction, 116
flow chutes and levees, 179
flow pathways, 199
jam downstream, 90
large woody debris (LWD), 89
megadebris flow, 236
Deltas, 10
Denver, Colorado (USA), 4, 211, 279
Auraria Campus, 278–280
winter 1959 issue of the *Georgia Review*, 278

Denver, Colorado (USA) (*cont.*)
creation of Metropolitan State College, 280
education development, 280
historic and legacy structures, 280
map of, 279
University of Colorado Denver, 280
Desert dust, 200
Desert geomorphic hazards, 195. *See also* Phoenix metropolitan region
alluvial fan flooding, 195
development burgeoned on alluvial-fan surfaces, 196
debris flows, 199
West-looking view, 199
haboobs and dust storms, 199, 200
street flooding
ballena, Fountain Hills built on, 195
cost-benefit trade-off of engineering, 195
development on pediments, 197
ENSOevent led to, 198
flooding being routed through, 196
in planned and unplanned housing developments, 195
Desertification, 199, 264
Desert pavements, 183
DGPS. *See* Digital geographic positioning systems (DGPS)
Digital cameras, 303, 310
low-cost, measure soil erosion using, 307
Digital geographic positioning systems (DGPS), 84
Digital photogrammetry, 304
Digital photography, 4, 84
along with topographic and spectral analysis to study lunar surface, 307
measurement of vegetation cover in the Arctic accurately, 307
typical sampling of sidewalk during photographic field survey, 308
Digital single lens reflex (SLR) camera, 310
Digital stereophotogrammetry, 307
Dirt cracking, 180, 181
Disaster-prone areas, 318
Dolomite, 272
Dominant rock decay process, 180
Doornfontein, 136
Droughts, 27
3-D system, 304
Dulag 121 Museum, 37
Durango-Silverton Railroad, 206

E

Earth components, 208
Earthquakes, 158, 236, 251, 318, 319
Earth science, 208
Ecological dynamics, 11
Eemian interglacial, 60
Electromagnetic current meter, 84
El Juf depression, 270
Environmental changes, 1
Environmental damages, 119
Environmental elements, 115
Environmental planning, 119
Erosion, 10, 18, 63, 104, 151, 264, 282, 287
accumulation of fine and medium sands, 61
accumulation terrace, 62
channel banks and bed, 96
and flooding in, 82
gullies, 71
lateral river valleys, 65
processes, 18
and rill, 104, 135
river bank, 232, 248
rocks of the Witwatersrand Supergroup, 130
soil, 151, 219
degradation led to, 10
surfaces, 261, 265
vulnerability to, 200
wind, 47, 183
Estuaries, 10, 14
Eureka, 211

F

Fe and Al oxides, 124
Federal Emergency Management Agency (FEMA), 195
Fe element, 124
Feldspar, 269
Ferralic arenosol, 121, 122
Ferruginous concretions, 118
Fish counts, and spawning activity, 84
Flood, 14, 27, 57, 70, 186, 318
Alluvial Fan Flooding, 195
Andalién River floodplain, 161
characteristics, 248
and disease risk, 138
flash, 10
floodplain in the bottom of the Warta River, 65
Floods Directive 2007/60/ EC, 152
frequency analysis, 90
in Johannesburg, 140
to landslides and earthquake liquefaction, 150
megafloods, 236, 245
to prevent, 97
Ramghat, 253
risk, 10, 25, 29
Salt River, 198
street flooding in, 195
urban area, 82

Florida, sinkholes formed, 253
Flow hydraulics, 90–93
depth-velocity data for 27 Kokanee redds in reach 3, 92
field measurements of flow depth and velocity, 90
Froude number, 90
HEC-RAS simulations, 90
longitudinal distribution
of modeled flow velocity Lower Vernon Creek, 91
modeled *Fr* in the main channel of Lower Vernon Creek, 92
maximum and minimum ranges for Okanagan River Sockeye salmon, 90
modeling results for mean flow velocity and, 90
"subcritical" or tranquil flow conditions, 90
Fluorine (F), 124
Fluvial systems, 10, 218, 318
Food waste, 116
Fossils, 191, 270
Fountain Hills, 181
Frontal Moraine Hills, 71
Fuel stations, 101

G

Gdańsk Bay, 45, 46
Geographical distribution, and TD types, 120, 121
Geographic information system (GIS) database, 304
Geomorphic
agents, 2
change, types of, 103
hazards, 1
map of Krygowski, 58
processes, 14
Geomorphometry, 222
application of ANNs in, 224
Georesources, 2
Geospatial information systems (GIS), 222
database, 4
Geotourism, 2
Gila Range, 199
Gila River, 187, 188
Glacial abrasion, 47
Glaciated mountains, 3
Glaciation, 2, 212
Global position system (GPS), 307
Global warming, 319
Główna River, 65
Gluszynka Stream valley, 65
Gneiss boulder, 269
Google Earth© map, 84, 103
Gorkha earthquake, 232, 250
GPS. *See* Global position system (GPS)
Granitic landforms generated by rock decay, 181
Granitic rocks, 190
Gravel
alluvial, 252
deposited by glacial water, 64
fluvial, 190
good spawning, 87, 89
grains, 49
material, 97
pit, 44
Pokhara, 236, 238, 246
of the Poznañ phase, 72
sized sediment, 96
Gretagmacbeth ColorChecker Color Rendition Chart, 306
Gros, 29

H

Haboobs, 199
Hazards, 1, 3
desert geomorphic, 195–200
environmental, 140
natural, 157
Heavy metals, 117, 118
Highlands, 279
map of, 279
High-resolution digital cameras, 304
High-resolution optical imagery, 307
Himalayan collisional tectonics, 232
Himalayan tectonics, 3
Holocene, 10
sediments, 60
Hong Kong Airport, 103, 109
Howardsville, 211
Human activities, 4, 10, 11
environmental alteration by, 115
footprint, 10
Human-built landforms, 218
Human civilization, 115
Human-created landscape, 1
Human-environment interactions, 1, 11, 224
Human-environment problems, 222
Human-environment relationships, 10
Human geomorphology, 1
Human influences, 4
"Human-made" creations, 4
Human mobility, 101
Human transformation, 1
Humidity, 131
Hurricanes, 319
thunderstorm, 199
Hydraulic modeling, 86
using HEC-RAS, 84
Hydraulic parameterization, 84, 85

I

IDIP method. *See* Integrated digital photography and image processing (IDIP) method
Ijebu-Ode in Nigeria, 269, 271
Illegal disposal, problem of, 125
In-channel flow hydraulics, 85
  parameterize conditions, of active Kokanee spawning redds, 85
  shallow flow depth, in most of Creek, 85
Incheon International Airport in Seoul, 106
Indian Ocean
  anticyclonic conditions, 131
  flowing Limpopo River system, 131
Industrial Revolution, 37
Instituto de Natureza do Tocantins (NATURATINS), 119
Instruments, selected as tools to measure surface roughness, 304
Integrated digital photography and image processing (IDIP) method, 305
  application of method outdoors through calibration, 307
  camera mounted facilitate movement, 305
  digital 3-D mapping, 307
  limestone sensors, exposed up to, 305
  mean versus median L for uncalibrated versus calibrated lightness values, 311
  measuring surface roughness, 309
    RMS roughness, 309
    SD *L* values, 309
  sampling height for photographs, 305
  uncalibrated versus calibrated lightness values, 312
  use Adobe Photoshop/Corel Draw, 306
Iron sulfides, 272
Island of Mallorca, 11

J

Johannesburg (South Africa), 129, 317
  annual precipitation, 131
  Atlantic Ocean-flowing Vaal/Orange River system, 131
  challenges of city today, 139
    food security, 142
    industrial site rehabilitation, 140–142
    mine pollution management, 140–142
    sustainable development, 143
    urban greening, 142
    urban water management, 140
  contemporary changes, 129
  development of city, 135
    the Apartheid era (1948-94), 138
      demographic patterns, 138
      Klip wetland, 138
      Soweto, 138
      suburb of Sophiatown, 138
    gold rush of the 1880s, 136
      Commissioner Street, 136
      Government Gazette, 136
      Rand Club, social club and stock exchange, 136
      spatial patterns, 137, 138
    post-1994 development, 139
      "ethnopolitical" dimension, 139
      international markets and investors, 139
      reinventing the purpose of the inner city, 139
    precolonial development, 136
  geology and mineral resources, 134
  Indian Ocean-flowing Limpopo River system, 131
  Jukskei River, 131
  key element driving the postcolonial development, 129
  Klip River, 131, 138
  landscape of Melville Koppies, 133
  map and regional topography, 131
  natural and urbanized catchments, 131
  physical environment of city, 130
    climate, 130
    ecosystems, 130
    topography, 130
  physical patterns and urban makeup, 129
  regional-scale Google Earth image of, 132
  regional topography, 131
Jonathan, 212
Jukskei River, 131
Juncus effusus, 140
Junikowski Stream, 56, 65

K

Kai Tak Airport, 109
Kalamalka Lake, 83, 90, 96
Kansai Airport, 110
Klip River, 131, 138
Knowledge, 3, 143, 150, 152, 222
  important gaps in, 31
  improving, 268
Kokanee salmon, 82, 84, 96
  spawning habitat, 82
Komorów reservoir, 43
Kunów sandstone, 49

L

Lagoons, 14, 116, 120
Lake Abay, 269
Lake City, 206
Lam Chan, 109
Landfills, instability in, 122
Land leveling, 106
Land reclamation, 104
Landscape urbanism (LU), concept of, 152

Landslides, 1, 156, 318. *See also* Flood
the Amphitheater, 215
contributions and beaver ponding, 94
hazardous, 189
of the San Juan Mountains, 214
Land subsidence, 318
Large woody debris, 89
Laser imaging detection and ranging (LIDAR), 308
Laser scanning, 304
Las Sendas, 196
Late Holocene, 10
Lead (Pb), 117
Leszno phase, 61
LIDAR. *See* Laser imaging detection and ranging (LIDAR)
LiDAR. *See* Light detection and ranging (LiDAR)
Light detection and ranging (LiDAR), 222
Limestone-cemented layers of mud bricks, 271
Limpopo River system
Lipowa Street, 49
Lower Vernon Creek, 83, 84, 98
engineered and channelized portion of, 89
habitat potential of, 84
reach division, showing reach boundaries and cross-sections, 85

M

Machhapuchhre Peak, 232
Ma Ha Tuak Range, 199
Majewski Pencil Factory, 37
Mallorca, 317
Manganese (Mn), 117
Mass wasting, 189
Mazovia Park of Culture and Recreation, 43
McIntyre Dam, 81, 83
Mediterranean
basin, 10
city. *See* Palma
coastal cities, 11
hydrologic cycle, 10
Megalopolises, 1
Megastorms Mercury (Hg), 117
Mesa, 196
Mesopotamia, 10
Metallic
corrosion, 272
elements, 117
ions, 118
structures, 122
Metamorphoses, 1, 74
Methane emissions, 140
Metropolises, 1
Metropolitan Regulatory Plan of Concepción (PRMC) of 1980, 154
Mineral Creeks, 214
Mine waste dumps, 318
Mining waste, 117
Mission Creek (near Kelowna), 83
Mn oxides, 124
Molybdenum (Mo), 117
Monsoonal climate, 3
Moraine plateau, 56, 57, 64, 65, 71
Morasko Hill, 64, 66, 71
spatial distribution of main impact craters, 66
Mudstones
bricks, 271
decay of, 262–264
Mummy Mountain, 199
Munich International Airport, 110
Museum exhibitions, 37

N

Na Bàrbara catchments, 18
Na Bàrbara rivers, 29
National Research Council, 208
National Science Foundation, 206
National Weather Service (NWS), 216
Natural systems, 11
Nazi occupation, 47
Neolithic architectural sites, 4
Neolithic palaeoenvironmental deterioration, 270
Neolithic period, 10
Nepal, 232
main places of tourism interest, 244
palaeoseismicity, 251
Pokhara. *See* Pokhara
recent birth of a major city in, 238–242
tourist city with major attractions, 242–243
New Kitakyushu Airport, 108
Nickel (Ni), 117
Nigeria, 271, 318
Niger River basin (Africa), 2, 4
Nutrients cycle, 117

O

Odra glaciation, 60
Ogun State, 271
O-IDIP method application, 305
in downtown Toronto, 305
limitations of photometric approach, 305
to measure surface roughness on flat surfaces, 313
contrast adjustments, 313
median *L* values, 313
objectives, 305
three-dimensional (3-D) quantification, 309
use Adobe Photoshop/Corel DRAW, 306
Okanagan basin, 82, 83, 317
Okanagan Falls, 81

Okanagan Lake, 82, 83, 89, 90, 96
Sockeye salmon into, 83
Okanagan main-stem lakes, 81
Okanagan Nation Alliance (ONA), 83
Okanagan River, 81, 82
Okanagan River Restoration Initiative (ORRI), 83
Okanagan River system, 83
Sockeye salmon, 83
Oldest Dryas, 63
Oligocene aquifer, 43
Oliver, BC, 81
One-dimensional hydraulic flow model (HEC-RAS 4.1), 86
Ophir, 226
Optical measurements, 308
Organic components, 272
Organic matter, 117, 118
degradation, 304
Osoyoos Lake, 81
Ostrówek, 67
Ostrowiec Świêtokrzyski, 49, 51
Ostrów Tumski Island, 57, 72
Ouray, 206, 211
Ouray Hydroelectric Dam, 219
Outwash plains, 71

P

Paleofloods, 3
Palma, 11, 317
basin, 15, 18
catchments, 14
Palma geography, 11
analyzing urban evolution and relationship with environment, 18
contemporary age, 28–29
Islamic period, 25–26
late Middle Ages, 27
modern age, 28
Roman and late ancient age, 24
Talayotic period (BC 3000-550), 18
ArcGIS's 3DAnalyst, 14
carbonate Miocene platform, 15
city location, 13
digital elevation model, 12
flat platform areas, 11
fluvial drainage network, 14
arrangement, 14
geomorphological reconstruction of Palma, Roman Age, 26
hardness of calcarenites and tabular reliefs, 15
hydrology of catchments in Palma basin, 14
hygienic conditions, 17
land use as crucial change of urban geomorphology, 29–31
between 1956 and 2012, 30
agricultural uses, 30
land-use transformation, 31
urban development, 29
linked to succession of three main walled enclosures, 17
Islamic (late 11th century), 17
Renaissance (1575), 17
Roman (BC1), 17
location of Mallorca Island, 12
map of the city of Palma, circa 1726, 12
Mediterranean Sea, 11
Mediterranean's Quaternary geodynamics, 11
morphometric analysis, Palma basin catchments, 14
mountain areas, 11
orographic effect, 15
paleohydrological studies, 14
parameters, 14
phase of sedimentation, 14
Postglacial Marine Transgression, 14
quaternary depositional process facilitates, 15
relief subunits, 14
Palma Alta (Upper in Catalan), 14
Palma Baixa (Lower in Catalan), 14
Prat de Sant Jordi, 14
relief units, 13
coastal area comprise, 13
the Marina, 13
Marratxí Hills, 13
Pla de Palma, 13, 15
Tramuntana Range, 13
semihorst geometry, 11
special relationship with water as a resource and, 18
topography, fluvial network, 12
torrentiality and clinometric variables, 15
Tramuntana Range, 13, 15
Upper Miocene carbonate platform, 15
Pastoralism, 188
Pedimentation, 190, 318
Penticton, 81, 83
Penticton Channel, 83
Periglacial processes, 213, 217
Petra, Jordan, 4
Phewa Lake, 245, 255
the Phoenician, 11
Phoenix metropolitan region, 3, 177
anthropogenic processes, 187
biological soil crusts (BSC), 179
common desert geomorphic processes, 179
Coolidge Dam, 188
core stones, 181
crustal extension resulted from, 179
debris-flow pathways, 199
debris flow, zones, 190

desert dust, 200
desertification, 199
dirt cracking, 180, 181
Federal Emergency Management Agency (FEMA), 195
Fountain Hills, 181
geological map of bedrock ranges in, 180
geomorphic landscapes, 179
Gila River, 187
aeolian transport from, 189
timeline of anthropogenic alteration and large flood events, 189
granitic forms, 179
interaction between aeolian, and fluvial processes, 187
excavation exposed, 190
grain size analysis of aeolian and fluvial sediments, 188
mass wasting, 189
debris flow levees, 193
debris flows interface with urbanization, 192
mountain slope, 193
potential danger areas as Camelback Mountain/Phoenix Mountains, 191
pressure-release joint of granitic surface, 193
rock slides, 190, 193
talus from rock falls, 193
mean annual precipitation, 177
metamorphic slopes tend to host debris-flow chutes and levees, 179
pediments
development on, 197
inselberg landscapes, 194
rock coatings, 180, 185
rock decay, 180
dominant rock decay process, 180
granitic landforms generated by, 181
desert pavements, 183
rock varnish, 186
Roosevelt Dam, 198
Salt River, 187
soils, 180
Sonoran desert setting of, 177–180
state of Arizona as seen in Google Earth, 178
tors, 181
valley fever, fungus Coccidiodes, 200
Photogrammetry, 304, 310
Laetoli footprints in Tanzania, 310
Photometric approach, 311
to apply digital photography and image processing for data acquisition, verification via, 313
laser profiling instrument, assessing soil roughness, 312
median *L* values, 311
multiple scattering to suppress shadowing effects, 312
surface roughness parameter, 312
total image size, 312
Photo of Camp Bird Mine tailings, 218
Physical activity alterations, 115
Physical landscape, 3
Plasticizers, 49
Platte River, 279
Pleistocene glaciations in Poland, 38
Pleistocene sediments, 56, 60
Pokhara, 3, 232
active mountain in a subtropical environment, 233
Annapurna Conservation Area Project, 242
Annapurna II, III and IV, 233, 235
anthropogenic-induced hazards, 253
sand mining, 253–254
Bhim Kali Dhunga, 236, 239, 246
catastrophic, geomorphic evolution, 234–239
cause of flood, 248
climate, 234
cross-section of the himalayas, 234
distinctive landforms, 237
Earthquake Disaster Preparedness and Response Framework Report for, 250
earthquake hazards, 250
environmental impacts
pollution, 255
urbanization versus geoheritage preservation, 255
flood hazards, 248–249
Gangapurna, 235
Gaunda-Gachok conglomerates, 236
geomorphological characteristics, 232
hot springs ("tatopani" in Nepalese) of Kharpani, 246
importance of tourism in the Pokhara economy, 244
Kali Khola, 235, 252
karstic features, 246
Kaski District, 233
Kawaguchi, Ekai, 242
lakes, legacies of catastrophic events, 245–246
Lamachaur, 235
Lamjung Himal, 235
landscape change, 243
land-use map of valley, 240
Lesser Himalayan metasedimentary rocks, 233
location map of valley, 233
long, dramatic, and complex history, 232
Main Central Thrust (MCT) Front, 236, 246
major attractions of valley, 245
major geosites of valley, 247
mountainous setting, 235
natural hazards and risks, 248
Nilgiri limestones, 236
nucleated settlements, 234
Phewa Lake, 255

Pokhara (*cont.*)
precipitation, 234
quaternary deposits, 238
risk sensitivity map of the Pokhara Submetropolitan City and, 252
savanna-type vegetation, 234
sedimentary rocks, 233
Seti Khola, 233–235, 255
flood, 249
Seti River, 233
floodplain, sand mining in, 254
sinkholes, and subsidence, 250, 253
tourist attractions, 232
UNDPon Earthquake Risk Reduction and Recovery Preparedness/ERRRP(2009), 250
vegetation, 234
xerophytic plants, 234
Poland, 317
capital, 37
human activity, 56
thickness of these sediments, 55
Polish Lowland, 56
Pollutants, 118, 119
Pollution, 1, 140, 255
sources, 272
Polson Park, 87, 96
Pomeranian phase, 63
Population growth, 318
Porosity, 304
Portland, 214
Power Plant, 219
Power spectral density (PSD), 308
Poznañ, 2, 56, 317
anthropogenic changes in morphological landscapes, 66–70
area, geomorphological map of, 62
digital elevation model and main elements of hydrographic network, 58
Fortress of Poznañ, 71
Genius Loci, 71
geosite, 72
geological and sedimentological setting, 60–64
geological cross-sections through, 72
geomorphological setting, 64–66
geosite of "Morasko Hill, ", 71
hydrographic network in 18th century, 59
left-bank tributaries, 56
lowest river terrace system in the Warta River valley, 61
in the Medieval Age, location of, 57
moraine plateau, 65
sedimentation of clay deposits, 61
surface waters, 56
the Szachty, 71
urban geosites, 71–73
urban sprawl, 68, 69
Warta Gap, 56, 60, 65, 71
Warta River, 56, 60
channel sediments, 63
maximum water levels, 60
Wielkopolska Lake District, 56
Progradation processes of coastline, 18
Pruszków, 37, 317
abiotic natural resources, 37
abiotic resources, 37
cultural heritage, 37
erratics, 43–49
Cambrian age, 47
central John Paul II Square, 49
commemorative inscription, 47
crescentic fractures, 46
gaize bed exposed in Lower Vistula Valley, 46
granite, probably from southeastern Sweden, 47
list of other erratics present in the urban space, 50
made from granite, 44
Novabrik elevation brick, 49
parallel-arranged fine crests and troughs, 48
piece of granite-gneiss, 44
Sandstone, 46
Småland granite, 46
in the town center, 47
georesources, 37, 39
and its surroundings, 39
geotourist values, 37
geovalues, of geotourism interest, 39
location, 38
Matzevahs in the Jewish cemetery, 51
municipal authorities, 37
natural and cultural resources, 37
objects of abiotic heritage, 40, 41
relief and deposits, 40–41
stones in open urban space
commemorative inscription, 47
Gabions: decorative reinforcement, 49, 50
Novabrik elevation brick, 49
tourist values, 37
water, 43
Pyrite (FeS_2), 140

Q

Quaternary aquifer, 43
Quaternary sediments, 55, 56
thickness, 55

R

Radar imaging, 304
Rainbow Trout, 96
Rains, 123

Rainwater infiltration, 140
RASI. *See* Rock Art Stability Index (RASI)
Red Sea, 269
Remote-sensing techniques, 304
Ridgway Dam, 219
Rill erosion, 104
Rio Grande, 207
  Railroad, 211
Rio Verde, 196, 197
River floodplains, 318
  modify, 226
River modification, 318
River mouths, 10
River Niger basin, 269
  warm wet climates of, 269–270
River restoration-rehabilitation (RR), 150
Rock Art Stability Index (RASI), 278, 281, 282, 298, 300
Rock coatings, 180, 185–186
Rock decay, 180
Rocky Mountains, 207
Roosevelt Dam, 198
RR. *See* River restoration-rehabilitation (RR)

S

Sahara Desert, 270
Sahel regions, 269
Saint Elizabeth of Hungary Roman Catholic Church, 278, 284
  added to National Register of Historic Places, 281
  collection food and money for, 281
  cultural stone stability index (CSSI), 282
    analysis, 283
    basics of, 281–282
    east-facing panels, 287–292
    north-facing panels, 283–288
    overall assessment, 297
    scoring system with accompanying qualitative interpretation, 282
    south-facing panels, 293–296
    west-facing panels, 296–298
  design by Jules Jacques Benoit Benedict, 280
  Heinrichs, Leo, 281
  located on Auraria Campus in downtown Denver (USA), 282
  May Bonfils Trust, 280
  spiritual home for Denver, 280
  structure follow Romanesque Revival style, 280
Salmonids, 83
  gravel material, suitable for spawning, 97
  typical flow conditions, 90
  valuable spawning habitat for, 87
Salt River, 187, 198
Sandstones, 263
  texture and mineralogical properties, 263
San Francisco Airport, 105, 109
San Juan Mountains, 3, 206, 318
  artificial neural network (ANN)
    parameters and categories, 224
    prediction of urban suitability, 228
  climograph of Telluride and Ouray, Colorado, 209
  comprising volcanic summits, 207
  critical zone
    graphical depiction, 210
    new awareness of, 208
  drained by, 207
  feature, 207
  generalized map of study area, 206
  geology, 208
  geomorphic processes, 212
    anthropogenic building, and modifications of landforms, 217–221
    fluvial, 213–214
    glaciation, 212
    mass movement, 214–217
    periglacial processes, 213
  mining town development, 210–211
  planner's dream, 221–222
  during the Pleistocene, 207
  population density, 208
  predicting urban suitability in, 223–225
  southernmost portion, 207
  Telluride (Upper San Miguel) mining district, 212
  towns
    distance to river, 227
    elevation, 227
    local relief, 227
    local relief, smoothest/roughest surfaces, 226, 227
    North-south orientation factor, 227
    summer solar irradiance, 227
    topographic shielding, 227
    winter solar irradiance, 227
  Uncompahgre mining district, 211–212
  variation in climatic conditions, 207–208
  volcanic nature of, 206
San Miguel Rivers, 207, 214
  valleys, 212
SAR. *See* Synthetic aperture radar (SAR)
Sa Riera, 18
Satellite imagery, 3
Scandinavian Pleistocene glaciation deposits, 55
Sea level, 10, 18
Sedimentation materials, 262
Sedimentation processes in valleys, 116
Sediment basins, 219
Seismic activity, 1
Selenium (S), 117
Seti Khola, 232
  flood, 253

Seti River, 249
  floodplain, sand mining in, 254
Sewer treatment, 118
Shaw Butte areas, 199
Shear stresses, 220
Shuttle Radar Topography Mission (SRTM), 58
Siberian continental zone, 269
Sieve analysis, 84
Silicates, 272
Silicosis, 140
Silver, 210, 211
Silverton, 206, 211, 220, 221, 226
Sinkhole collapse, 318
Skaha Lake, 81
Slope stabilization, 318
Small-scale alteration, of relief, 109
Socioeconomic aspects, 117
Socioeconomic organization, 18
Sockeye salmon, 82, 98
  into Skaha Lake, 83
Soils, 180
  contamination, 117–119
    by TDs In the urban area of Araguaína, 119
  degradation, 10
  depletion, 110
Solar radiation, 226
Solid waste, 119
  landfills, 124
Solifluction, 216
Sonoran Desert, 3, 177, 199
  distinct biogeography, typical of, 177
    Barrel *(Ferocactus cylindraceus)*, 177
    Brittlebush *(Encelia farinosa)*, 177
    Catclaw acacia *(Acacia greggii)*, 177
    Creosote bush *(Larrea tridentata)*, 177
    Desert globe mallow *(Sphaeralcia ambigua)*, 177
    Elephant tree *(Bursera microphylla)*, 177
    Hedgehog *(Echniocereus engelmannii)*, 177
    Iconic saguaro *(Carnegiea gigantea)*, 177
    Ironwood *(Olneya tesota)*, 177
    Juniper osteosperma, 178
    Ocotillo *(Fouquieria splendens)*, 177
    Palo verde *(Parkinsonia microphylla)*, 177
    Pinus monophylla, 178
    Triangle-leaf bursage *(Ambrosia deltoidea)*, 177
Southern Sahara, 269
South Mountain, 199
Spatial-temporal frameworks, 2
Spawning habitat assessment, 93. *See also* Salmonids
  hydraulic suitability, 94
  integrated assessment, 95
  peak period of, 84
  substrate suitability, 93–94
Stone-built urban cultural heritage, 261
Stone decay, 278
  patterns, 4
Stream walk
  reach and subreach characterization, 84
  reconnaissance, 84
Submerged channels, 18
Substrate size, 89
  channelized portion of Lower Vernon Creek through Polson Park, 89
  cross-sections, 89
  portions of the Creek, heavily modified by, 89
Sulfuric acid, 271
SungboEredo, 269
Sunnyside Mine, 211
Superimposed Holocene deposits, 55
Surface roughness, 304
Synthetic aperture radar (SAR), 222

T

Take-off operations, 110
TDs. *See* Technogenic deposits (TDs)
Technogenic deposits (TDs), 3, 116–118
  soil contamination, 119
Technogenic soils, 118
Telluride, 206, 207, 211, 212, 219, 220, 226
Terrestrial laser scanners, 304
Terrestrial (ground-based or close-range) photogrammetry, 310
Tibesti massifs, 270
Toncontín Airport in Tegucigalpa, Honduras, 107
Topcon RTK-DGPS system, 84
Topographic surveys, 84
  cross-sectional, 84
  reconnaissance stream walk, 84
Tors, 181
Tourism, 2, 31, 39, 52
  abiotic, 2
Toxicity, 119
  for Cd and As, 124
  materials in Araguaína TDs, 124
  and permanence in the environment, 118
  risk of the elements, 116
  toxic elements of paint, 119
  in urban areas, 116
Transmission X-ray microscope
  with 3-D cohesive zone modeling, 308
Triangulated meshes, 304
Turffontein, 136

U

Uncompahgre River, 207, 219, 220
  valley, 211
United States Forest Service, 208
Uranium (U), 118

Urban agglomerations, 10
Urban architecture, 262
Urban areas, 115
Urban-built environments, 2, 3, 261, 270
  architecture, 262
  and prevailing atmospheric pollution, 270–272
Urban climate, 1
Urban evolution, 18
Urban geomorphology, 1
Urban interventions, 18
Urbanization, 13, 28, 83, 84, 239, 241, 252, 255, 261, 317–319
Urban stone decay, evidence of weathering consequent to, 264–269
  chemical weathering
    and sediment production, 266–269
  geology, materials, and deep weathering, 264–266
Urban transformation, 18
  of Palma, 11
U-shaped valleys, 212
Utrata River, 43

V
Valley fever, fungus Coccidiodes, 200
Vernon Creek watershed, location of, 82
Visual descriptors
  ideal slope for Kokanee spawning and, 87
  Reach 1-5, 87–89
  Reach differentiation based on, 87
  riffle-pool sequences, 87

W
Wading rod, 84
WadiSaoura, 270
Wanakah Mines, 212
Warsaw, 37, 56
Warta River, 2, 56, 57, 72, 317
  floodplain, 73
  terraces system concept, 65
  tributaries, 63
  valley, 57, 58, 61, 63
    hypsometric changes of terrain surface, 61
    relief, 56
Waste-based pollution, 318
Waste contamination, 140
Waste disposal, 119
Water Survey of Canada, 86, 90
Weathering, 4, 104, 261, 262, 278
  ambiguous behavior of alumina, 263
  around the Ikorodu area of Lagos State in Nigeria, 262
  chemical weathering
    and sediment production, 266–269
  contemporary research into sandstone, 263
    climatic theory, 263
    structural theory, 263
  3-D quantification of, 305
  geology, materials, and deep weathering, 264–266
  geomorphological research, 263
  primary minerals into secondary ones, 262
    chemical composition due to, 262
  sandstone, 262
Web-based resources, 103
Wellington airport on a tombolo, location of, 107
West Africa, 271, 272
Wielkopolska Fossil Valley, 60
Wind abrasion, 47
Wolman pebble counts, 84
  using GRADISTAT, grain size statistics for, 89
World Reference Base for Soil Resources (WRB) (2014), 119
World War II, 37

X
X-ray CT with ground-penetrating radar (GPR), 304
X-ray imaging apparatus, 304
X-Rite ColorChecker, 306
X-Rite SP68 Sphere Spectrophotometer, 306

Z
Zinc (Zn), 117

Printed in the United States
By Bookmasters